PARASCIENTIFIC REVOLUTIONS

Proximities
Experiments in Nearness
David Cecchetto and Arielle Saiber, Series Editors

Published in association with the Society for Literature, Science, and the Arts

---

Microbial Resolution: Visualization and Security in the War against Emerging Microbes
Gloria Chan-Sook Kim

Parascientific Revolutions: The Science and Culture of the Paranormal
Derek Lee

PARASCIENTIFIC REVOLUTIONS

The Science and Culture of the Paranormal

DEREK LEE

Proximities

University of Minnesota Press

Minneapolis

London

Every effort was made to obtain permission to reproduce material in this book. If any proper acknowledgment has not been included here, we encourage copyright holders to notify the publisher.

Portions of the introduction are adapted from "The Man in the Macintosh and the Science of the Occult," originally published in *James Joyce Quarterly* 55, no. 3–4 (2018): 347–69. Portions of chapter 5 are adapted from "Postquantum: *A Tale for the Time Being, Atomik Aztex,* and Hacking Modern Space-time," originally published in *Multi-Ethnic Literature of the United States* 45, no. 1 (2020): 1–26, https://doi.org/10.1093/melus/mlz057.

Published by the University of Minnesota Press
111 Third Avenue South, Suite 290
Minneapolis, MN 55401-2520
http://www.upress.umn.edu

ISBN 978-1-5179-1888-0 (hc)
ISBN 978-1-5179-1889-7 (pb)

A Cataloging-in-Publication record for this book is available from the Library of Congress.

The University of Minnesota is an equal-opportunity educator and employer.

CONTENTS

Introduction: From Paradigms to Parascience   1

1 Toward a (Meta)physics of Precognition:
   The Science and Aesthetics of the Psitron   33

2 Gaia, the Paranormal Planet   67

3 Remote Viewing: Ingo Swann, Project Stargate, and
   Weaponizing the Paranormal   99

4 On Ghosts and Ghost Vision: *The Hundred Secret Senses,*
   Comfort Woman, and the Asian American Spirit Medium   139

5 The Multiverse and the Mind: The Alternative Chronologies of
   Atomik Aztex and *A Tale for the Time Being*   175

6 The Structure of Parascientific Revolutions   211

Acknowledgments   221

Notes   225

Index   259

INTRODUCTION

From Paradigms to Parascience

On January 12, 2013, biologist and best-selling author Rupert Sheldrake took the darkened stage of TEDx Whitechapel in London's East End. Wearing a crisp black suit, a white shirt, and no shoes, the scientific iconoclast quietly surveyed the capacity crowd. The theme for the evening—Visions for Transition: Challenging Existing Paradigms and Existing Values—had solicited several freethinkers to "explore new and innovative solutions in response to the political, economic, social, and spiritual challenges that are converging on our planet."[1] Many speakers took this proposition to heart. Journalist Graham Hancock encouraged the audience to try ayahuasca and transcend the gridlock of contemporary culture. Satish Kumar, a Jain monk, told the tech-savvy crowd actual well-being came from physically working the land. Tim "Mac" Macartney said trees were our best teachers. For many, though, it was Sheldrake's presentation that crossed the line from provocative to dangerous.

Sheldrake's lecture drew from his recently published book, *The Science Delusion* (2012), which argued true scientific inquiry was constrained by ten dogmas that had ossified into facts; only by letting go of these false assumptions could science become more accurate, more free. During his talk, he criticized the mechanistic view of nature, the immutability of mathematical constants, and the superiority of biomedicine over spiritual healing among other tenets. Perhaps the most unorthodox idea raised was morphic resonance (also known as morphic fields), his theory of mind that "everything in nature has a kind of collective memory."[2] If rats in a London laboratory were trained to perform a new trick, he claimed, rats on the other side of the world would learn the same trick faster because of a shared, dynamic species memory. "Surprisingly, there's already evidence that this actually happens," he added.[3] In other words, Sheldrake declared that species-wide telepathy existed and that it could be proven. He added that such radical knowledge transmission was not limited to animals; it could also be located in plants, bacteria, and even abiotic entities like crystals.

Sheldrake's talk received polite applause but soon came under harsh rebuke by P. Z. Myers, Jerry Coyne, and Kylie Sturgess, members of the Skeptical movement, who argued that his presentation had no place in a forum dedicated to the best in technology, engineering, and design—the key terms of TED's acronym.[4] TED agreed. The organization promptly removed Sheldrake's presentation from its video channel, citing its "factual errors."[5] At this point, things got ugly. Sheldrake's supporters began posting hundreds, then thousands, of angry comments blasting the organization's censorship practices and scientific dogmatism. The TEDx Whitechapel event organizers defended Sheldrake's right to question mainstream science. Not lost in the hubbub was the hypocrisy of removing an antiestablishment talk when the explicit goal of the event was to challenge existing paradigms. Sheldrake's critics responded that TED was well within its rights to remove a talk misaligned with the motto "Ideas Worth Spreading." To question the validity of science and spread testimonials about telepathy was socially reckless, even hazardous. After weeks of debate, TED settled on a compromise. Sheldrake's presentation would return online, but not on TED's primary video archive. Instead, it would reside in a blog section where readers could continue the debate, neither banished nor embraced but instead relegated to a third space outside the standard venues of thought.

From this marginalized position, Sheldrake's talk nevertheless continues to haunt TED. As of this writing, the number of comments on his outcast video approaches ten thousand. "The Debate about Rupert Sheldrake's Talk" TED page not only highlights the cultural politics of what counts as acceptable scientific knowledge but also speaks to the extraordinary power of psychical discourse and the enduring afterlife of "bad ideas" in our supposedly rational world. The Sheldrake incident reminds me of Sigmund Freud's return of the repressed. In studies like "The Uncanny," Freud argues that primitive and instinctive desires repressed from consciousness cannot be eradicated, for they linger in the unconscious and continually hector us as slips of the tongue, fantasies, and personality disorders, permanent reminders of their presence and indestructibility.[6] In many ways, Sheldrake's talk reenacts this drama for the twenty-first century, with his scientific heresies not quite visible and not quite vanquished, skulking at the borders, a persistent specter of our pseudoscientific past and future.

In its political, cultural, and even epistemological aspects, this episode captures the subaltern status of the paranormal mind in contemporary life while also demonstrating its residual power. It also solicits several larger questions about the existence of the paranormal in contemporary life. How does a model of consciousness that science has been unable to verify for well over a century cultivate new theorists and redeemers? What are the aesthetic fields, scientific disciplines, cultural agents, institutions, and Western and non-Western epistemes that allow concepts like telepathy to keep circulating throughout mass culture? What other forms has the paranormal mind taken, and how have they evolved over time? What does this unpredictable ecosystem of information permutation and dissemination reveal about entanglements between science, art, and myth? This study is an attempt to shape a conversation around such vexing questions.

The paranormal mind can be broadly defined as a mind (or a concept of mind) capable of extraordinary abilities like telepathy, clairvoyance, and precognition that defy traditional scientific explanation. Religion scholar Jeffrey Kripal describes the paranormal as "the unacknowledged, unassimilated Other of modern thought" excluded from mainstream science and religion,[7] and in this respect, the paranormal mind covers a diverse range of cognitive abilities that modern society considers supernatural, inexplicable, or just plain weird: telekinesis, spirit communication, astral travel, divination, phylogenetic memory, cosmic consciousness, and psychometry, just to name a few. While the paranormal mind may elicit eye rolls or incredulous laughter from the contemporary reader, it actually draws from a rich cultural history at the intersection of the psychological sciences, spiritualism, literature, technology, and the occult going back over a century. We must also acknowledge that its entire genealogy is pockmarked by controversy and doubt. Ever since the first scientific models of the paranormal mind first arose in the Victorian era, critics have viewed it with suspicion, considering it wishful thinking, metaphysics masquerading as materialism, or religious bunk conning its way into scientific respectability.

Despite all the disciplinary forces arrayed against it, one could argue the paranormal now enjoys greater media exposure, market infiltration, and resiliency to scientific critique than in any other period in history. In film and television, franchises like Star Wars, Paranormal Activity, and Stranger Things have normalized psychical phenomena as higher manifestations of consciousness. In comics, telepathy is a common superpower in the most popular titles, be it Wonder Woman or X-Men. The genre of supernatural romance has its own aisle in Barnes & Noble bookstores. A 2009 Pew Research survey revealed 18 percent of the U.S. adult population believes they have been in the presence of a ghost, and 29 percent have felt in touch with the dead.[8] Paranormal research and applications have continued deep into the twentieth century, such as Russia's mass hypnosis television broadcasts in the 1980s.[9] For all of modernity's misgivings, the paranormal mind is very much alive in the twenty-first century. Like Rupert Sheldrake's outcast TED talk, it is a haunting presence, one beyond the borders of establishment science yet never too far away. The best and brightest scientific minds continue to condemn it and the forces of intellectual history are arrayed against it, but the paranormal mind always seems to survive. It evolves. It thrives.

This book considers the cultural and theoretical development of the paranormal mind over the twentieth and twenty-first centuries with a strong emphasis on its primary drivers in the post-1945 era: literature and science.[10] In recent decades, there has arisen keen literary interest in occultism, though most studies have focused on the Victorian psychological sciences and their connections with modernist writers like W. B. Yeats, T. S. Eliot, and James Joyce.[11] This study goes deeper into the twentieth century and farther afield from its roots in the Occult Revival. One of my central theses is that the paranormal mind has evolved considerably through contact with new Western and non-Western sciences; consequently, this study will touch on quantum physics, neurology, systems biology, microbiology, Mesoamerican chronology, Korean shamanism, and other indigenous sciences to see how these disparate fields have renewed paranormal logics. At the same time, it traverses the literary lineages of occult mentation beyond modernism. While the fin de siècle (1880s to 1890s) and modernist eras (1890s to 1940s) are the literary periods most often identified with the paranormal mind, they are hardly the only ones.[12] Twentieth- and twenty-first-century fiction has a fascinating history of engaging with the paranormal,[13] and I focus here on science fiction and ethnic fiction as new kinds of literatures of consciousness that, just like modernist novels, attempt to explore the uncharted landscapes of the human mind. In addition, the book examines several overlooked literary genres where the paranormal has been frequently retheorized and recirculated, such as New Age memoirs, psychic how-to guides, government training manuals, remote viewing training protocols, parapsychological journals, and declassified CIA intelligence reports. Many of these areas have been wholly ignored by scholars because of their pseudoscientific connotations, but I contend that a true history of the paranormal mind must account for its permutations across the full aesthetic landscape of high-, low-, and middlebrow cultures.

Parascientific Revolutions also provides an epistemological theory of the paranormal. My aim is not to prove the existence of the paranormal mind but to gain insight into its circulation, resilience, and architectonics. The history of ideas is littered with outrageous scientific concepts, most of which are soon forgotten. Nevertheless, why do certain ideas like clairvoyance continue to persevere, even spread, when so many others have disappeared? How is pseudoscience reproduced and reinvented?[14] What social and literary forces power the dissemination of the supernatural? Much has been written on the production and gatekeeping of mainstream scientific knowledge (physics, psychology, etc.), but comparatively little research has gone into the structures, processes, and flows of pseudoscience. Indeed, one of the book's primary arguments is that pseudoscience follows patterns of rejection and regeneration that can, and should, be theorized. The durability of the paranormal mind—the ultimate pseudoscientific object—flies in the face of post-Enlightenment narratives of epistemic progress. This suggests to me the presence of an alternative, and largely unrecognized, mode of (pseudo)scientific knowledge creation deserving of attention, not only for what it reveals about our esoteric past but also for what it portends for our epistemological future.

Before falling too deeply down the rabbit hole of paranormal history and theory, it is worth addressing the elephant in the room: Why bother studying the paranormal mind? Why investigate a crazy concept that most people do not even believe exists? My first answer is, the paranormal mind is not crazy at all. Countless philosophers, scientists, and artists have engaged with it over the years. As Egil Asprem notes, "people working at the cutting edge of fields as diverse as physics, chemistry, physiology, and literature" have sought to comprehend it.[15] From Nobel laureates like J. J. Thompson and Charles Richet to literary heavyweights like Arthur Conan Doyle to ordinary folks conducting séances in their parlors, the paranormal mind has represented an essential aspect of the human to multitudes. For the Society for Psychical Research (SPR), an English scientific organization dedicated to studying spiritualistic phenomena, the paranormal brought together some of the greatest questions facing mankind: What happens to us after death? Why do we dream? Is there some greater connection between all living things? The sheer volume of scientific and aesthetic inquiry dedicated to answering these kinds of questions over decades is testament to its import. Historian Seymour Mauskopf has argued that pseudosciences are important because they reveal more about the process of scientific development (e.g., demarcation, credibility, institutionalization) than mainstream sciences.[16] As such, the paranormal mind is a privileged site because the parapsychological "comes as close as any example to being paradigmatic for modern marginal science."[17] Perhaps the best, and bluntest, answer is, pseudoscience matters. It affects public discourse, medical policy, and social perceptions of science. If the contemporary antivaccination movement and climate change denialism have revealed nothing else, it is that quasi-scientific information has profound material effects on political life. The paranormal may operate on lower frequencies, but it still participates in the same epistemic fracas as information, misinformation, and the will to believe. In *Outsider Theory*, Jonathan Eburne writes, "Even if we wish to protect ourselves from the dangerous effects of 'bad' ideas, it is all the more important to know where and how they function, how they impress themselves upon what we think and how we know."[18] I wholeheartedly agree. To be true to itself, a philosophy of science must recognize that the development of facts and non-facts is crucial for understanding the scientific system in its entirety. Exploring the paranormal mind, then, is a way of studying the overlooked technics of knowledge and pseudo-knowledge circulation in the modern age.

One of the reasons scholars ignore the paranormal is its failure to align with existing scientific narratives. Auguste Comte famously described science as the last and highest stage of human intellect, where the theological and metaphysical are finally discarded in favor of true knowledge.[19] Science here is the implicit light of the Enlightenment banishing the darkness of folklore, myth, and superstition. Though philosophy of science has long since curtailed Comte's grand vision, his metanarrative of science as a perpetual march toward a perfected and complete epistemology still underwrites contemporary conceptions of science as a force for truth writ large. This has consequences for anomalous areas of inquiry, like the paranormal, as well as for projects seeking to analyze them. Studies of early twentieth-century occultism have critical purchase because the Victorian sciences supporting them were once viewed as legitimate. The same cannot be said for paranormal phenomena in the late twentieth and early twenty-first centuries, the temporal localization of this book. The calcification of disciplinary field lines and improvement of scientific practices has cast psychical research out as a sanctioned research area. Andreas Sommer, noting even starker realities, writes that "there is a genuine danger of attracting the ire of certain influential groups and figures who have been quick to level charges of 'relativism' and even 'anti-science'" when addressing the paranormal.[20] In sum, the declining legitimacy of the paranormal mind as a scientific object over the course of the long twentieth century has hampered its study as a cultural object as well.

With all due respect to modern science, however, I do not believe the paranormal is an illegitimate subject for meaningful inquiry. There is still much to learn about the history, development, and dissemination of unorthodox science in the present age. I would further argue that science is not a unilateral force for eradicating heterodoxy. In fact, a core premise of this book is that science and literature are principal agents in the dynamic production, development, and evolution of the paranormal mind. As modernist scholars like Alex Owen, Mark Micale, and Mark Morrisson point out, occult research and narratives have informed each other from the very start.[21] My study uses the paranormal mind to unveil structural aspects of this epistemic synthesis, the unseen means by which science and literature have collectively acted as engines for paranormal theorization, production, dissemination, and recalibration. *Parascientific Revolutions* elucidates the pathways beyond standard science where rejected scientific ideas continue to linger. So much of modern science studies, from Karl Popper to Thomas Gieryn, has explored how scientific truths come into being. This is, and continues to be, an important task. But we cannot ignore the fact that ideas at the margins of the scientific imaginary, like morphic fields, continue to exist uneasily between fact and fiction. A primary goal of this book is to illuminate the epistemic processes—the scientific, literary, philosophical, and cultural forces—shaping the paranormal over the arc of the twentieth and twenty-first centuries.

Central to this project is what I call *parascience,* the epistemic space on the boundaries of establishment science where its unconfirmed ideas reside. While this term will be explored in greater detail later, the basic premise is that "failed" scientific concepts like telepathy do not immediately disappear when pushed to the margins of scientific discourse. Parascience is not an intellectual graveyard, a final resting place for bad science. Rather, I view it as a vibrant, interdisciplinary arena where unorthodox science amalgamates with other literary, mythic, philosophical, Western, and non-Western modes of thought, and it is here that pseudoscientific ideas are often reformulated and returned into mainstream culture. Literature plays a dominant role in parascience because it serves as a primary means through which untraditional ideas like the paranormal mind are taken up, theorized, and circulated to new publics. For instance, as we shall soon see, science fiction has played a seminal role in maintaining telekinesis in the cultural consciousness, regularly offering new explanations for an unexplainable power. Ironically, science is the other principal agent abetting paranormal evolution. As psychical research has waned, new scientific frameworks and programs have emerged to substantiate the paranormal mind. *Parascientific Revolutions* examines several of these idiosyncratic pathways. It follows how clairvoyance hybridized with cartography and geodesy to produce the Cold War surveillance technology known as remote viewing. It reveals how quantum physics fused with precognition to conjure faster-than-light thought particles. It traces how Chinese metaphysical theories of qi authorized communication with ghosts. This book looks past the standard psychical models, literary genres, and epistemological frameworks historically used for understanding the paranormal and instead follows its peculiar multiplication and propagation across global culture. By focusing on parascience rather than science alone, this study captures and expands on the spectrum of processes, forces, disciplines, and ideologies constituting the paranormal in the contemporary era.

Psychical Research and the Paranormal Mind

The notion of a human mind capable of marvelous gifts of prophecy, enlightenment, and communication with the dead can be found in nearly every Western and non-Western culture throughout recorded history. In order to appreciate the modern trajectory of the paranormal mind as a scientific and literary object, though, we must first acquaint ourselves with the Occult Revival, the SPR, and the scientification of the supernatural leading up to the 1940s.

The Occult Revival is an umbrella term for a late nineteenth- and early twentieth-century cultural movement centered in Britain involving the parallel rise of esoteric groups with widespread public interest in séances, mesmerism, astrology, sorcery, divination, and necromancy. In contrast with the first wave of Victorian spiritualism that began in 1848 with the Fox sisters in Hydesville, New York, the Occult Revival was primarily a middle- and upper-class movement driven by artists and intellectuals attracted by the promise of hidden knowledge and ancient wisdom.[22] Although spiritualist in principle, the Occult Revival was also highly scientific in its aims and methods because its practitioners desired deep insight into nature. Groups like Madame Helena Blavatsky's Theosophical Society claimed the "Secret Doctrine of the East" could unlock the unconscious, the fundamental laws of the universe, and the truth of the soul.[23] Aleister Crowley's Thelema aspired to harness the true powers of the will through sexual magic and demonology. The Hermetic Order of the Golden Dawn sought knowledge of astral travel, scrying, and divination, among other arcane practices. Collectively, these organizations support Jason Josephson Storm's claim that the myth of disenchantment (the idea that modernity banished supernaturalism from its world view) is itself a myth.[24] According to historian Alex Owen, these groups all shared the belief that "reality as we are taught to understand it only accounts for a fraction of the ultimate reality which is just beyond our immediate senses."[25] It was in this hidden universe that occultists sought knowledge. Given their arcane engagement with scientific aims, the Occult Revival can be understood as a simultaneously premodern and modern movement.[26]

The SPR was one of the smallest Occult Revival organizations, but it was easily the most prestigious thanks to its dedication to scientific principles and methods. Founded in 1882 by several Cambridge scholars intrigued by the occult, the SPR had clear goals: "to investigate that large body of debatable phenomena designated by such terms as mesmeric, psychical and spiritualistic," and to do so "without prejudice or prepossession of any kind, and in the same spirit of exact and unimpassioned enquiry which has enabled Science to solve so many problems, once not less obscure nor less hotly debated."[27] For such a small association, the SPR immediately distinguished itself with a formidable list of members, including prominent scholars (Henry Sidgwick, professor of moral philosophy at Cambridge and SPR's first president), politicians (prime ministers William Ewart Gladstone and Arthur Balfour), scientists (physicists Balfour Stewart, Oliver Lodge, and William Crookes), aristocrats (the earl of Carnarvon, the marquis of Bute), and writers (Alfred Lord Tennyson, John Ruskin, Lewis Carroll).[28] For years, the British public marveled at the incredible phenomena associated with occult circles: speaking to the dead, reaching higher planes of reality, beholding visions of distant events, witnessing ectoplasmic emanations, and so on. The SPR took these claims seriously and sought to elucidate their scientific principles.[29]

The SPR's original methodology involved sending teams of investigators across the English countryside to visit haunted houses, observe mediums, and collect stories of paranormal phenomena. Much like eighteenth-century naturalists scouring the world for rare specimens, the SPR collected the supernatural in all its forms, be they holy visions or feelings of dread. From this raw material, the SPR categorized the many varieties of occult phenomena in a practice Seymour Mauskopf and Michael McVaugh have since dubbed the "natural historical" approach.[30] Early psychical researchers recognized the limits of their field. The arbitrary and spontaneous nature of spiritualist phenomena prevented its study as a laboratory science (at least in the beginning), but the SPR could apply a level of analytical rigor to experiences typically dismissed as beyond science. More conservative scientists of the period disregarded the supernatural, while occultists often thought that overrationalizing would impede their mystic endeavors.[31] Psychical researchers split the difference, for they "accepted the reality of these experiences . . . and they posited causal mechanisms that often relied upon an innovative understanding of the human mind."[32] The lasting contribution of the SPR was reframing the paranormal from spiritual curiosities into scientific objects worthy of study.

For the SPR, the human mind was the linchpin for explaining a wide range of occult abilities. The late nineteenth and early twentieth centuries were marked by a cultural turn inward to understand the human subject. According to Mark Micale, "During these years, artists, philosophers, and scientists probed beneath the surface identity of reason in order to uncover deeper irrational or nonrational levels of human experience and cognition."[33] The paranormal mind was essential for comprehending these depths. What are its general characteristics? What are its underlying properties and physiological mechanisms? What are the range of its powers, and how can they be assembled into a comprehensive system of knowledge? In time, the SPR identified four major categories of interrelated phenomena that now characterize the paranormal mind: telepathy (mind-to-mind communication), clairvoyance (knowledge of distant objects or past events), precognition (knowledge of future events), and telekinesis (mental control over physical matter). From these powers spring other inexplicable experiences. For example, in *Phantasms of the Living*, psychical researcher Edmund Gurney argues that ghosts are not actually undead beings but rather telepathic projections unconsciously sent from a person in distress to a percipient who "sees" it.[34] In this manner, the SPR developed the vocabulary and discourse that still frame the paranormal today.

In addition, the SPR produced the first scientific model of the paranormal mind in 1901 with the posthumous publication of Frederic W. H. Myers's *Human Personality and Its Survival of Bodily Death*. Myers argued that the human mind consists of two parts: the superliminal self and the subliminal self. The former represents the conscious mind, comprising everything we are phenomenologically aware of. The latter represents a pre-Freudian theory of the unconscious collecting "*all* that takes place beneath the threshold, or say, if preferred, outside the ordinary means of consciousness."[35] According to Myers, the subliminal self could communicate directly with other minds, project ghosts, receive prophetic visions, and give rise to strokes of genius. Unlike Freud's theory of the unconscious, the subliminal self was "porous"; it could leak into the minds of others at an unconscious level that could sometimes, but not always, rise to the level of conscious awareness. Through this mental mechanism, talented individuals with highly developed subliminal selves (such as spirit mediums) could glean information about complete strangers and speak with the dead. Perhaps the most astonishing of Myers's claims was the persistence of the subliminal self beyond the expiration of the material body. Myers noted that "a manifestation of persistent energy" connected to a deceased person could exist as a force that continues to communicate with mankind.[36] This explains at least one technical aspect of mediumship: the eternal subliminal selves of the dead can contact the subliminal selves of the living if someone is sensitive enough to hear it. In short, Myers scientificized the soul.

The subliminal self is the first Western scientific model of the paranormal mind insofar as it provides a unified framework (or a "measured base," as Myers put it) explaining the nature and mechanisms underlying occult phenomena.[37] Its powers may seem fantastical to modern readers, but the subliminal self has proven highly influential to twentieth-century thought. According to psychologist and former president of the American SPR (ASPR) William James, "to Britain, Frederic Myers was the greatest living systematizer of the notion of the unconscious mind," and "Myers's subliminal theory of self would provide a foundation for the most important psychology of the twentieth century."[38]

The SPR wielded considerable literary influence throughout the early twentieth century. Arthur Conan Doyle was an SPR member, and his experiences with mediums and séances informed much of his Professor Challenger series. May Sinclair's ghost story collections *Uncanny Stories* and *The Intercessor and Other Stories* were inspired by psychical research. Virginia Woolf was similarly fascinated by the psychical, and the telepathic connection between Clarissa Dalloway and Septimus Smith in *Mrs. Dalloway* is perhaps the most famous example of occult science informing her work.[39] Pulp author Sax Rohmer drew from his psychical interests to conjure the villainous Dr. Fu Manchu as a master of esoteric science. Even James Joyce, who was famously skeptical of the occult, did not hesitate to include psychical research in his own works.[40] While paranormal phenomena may be spontaneous, it played an ongoing role in modernist aesthetic production.[41] These writers shared with occultists an intense desire to understand the unseen inner workings of the mind. The psychological sciences were burgeoning fields in the early twentieth century, and the vistas of consciousness remained wide open for scientists and artists alike. Similar to any scientific object before disciplinary borders close, the human mind was a blank state for articulating diverse theories. As Roger Luckhurst writes, psychical concepts were "theorized at vanishing points—just where confident demarcations between truth and error, science and pseudoscience, could not at the time be determined."[42] Psychical research played a critical role in such aesthetic and scientific hybridizations.

The influence of the SPR and the ASPR waned over the course of the 1920s. One reason was the internecine conflict between spiritualist-leaning and science-leaning factions of these organizations.[43] According to historian Brian Inglis, the latter hoped to "maintain the most rigorous of scientific standards," which resulted in an agenda focused on disproving mediums and receiving academic plaudits.[44] The first activity was a negative endeavor that only served to highlight the hucksterism endemic to the occult. The second was simply a lost cause; no smoking gun could ever be found that would convince traditional scientists to accept psychical theories.[45] As a biologist once told James, "Even if telepathy were proved to be true, savants ought to band together to suppress and conceal it."[46] Psychical research was an affront to positivistic science; its connections to spiritualism, Theosophy, and Magick placed it outside the prevailing paradigms of the human mind. Many citizen science experiments took place during this period. As Alicia Puglionesi observes, many normal Americans informally contributed their paranormal experiences to the ASPR to do their part for the field, most famously Mary Craig and Upton Sinclair's telepathic image transmissions in *Mental Radio* (1930).[47] Such activities highlight the tenacity of populist esotericism. Nevertheless, psychical research never gained entry to mainstream scientific discourse, save for the following exception.

In the 1930s, J. B. Rhine's Parapsychology Laboratory at Duke University emerged as the new nexus of psychical research. Under the auspices of William McDougal, chair of Duke's psychology department and an ASPR veteran, Rhine established the first sustained academic outpost for psychical research in 1927.[48] Rhine's work was revolutionary because of his laboratory-based methodology. Believing psychical research needed to evolve past the natural historical approach toward controlled experiments "under conditions of systematic observation," Rhine conducted his work in laboratories and used the newfound power of statistics.[49]

A typical experiment in Rhine's laboratory involved testing subjects who could perceive or predict images, numbers, or occurrences at rates that were statistically anomalous. For example, Rhine had Duke undergraduate students guess a series of Zener cards. These specialized decks of twenty-five cards feature five runs of five images: a circle, a plus sign, wavy lines, a square, and a star (Figure 1). Test subjects were asked to mentally visualize cards drawn by the experimenter hundreds of times. Basic mathematics suggests that an ordinary person could only correctly guess 20 percent of the time, or a one in five chance. This was true for most subjects, who clearly did not have what Rhine called "extrasensory perception" (ESP), which denotes telepathy, clairvoyance, and precognition.[50] (The more general term *psi* refers to ESP plus telekinesis.) Some consistently guessed above average, however. For example, one student, A. J. Linzmeyer, achieved stretches in which he guessed 15 of 25 and 21 of 25 cards correctly.[51] Statistical analysis indicates that the possibility of such runs occurring by chance was 1 in 30 billion, which is to say impossible. According to Rhine, subjects like Linzmeyer possessed a "sixth sense" unknown to current science that allowed him insight into distant objects or distant minds.[52]

Figure 1. Zener cards (also known as ESP cards). "File:Zener cards (color).svg," by Mikhail Ryazanov, from a talk delivered April 1, 2014, 01:30, licensed under CC BY-SA 3.0.

With laboratory conditions, modern scientific methods, reproducibility, and statistical evidence bolstering his results, Rhine and his collaborators brought new levels of credibility to *parapsychology*, his updated term for psychical research. His 1934 text *Extrasensory Perception* riveted the nation and sparked debate about the paranormal mind from scientists and laypeople alike. Although they could not refute Rhine's mathematics, many scientists could not bring themselves to believe in supernatural abilities that defied the rules of physics and biology. Others reveled in data now proving what they had always suspected: the mind possessed amazing powers of telepathy, clairvoyance, and precognition. Rhine's success shifted the intellectual leadership of psychical research from Europe to America and confirmed the ascendance of empirical methods over anecdotal ones in psychical studies.[53]

During this time, science fiction replaced modernism as the literary field most engaged with parapsychology. One of the undergraduates Rhine tested at the Parapsychology Lab was the future editor of *Astounding Science Fiction*, John Campbell, who became a lifelong convert to what he called "psionics."[54] One of the greatest editors of the golden age of science fiction, Campbell assisted the careers of famed writers like Isaac Asimov, L. Ron Hubbard, and A. E. van Vogt. Many of these authors consequently took up psionics as a major component of their artistic visions (while also guaranteeing sales to *Astounding*). Novels like van Vogt's *Slan,* Henry Kuttner's *Mutant,* and Alfred Bester's *Demolished Man* envisioned new telepathic races and their place within society. As any science fiction reader knows, this paranormal legacy is now entrenched and ubiquitous to the genre. Campbell recognized science fiction could play an ideological role by pushing mass culture where science refused to explore. As Campbell wrote in a 1953 letter to Rhine, "I am using fiction to induce competent thinkers to attack such problems as psi-effects."[55]

While Campbell's advocacy for psi-fi worked extraordinarily well,[56] Rhine's success in the academy proved shorter lived. Despite some interest and repeated attempts, no equivalent to Duke's Parapsychology Lab appeared at other universities. According to Mauskopf and McVaugh, parapsychology never developed a strong enough case for mainstream psychology departments to accommodate such a controversial research program.[57] In addition to such disciplinary qualms, there was a growing sense that ESP was founded on faulty science. Famed psychologist B. F. Skinner discovered that images on the front of Zener cards, under certain light conditions, could be seen from the back—an oversight suggesting cheating on the part of subjects and incompetence on the part of experimenters.[58] Chester Kellogg attacked Rhine's use of statistical methods as flawed.[59] C. E. M. Hansel argued that parapsychologists usually wanted to believe in ESP, which led to conscious and unconscious bias.[60] John Kennedy hypothesized that researchers made recording errors in their data to favor ESP—a veiled accusation of outright fraud.[61] Yet another branch of criticism emerged in the broader failure to replicate experiments. Universities like Colorado, Minnesota, Kansas, and Antioch all conducted ESP experiments in the 1930s and 1940s without achieving the incredible results found at Duke.[62] Rhine and his associates claimed test subjects found at other universities were simply not as talented as the ones they themselves had located, but the inability to reproduce his experimental success proved damning. It did not help that the Parapsychology Lab eventually secured enough funding to detach itself from Duke in 1947. Such a maneuver may have made financial sense, but it proved disastrous from a disciplinary perspective because parapsychology no longer enjoyed Duke's academic prestige, institutional connections, or ability to confer degrees.[63]

By the midpoint of the twentieth century, the figurative writing was on the wall: the prospects of psychical research had plateaued, and the field was slowly dying. According to Inglis, the London SPR had half as many members in the 1930s than it did in the 1920s—a shrinking that would only worsen over time.[64] Montague Ullman's Dream Laboratory at Maimonides Medical Center performed innovative research on telepathically sending images into sleeping subjects' dreams, but this work also failed to be replicated.[65] There were also fewer academic psychologists willing to risk their careers by associating with "fields of research whose subject matter public opinion has equated with quackery, folly, intellectual vulgarity, and mental illness."[66] Compounding these issues was the lack of a standardized psi theory. "We have no science of parapsychology," SPR veteran Gardner Murphy admitted. "No theoretical system tightly and beautifully organized in the manner of an architect; no solid beams of repeatedly confirmed findings, reproducible by a careful experimenter who can exactly follow the specifications, sure of the general trend of the results he will get."[67] While I contend that Myers's subliminal self is a theory, I sympathize with Murphy's larger point that psychical research had shockingly few principles to explain the physical or psychological aspects of the paranormal mind after decades of data collection. This is why Asprem calls psychical research a history of failure. "Researchers could not agree between themselves on experimental protocols, fundamental hypotheses, or even whether or not one should seek acceptance from the scientific establishment in the first place," he writes. "A number of 'paradigms' for research were proposed, but none of them won general acceptance, and none of them managed to produce results that were convincing to outsiders."[68]

If the narrative of science is a steady, inexorable march toward greater truth, then the paranormal mind has no apparent place in that march. In the marketplace of ideas, concepts like psi have hardly proven themselves rigorous enough to exist within science's ranks. The irony, of course, is that psychical researchers have always championed science. The SPR's method is "the method which our race has found most effective in acquiring knowledge," Myers wrote in *Human Personality*. "It is the method of modern Science."[69] From the natural historical method to the statistical method to the development of academic publications like the *Proceedings of the Society for Psychical Research,* psychical researchers have always craved scientific acceptance. But modern science has never championed the psychical—and in fact has happily served as its foil. Psychical research may have viewed itself scientifically, but it was "real science" that showed just how far the esoteric discipline had to go.

Beyond issues of rigor and replication, the main problem psychical research faced was its misalignment with the dominant ideologies of science. As Inglis notes, "The reality of paranormal phenomena would mean making nonsense of much that is still being taken for granted, and taught, in many other disciplines. In biology, psychology, anthropology, medicine, and even history; most of all, perhaps, in philosophy."[70] In 1939, Jules Romains argued that psychical research amounted to a new scientific revolution: "Some of the most important results obtained through psychic experiments, as soon as they are confirmed, (if they have to be) and officially recognized as 'true' will be a threat to positive science *within its own frontiers.*"[71] For decades, psychical researchers had argued the human body functioned in strange ways involving undiscovered senses and physics-defying mechanisms. If even a fraction of these claims proved true, they would topple existing paradigms. Epistemologically speaking, psychical research represented a fundamental threat to modern science. Was it a surprise, then, that establishment scientists deployed a host of tactics (e.g., desk rejections from mainstream journals, refusal to believe evidence, claims of triviality) that continually dismissed the findings of paranormal investigators?[72] Individually, these strategies were frustrating to parapsychologists; collectively, they speak to disciplinary headwinds hostile to their continued existence.

The methodological and ideological attacks on psychical research highlight another narrative crucial for the pages ahead, namely that parapsychology has effectively—and permanently—been proven pseudoscientific. The argument here is that normal science has done its job. As the work of Kennedy, Skinner, Kellogg, and Hansel suggests, parapsychologists were an unreliable bunch trafficking in poor technique, fallacious reasoning, and fraud. By extension, the paranormal mind becomes irrelevant as an object of study. During the fin de siècle, psychical researchers may have exploited uncertainties in the psychological sciences, but since then, science had systematically exposed their outrageous claims. It had fulfilled its positivistic purpose as an instrument of truth and banished psychical research from its existing paradigms. Without practitioners, scholarly prestige, or even valid research questions to pursue, the paranormal mind was left outside the boundaries of science, where it would surely fade away. The problem with this story is that the paranormal mind never disappeared. The rest of this introduction and the chapters that follow will attempt to explain why.

Science and Parascience

In his classic work *The Structure of Scientific Revolutions*, philosopher of science Thomas Kuhn popularized his now-famous paradigm theory of scientific progress. Building on the work of Ludwik Fleck, Kuhn argued that all scientific progress takes place within a "paradigm," or the underlying assumptions, beliefs, and methods of a given scientific era.[73] A paradigm like Newtonian physics is powerful because it determines not only the knowledge scientists can create but also the questions they are authorized to ask.[74] Like Michel Foucault's concept of the episteme, it contains the conditions of its own possibilities. While Kuhn's theory has been criticized over the years, it nevertheless remains a dominant theory of scientific epistemology in science studies.[75] Indeed, in our post-Kuhnian era, references to "the current paradigm" and "paradigm shifts" across politics, social science, and culture speak to its outsize influence.

Kuhn's paradigms play a prominent role in this study because they are a preeminent model of knowledge production that is irreconcilable with pseudoscience. Kuhn contends that scientific ideas that fail to achieve mainstream acceptance are essentially thrust out of the paradigm, where they "disappear to a very considerable extent and then apparently once and for all."[76] This makes intuitive sense. If a medical concept like hysteria no longer fits within the principal knowledge frameworks of an era or fails to cultivate new proponents, then its long-term prospects are bleak. Yet we cannot deny that formerly scientific ideas like telepathy continue to thrive deep into a present age dominated by reason and STEM (science, technology, engineering, and mathematics) education. The Sheldrake TED controversy is only one example of this. Despite what Kuhn and other philosophers of science might say, the fact of the matter is that some pseudosciences refuse to disappear from either scientific discourse or the culture at large. In contrast to Comptean narratives of scientific progress and Kuhnian models of epistemic policing, heterodox science always seems to persist.

One of my fundamental tasks in this book is challenging the glorified mythology of science as a tireless force that has consistently and effectively expunged pseudoscience from scientific and popular culture. Both literature and science play structural roles in the intellectual advancement, perpetuation, and dissemination of pseudoscientific concepts like the paranormal mind. For literary critics, destabilizing this narrative helps open a vibrant new space for understanding the authors, genres, and historical contexts that have informed supernatural discourse. For science and technology studies (STS) scholars, it similarly broadens the range of knowledge production to include occult sciences, border sciences, failed sciences, and pseudosciences contra the establishment fields historically central to the discipline.

The standard way of rationalizing the persistence of heterodoxy is to view it as a relic of a bygone era.[77] According to philosopher of science Bruno Latour, modernity is widely (mis)understood as an historical rupture separating the contemporary regime of science (empirical knowledge) from the previous regime of the social (cultural, mythic, or theological knowledge).[78] Accordingly, pseudoscientific ideas can be understood as remnants of an archaic, irrational era persevering into the scientific present. Anthropologist Edward Burnett Tylor might describe them as "cultural survivals"—mystic residues that, like superstitions, have persisted into modern society.[79] This is ostensibly why dowsing and astrology still exist; they are carry-overs from a less scientific epoch.

However, the sheer prevalence of heterodox ideas suggests there exist epistemological processes beyond the dominant scientific paradigms that play a role in modern knowledge circulation. Two epistemic models in particular warrant attention. Religion scholar Christopher Partridge argues that nontraditional ideas exist in a third space between organized religion and secular knowledge called "occulture."[80] Occulture is a religious-cultural milieu that contains the "vast spectrum of beliefs and practices sourced by Eastern spirituality, Paganism, Spiritualism, Theosophy, alternative science and medicine, popular psychology, and a range of beliefs emanating out of a general interest in the paranormal."[81] For Partridge, occulture acts as a reservoir for rejected and oppositional beliefs divorced from scientific or religious institutions. Crucially, he contends that occulture's primary medium is pop culture; cinema, television, and mass-market literature keep heterodoxy alive in modern society.

Philosopher Massimo Pigliucci has a variation of this concept he terms "almost science."[82] Defined as the "middle territory between science and non-science," it is the episteme of scientific programs that have not yet achieved mainstream recognition due to lack of evidence or the inability to investigate. One example of almost science is the Search for Extraterrestrial Intelligence (SETI), which remains indeterminable because we have not yet received alien contact.[83] Another is string theory, a powerful theory in physics that exceeds the current technological ability to verify it.[84] Almost science is politically important because it serves as a battleground for distinguishing real science from pseudoscience. In contrast to Partridge, Pigliucci claims philosophy is the key medium for cogitating pseudoscience. "Philosophy has often been the placeholder for areas of intellectual inquiry that have subsequently moved to the domain of science," he observes, with psychology and physics being notable examples.[85] Philosophy is therefore the site where not-quite-scientific ideas like ufology are rightfully analyzed and theorized. It is, in his words, "the continuation of science by other means."[86]

Partridge and Pigliucci address a serious gap in post-Kuhnian science studies, namely the zones of knowledge on the outskirts of accepted science. Their work is important precisely because they see marginal ideas as central to modern thought and therefore worthy of serious exploration. We *moderns* (to borrow Latour's term for post-Enlightenment citizens) often think we exist in a world of purely scientific knowledge distinct from socialized belief, but such an assumption is naïve and false. Concepts like the subliminal self, hypnotism, and telepathy all have scientific roots, and they still belong to scientific and pseudoscientific discursive networks. Neither literary nor science studies scholars have paid enough attention to the processes, structures, and realms of knowledge on the fringes of mainstream science even though they play a powerful role in modern culture. As Eburne contends, "We need a sustained reckoning with the circulation of outlandish ideas."[87]

To this end, I propose that the paranormal mind is not just a relic of ancient or outmoded thought but rather a dynamic product of a scientificocultural third space I call *parascience*. I define it as the epistemic realm on the borders of establishment science where heterodox concepts can hybridize with other areas like literature, myth, philosophy, theology, non-Western thought, and novel research programs to produce new knowledge forms that persist across modern science and culture. Parascience thus identifies the site for the production *and* circulation of pseudoscientific ideas like the paranormal mind. *Para* in Greek translates as "besides," "alongside," or "beyond." Like the *para* in paratext, it describes the forces, structures, and cultural contexts that stand astride the object in question without necessarily demoting them beneath that object. It also operates like the *para* in paraliteracy, which Nahum Chandler describes as "a more universal inhabitation of thought and culture [arising] from a so-called partial position by way of its double position as simultaneously both within and outside of the mainstream."[88] Such an emphasis on parallelity, on questioning the relation of majority and minority discourses, appropriately reflects how I see parascience. It describes the circulation of knowledge alongside establishment science—neither less than nor greater than it, but at all times surrounding and coexisting with it. In the same way that paratext and paraliteracy invoke all the variables of authorship, history, genre, power, privilege, and politics surrounding a specific text, so too does parascience give name to the space enveloping orthodox scientific knowledge.

Parascience must be distinguished from *pseudoscience,* which has a different meaning and function. The term *pseudo* derives from the Greek word *pseudos,* "falsehood," and pseudoscience consequently means a false or sham science. This is problematic both denotatively and connotatively. In scientific discourse, pseudoscience is the ultimate insult because it is the opposite of good science; it suggests faulty thinking, and its psychological, cultural, and epistemological effects are deleterious.[89] It also suggests that the ideas under consideration do not rise to the basic requirements of real science and can be safely ignored. To attach yourself to pseudoscientific fields like phrenology would be to undercut yourself before you even have a chance to speak. In fact, the true power of pseudoscience is rhetorical—its implicit and potent ability *to suppress conversation*. The negative connotations of pseudoscience foreclose any possibility for meaningful conversation or critique; such a muzzling of ideas is antithetical to the spirit of this project, which aims to explore the intellectual history and theoretical processes behind one of the most contro-versial pseudosciences in modern culture.

In contrast, the function of *parascience* as a theoretical term is to open up a new critical space around those very scientific objects deemed irrelevant, illogical, or pseudoscientific. R. A. Judy argues that paraliteracy carves out a discursive space where there was previously none because of its ideological discordance.[90] Parascience follows the same logic. My aim is not to silence the conversation around telepathy and precognition because they are dangerous; rather, my aim is to explore the epistemic pathways and aesthetic fields central to their cultural endurance. In *Cultural Boundaries of Science,* Thomas Gieryn suggests we approach science cartographically because so much of its prac-tice depends on mapping out terrain, acknowledging borders, and identifying the agents of its cultural reification.[91] If we allow ourselves to think spatially, I envision parascience as the areas surrounding and suffusing paradigmatic science. In this dynamic expanse, ideas marginalized by establishment science interact with other discourses, including literature, art, philosophy, theology, traditional knowledge, folklore, magic, and new varieties of Western and non-Western science. These are realms of knowledge we often partition from mod-ern science. *Parascience* gives name to the rhizomatic space between these various epistemes, the bleeding borders between scientific and nonscientific discourse. In this regard, we can also consider parascience a Foucauldian counterdiscourse because, by giving renegade ideas a platform, we also allow them to "speak."[92] Parascience foregrounds outsider knowledge as well as the politics of scientific power.

Parapsychological circles have previously used the term *parascience* as a blanket label for all psychical-related areas like ESP, astral travel, and paraphys-ics.[93] This broad designation for untraditional scientific disciplines overlaps with my own definition, but I see parascience as more specific and capacious. It is more specific because it identifies the epistemic expanse where pseudo-science circulates rather than servicing as a catchall term for paranormal con-tent. It is more capacious because it includes discourses beyond the occult (e.g., theological, mythical, non-Western, speculative) as well as the processes allowing them to engage and renew concepts like the paranormal mind. My concept of parascience also distinguishes itself from Partridge's occulture and Pigliucci's almost science because it operates less as a cultural reservoir than as an epistemological engine. By this, I mean that parascience produces new types of knowledge. It is an active, evolving space where marginalized scientific ideas constantly engage with literary concepts, past and present ideologies, new research models, religious beliefs, age-old narratives, and non-Western epistemes to develop into new knowledge forms. Such transformations are not residues of the irrational past but novel hybrids in the Latourian sense, wherein modern, premodern, secular, sacred, and cultural elements recombine to instantiate concepts that are at once recognizable and estranged. Put differently, parascience represents the epistemic borderlands, a tertium quid between scientific and nonscientific imaginaries where unorthodox ideas are perpetually reconceived, retheorized, and rejuvenated through contact with a multiplicity of discourses.

A familiar example will help highlight this point. When the Fox sisters initiated the first spiritualist movement via their rappings with the dead, newspapers described their mysterious communication as a form of "spiritual telegraph."[94] By associating an ancient concept (speaking with ghosts) with a new one (Samuel Morse's wondrous new invention), the Fox sisters' mediumship emerged as a novelty acceptable to a rational, technophilic audience. Similarly, in chapter 1, we will see how the outmoded psychical power of precognition was given new life as faster-than-light thought thanks to the emergence of quantum mechanics. The process of hybridizing premodern and modern intellectual histories is a theme that repeats throughout the twentieth and twenty-first centuries. Parascience thus names a place and a process for ideational renewal, which is to say it identifies a vital arena where pseudoscience continually resides and reproduces itself.

The means by which unproven scientific concepts are renewed for cultural consumption and cogitation is the parascientific process, or *parascientification*. Understanding how this process unfolds is critical for tracking the circuitous development of the paranormal mind over the last several decades as well as for recognizing the larger patterns of pseudoscientific knowledge production in modern times. Central to this overlooked pathway are *parascientific objects,* my general term for ideas rejected by mainstream science that consequently circulate the interdisciplinary expanse of parascience. This contrasts directly with traditional scientific objects entrenched in the Western scientific establishment. Parascientific objects need not be material things but can exist as phenomena, theories, or models incompatible within dominant paradigms. Such objects may not align within existing ideologies (e.g., the subliminal self), have produced little evidence (e.g., cold fusion), cannot be tested with existing technologies (e.g., UFOs), or some combination of these. What unites parascientific objects is their epistemic exclusion. They are not, by definition, scientific objects; they do not cohere with Western science, they do not belong to any prestigious academic agendas, and they do not produce discoveries most researchers would welcome. They are the pariahs of modern science. While parascientific objects are not legitimate areas for scientific inquiry, this does not foreclose their continued existence as subjects of cultural fascination. In fact, they often receive an influx of fresh analyses because they are liberated from the strictures of the scientific method. Parascientific objects, no longer bound by empiricism or institutional funding, are often radically reformulated by artists, philosophers, and citizen scientists who draw from any number of scientific or nonscientific discourses to reimagine them. In this way, disparaged subjects like telekinesis can be mythologized, Orientalized, parapsychologized, and biologized into something else—something new. Transformed over the years and reformulated by outsider bodies of knowledge, the refurbished parascientific object can bear little semblance to its original form and may even work its way back into mainstream culture and science.

Science and literature are two primary agents of parascientification. That science resides at the root of pseudoscientific knowledge production might seem confusing, even heretical, but a quick glance at scientific history indicates otherwise. Phrenology arose from the confluence of criminology, physiology, Darwinism, and heredity. The modern anti-vax movement has roots in Dr. Andrew Wakefield's now-retracted 1998 *Lancet* article. Telepathy hinged on telegraphy, psychology, and statistics. In all these cases, the application (or misapplication) of novel scientific developments allowed parascientific objects to be deliberated in new ways. In short, science can, and has, made the illogical highly logical. It does so by offering rational frameworks and tools for legitimizing ideas at the borders of science and nonscience. *Parascientific Revolutions* illustrates how this process of scientific authorization has played out time and time again with psi.

Perhaps more interesting is the role that literature plays in the theorization and dissemination of the paranormal. I contend that it is the primary means through which parascientific objects have persisted across twentieth-century culture. According to Susan Squier and Sherryl Vint, literature plays a central role in the ideological—and therefore material—development of a culture. In *Babies in Bottles* and *Liminal Lives,* Squier claims literary forms are "technologies" that directly affect the world. Rather than merely reflecting scientific or technological progress, literature is an agent in the consolidation and construction of scientific fields; by deploying institutional discourses, epistemologies, and practices, fiction "defines what is knowable and brings those objects into being."[95] For Squier, literary tools like narrative and metaphor shape the interpretation of scientific findings as well as the findings themselves. Literature consequently plays a tangible role in how science is socially understood, practiced, and disseminated. Vint makes a similar claim about science fiction in particular. In *Bodies of Tomorrow,* she argues that the genre is a privileged site for inquiry because it critiques the world while also altering it. "SF [science fiction] is part of the field of ideology," she writes, "and as such can work to not only comment on cultural politics of the current moment but also to intervene and change this moment."[96] Campbell also emphasized the material effects of science fiction through its socioethical function. By exploring the political and sociological consequences of scientific innovation (space travel, artificial intelligence, etc.), science fiction plays a vital role in helping society consider the morality of scientific progress.[97] Science fiction performs an ethical function that science often does not, and in doing so, it changes the world.

While Squier, Vint, and Campbell all emphasize the power of literature to change scientific ideology, I see literature as crucial for shaping scientific epistemology. It catalyzes the theoretical production, advancement, and distribution of the sciences. This is particularly significant for ideas orthodox science has dismissed. When outmoded concepts like phlogiston are abandoned by normal science, their empirical study comes to an end; this is why Kuhn predicted their quick and inevitable demise. But on transitioning into parascience, ideas like ESP can engage with texts that perform alternative modes of hypothesizing, experimentation, and analysis. Broadly speaking, literature can pick up where science left off. In fiction, for example, offbeat theories of precognitive mechanics are posited and debated. Characters become test subjects for experiments that science will not or cannot perform. Plot emerges as a dataset where the impacts of precognition are further analyzed, leading to deeper speculation on the nature of things. In sum, texts are not limited to Campbell's brand of ethical reflection, for they also participate in the epistemological processing of scientific ideas through culture. Parascientific objects like precognition are not valid subjects of scientific inquiry, but a Philip K. Dick novel can treat it as a genuine object of concern; in teasing out its physicalist characteristics and biological effects, the novel can simultaneously advance the concept while broadening its epistemic reach. In its own way, literature carries out the theoretical and experimental labor modern science has disavowed, which is to say it is a principal actor for distributing pseudoscientific ideas into culture.

Parascientification thus follows a predictable pattern. A scientific idea may fail to meet the empirical or paradigmatic standards of modern science and consequently falls out of practice. This concept may not vanish from scientific discourse as a matter of course, though. Instead, it can migrate to the epistemic fringes where it engages with mythic, philosophical, and other scientific (Western and non-Western) discourses, most often through the medium of literature. The end products of such intermingling are epistemic hybrids, amalgamations of modern and premodern intellectual histories forged into something new. As I argue in the following chapters, heterodoxy infused by novel scientific models and alternative ideological platforms can, and have, migrated back to the cultural center. Since it no longer depends on bad science but on a range of alternative knowledges, a parascientific object becomes, or at least seems to become, legitimate. This renewed authorization grants its circulation back into mass culture and scientific conversations, where it can find different audiences and champions to keep the formerly neglected idea in circulation. To be clear, parascientification is never teleological. It is an ongoing scrum of forces with no prescribed end point. Such rampant epistemic circulation means no aberrant scientific concept is ever truly sidelined. It may exist outside paradigmatic thought, but it survives by continually shape-shifting and adapting into other forms.

The cyclical quality of parascientific objects informs one of the "revolutionary" aspects of *Parascientific Revolutions.* There are several distinct meanings at play here. Parascience is revolutionary insofar as it identifies the intellectual recycling of older concepts into new formations. Parascience may take up clairvoyance and recontextualize it with cartography to make it scientifically permissible, or it may take up telekinesis and reanimate it via speculative fictions so it once again becomes relevant in contemporary discourse. Parascientific objects remain outside traditional science, but they do not fade away as predicted because they exist as re-evolutions, perpetually circling through fresh networks of myth, literature, and science as hybrids of the ancient past and technoscientific future. "Revolutions" also acknowledges *The Structure of Scientific Revolutions.* Kuhn reconceived the relationship between science and culture, and this study builds on his work by focusing on epistemic processes beyond paradigms, especially the role of literature in scientific poiesis. Last, I do see *Parascientific Revolutions* as a radical way of approaching literary STS. I view parascience as a new frontier in the discipline where the forgotten, marginalized, and discarded objects of scientific history can be recuperated as vital entities in their own right. The paranormal is often dismissed as unworthy of academic critique, but its bizarre history and longevity demand a thorough examination about how and why such scientific anomalies persist. This project is a call to engage with the very subjects that science demands we ignore.

Fortunately, *Parascientific Revolutions* is not alone in probing the boundaries of science, pseudoscience, and literature. Scholars like Michel Serres, Paul Feyerabend, and Jeffrey Kripal have all made the case for exploring these neglected areas. Like Latour, Serres has long challenged the traditional ideology—and valorization—of modern science as a system purged of its archaic and social elements. In his *Hermes* writings (1968-1974), Serres notes "the 'rational' is a tiny island of reality. . . . All knowledge is bordered by that about which we have no information."[98] A proper science studies must admit this limitation and address the irrational and pseudoscientific in order to achieve a more comprehensive epistemology of the real. Parascience also taps into the contrarian spirit of Feyerabend's "anarchic epistemology." In his agonistic manifesto against positivism, *Against Method* (1975), Feyerabend claims science requires a "new conceptual system . . . that clashes with the most carefully established observational results and confounds the most plausible theoretical principles, or to import from outside science, from religion, from mythology, from the ideas of incompetents, or the ramblings of madmen."[99] Anarchic epistemology readily accepts outsider perspectives as part of a more expansive interrogation of science, and in doing so, it sets the stage for parascience. Kripal similarly makes a compelling case for literature as the nexus of paranormal discourse. Like Partridge, he sees mass-market forms like comics, pulp fiction, and UFO narratives as occultural reservoirs for paranormal thought.[100] Collectively, these intellectual rebels highlight the nonscientific, the quasi-scientific, and the pseudoscientific as important elements of modern culture whose time for exploration is overdue. Theirs is a vision of interdisciplinary science and literary critique I wholly share.

Parascientific Implications

Parascientific Revolutions expands the literary bounds, scientific histories, and methodological frames of the paranormal in humanities criticism and science studies. In terms of literary history, this project reconfigures the major narrative around literature and the sciences of the mind. As noted earlier, most scholarship on fiction and the psychological sciences has focused on modernism. The rise of psychology, psychoanalysis, and psychical research during the late nineteenth and early twentieth centuries had a tremendous impact on aesthetic production, and the amount of research undertaking Freudian and Jamesian approaches to modern writers over the decades is considerable.[101] Consequently, modernism is widely viewed as the literary epoch most engaged with the mind as a scientific and aesthetic object. Occult criticism has reinforced the phenomenological primacy of modernism, with critics from George Harper Mills to John Bramble arguing for the centrality of spiritualism to modernist literature.[102] Such an intense focus on the early decades of the twentieth century has essentially periodized the paranormal mind within the academy as a modernist phenomenon. My study challenges this story line. *Parascientific Revolutions* builds on the existing body of occult modernism by tracking the paranormal mind deeper into the twentieth and twenty-first centuries and expanding the scientific and literary fields where it resides. By illustrating how fields like quantum physics, neurology, microbiology, systems theory, Cold War technologies, Chinese medicine, shamanism, and other areas helped rationalize the paranormal for new audiences, this book highlights how literature in the post-1945 era (i.e., after modernism) engaged with a multitude of scientific frameworks to transmit heterodox knowledge.

Such epistemic diversification naturally expands the literary genres where critics can seek out the paranormal. Some, like speculative fiction, are well-established areas. Others, like psychic manuals, have rarely, if ever, been examined in literary criticism before. Whether a traditional or untraditional genre, my entreaty is the same: only by following the paranormal's wayward paths can we truly understand its idiosyncrasies, evolutions, and cultural resonances. This book does not attempt to survey all the paranormal science fiction of the late twentieth century. What is does do is stake a claim for science fiction as an intellectual successor to modernism in terms of literary phenomenology. Investigating the human mind did not necessarily peak with the modern novel, and it can be found in posthuman narratives, postapocalyptic futures, and dystopian tales that reimagine how the mind works, what its powers are, and what it all means.

Perhaps more unusual is ethnic fiction as a burgeoning medium for parascientification. For much of the twentieth century, ethnic literature has been viewed through the lens of realism—a culture manual for conveying authentic experiences with racism, oppression, and cultural otherness.[103] This otherness frequently draws on the supernatural: ancestral ghosts, spirit quests, prophetic dreams, visions of the distant past, mediums, and medicine men are classic hallmarks of the genre.[104] If nothing else, these motifs suggest the paranormal is culturally ubiquitous. This study interprets ethnic fiction as a crucial space for exploring the historical and futural possibilities of the paranormal. In doing so, it produces several benefits. First, it expands the paranormal sciences to include non-Western epistemes. In recent years, there has been an explosion of academic interest in indigenous sciences, such as Sean Hsiang-lin Lei's *Neither Donkey nor Horse* (2014), Douglas Falen's *African Science* (2020), and Kelly McDonough's *Indigenous Science and Technology* (2024). Different cultures throughout history have developed different sciences and technologies to address their specific concerns, which in turn produces varying interpretations of nature's operations (including supernatural phenomena). Just as science fiction extrapolates Western sciences to theorize the paranormal, so too does multicultural fiction deploy ethnoscience to reinterpret the awesome potential of the mind.[105] In chapter 4, for example, I discuss how Amy Tan uses qi cosmology to naturalize both the existence of spirits and our ability to speak with them. While my study primarily focuses on Asian sciences, this approach also models how Middle Eastern, African, indigenous, and other global scientific traditions can be applied in future scholarship.

A second intervention is bringing the gender politics of science and parascience into sharp focus. As feminist STS scholars have long observed, modern science is an historically androgenic endeavor that has systematically excluded women from its professional ranks, pathologized female biology, and naturalized gender stereotypes.[106] Puglionesi sees psychical research as an exception: thanks to the importance of female mediums to the discipline, it has accepted and appreciated women's labor in ways few other mainstream sciences ever do.[107] With border sciences, "everyone involved was in some way marginal," which leveled the playing field.[108] The women we encounter in paranormal ethnic fiction, like psychic explorers Kwan Li and Nao Yasutani, continue along a path blazed by Blavatsky, Annie Besant, and other forceful women in occult science. Ethnic fiction showcases parascience as a topsy-turvy arena where science *and* gender politics get turned upside down.

Third, ethnic fiction estranges the paranormal itself. The paranormal has been epistemologically marginalized because it stands outside Western science. However, if we follow the lead of scholars like Sandra Harding and view Western science as just one part of the larger world of multicultural science, then the paranormal is no longer intrinsically abnormal.[109] Furthermore, if we can accept decolonial claims that we exist in a *pluriverse* of global knowledge rather than a *uni-verse* of European rationalism, then the paranormal mind in ethnic literature emerges as an effective site for understanding how different cultural groups understand the natural and supernatural worlds.[110] In *Under Representation,* David Lloyd describes subaltern narratives not as progressive aesthetics but as disruptive ones; minority viewpoints discombobulate standard Western logic. Paranormal ethnic fiction unsettles the meaning of science, consciousness, time, space, death, and ghosts, and this has serious implications for the paranormal from a global perspective.

Within the interdisciplinary field of science studies, *Parascientific Revolutions* makes several interventions around heterodoxy, the relationship between science and pseudoscience, and epistemic production. This study underscores the significance of esoteric science as a subject of serious inquiry rather than an historical dead end to explain away. The history, sociology, and philosophy of science have been dominated by facticity. In the 1930s, Karl Popper famously demarcated science from nonscience via falsifiability,[111] with only the former being a matter of concern for science studies.[112] In the second half of the twentieth century, Kuhn's paradigm model emphasized the social production of facts from their emergence to their institutionalization. In the post-Kuhnian era, refinements like Imre Lakatos's "research program" model acknowledge the existence of scientific anomalies while still emphasizing fact generation. Even Latour, who has done so much for deconstructing the mythology of Western science, focused on black boxes and scientific truths in his early work.[113] Such a long-standing focus on facticity is important, but it also obscures the material impact of the not quite true, the esoteric, the unexplainable, and, yes, the supernatural. We cannot deny that scientific heterodoxies exist in contemporary society, yet we still do not have a philosophy of science to properly account for them. My goal is not to prove the validity of ideas like morphic resonance; rather, it is to acknowledge their presence and to investigate how they persist. This analysis brings marginal science into the fold of modern STS.

As a concept, parascience illuminates the nature of pseudoscientific knowledge production outside dominant paradigms. Kuhn's paradigm highlighted the cultural production of scientific facts, and Gieryn's boundary work reinforced the roles that media, government, and educational institutions play in policing scientificity. This book follows in their spirit, with the twist that it tracks pseudoscientific circulation, evolution, and dissemination. This approach draws several disciplines into the process of epistemic production usually disregarded in STS, including mass-market literature, occultism, and non-Western science. The overarching political intervention of parascience is centralizing the marginal: while modern science discounts occult, indigenous, female, and outsider perspectives, parascience demands we examine them anew.

Last, parascience complicates the relations between science and pseudoscience. As generally understood, the latter is a misapplication of the former; conversely, the former is deployed to disprove the latter. *Parascientific Revolutions* challenges this binary. As it draws on a host of historical, literary, and scientific examples, this project reveals time and again that science itself plays a role in the perpetuation of pseudoscientific ideas. By providing new models for interpreting the natural world, science conjures fresh logics for ideas like ESP that fail to cohere within dominant paradigms. This leads to two key insights. First, it suggests that more science cannot stamp out pseudoscience. According to the knowledge deficit model of science communication, the public makes better (i.e., more rational) decisions by gaining scientific knowledge.[114] The assumption is that ubiquitous STEM education will eradicate pseudoscientific belief as a matter of course. Clearly this has not happened, and the revitalization of pseudoscience from science itself explains why. Second, this study suggests that the survival of pseudoscientific discourse is not an aberration; it is, in fact, endemic to the structure to which it belongs, the parascientific space that, at all points, surrounds and feeds off scientific inquiry. Putting these two observations together, we might say pseudoscience is part and parcel of the scientific world. The following chapters use the paranormal mind to demonstrate how and why this keeps happening.

Chapter Summaries

Each of the five body chapters focuses on a different power of the paranormal mind: precognition (chapter 1), telekinesis (chapter 2), clairvoyance (chapter 3), spectral communication (chapter 4), and telepathy (chapter 5). The first three of these explore how literature has engaged with different emerging or marginalized Western sciences to reimagine psi. The last two show how multiethnic literature has drawn from non-Western sciences to reinvigorate the paranormal mind for modern culture. Throughout, there is a forward sweep in time from the mid-twentieth century to the early twenty-first century, so by the conclusion, we can reconcile our idiosyncratic past with the parascientific future.

Chapter 1 focuses on a hypothetical thought particle of precognition that emerged through the confluence of psychical research and 1950s-era quantum physics. In 1965, psychical researcher Adrian Dobbs proposed the existence of a faster-than-light thought particle—the psitron—as a fundamental unit of consciousness and the master key for explaining divination. Drawing from mid-twentieth-century quantum physics, neurology, and psychology textbooks, he developed a remarkable physicalist model of cognition to explain humans' ability to see the future. No evidence for this particle has ever been found, but the psitron nevertheless remains a parascientific object firmly embedded within our cultural discourse. By analyzing Philip K. Dick's *VALIS* and Alan Moore and Dave Gibbon's *Watchmen* as two texts advancing psitron discourse, I explore how unproven scientific ideas can adopt new forms thriving outside paradigmatic science.

Chapter 2 examines the intellectual development of planetary consciousness and telekinesis through systems theory and science fiction. First popularized in the 1970s by scientists James Lovelock and Lynn Margulis, the Gaia hypothesis interprets Earth as a living entity possessing both agency and intelligence. This chapter considers Gaia as an overlooked manifestation of the paranormal mind, a posthuman superorganism endowed with telekinetic control over life itself. It begins by tracking the rise of the "living Earth" in ancient Greece to its decline during the scientific revolution and its eventual resurgence in New Age thought. Fundamental to Gaia's second life is the rise of systems biology, with theorists like Ludwig von Bertalanffy, Humberto Maturana, and Ilya Prigogine attesting to the complexity of nonbiological systems and New Age critics exploring the interconnectedness of all things. By reading the Gaian works of Lovelock and Margulis alongside these scientists, this chapter shows how living matter and disembodied cognition have emerged as system properties of the Earth. These developments are interwoven with the apocalyptic fiction (*Vitals* and *Darwin's Radio*) of Greg Bear, who amalgamates systems theory and microbiology to reinterpret telekinesis on a global scale.

Chapter 3 examines the literary, scientific, and political evolution of clairvoyance during the 1970s to 1990s, when the U.S. government established the psychic spy program known as Project Stargate. For over two decades, the United States developed a series of clandestine programs that trained psychics to telepathically spy on Cold War targets using a novel mode of clairvoyance known as *remote viewing*. The creator of remote viewing and the intellectual leader of Project Stargate was Ingo Swann, a prolific writer of parapsychological fiction and nonfiction. Drawing from archival research of Swann's personal files, correspondence, and declassified CIA manuals, this chapter reveals how Swann altered the trajectory of twentieth-century psychical research by transforming it from an outsider science into the ultimate insider geopolitical technology. It begins with Swann's offbeat biography and delves into his government work with defense contractor Stanford Research Institute, where he fused clairvoyance with cartography to conjure remote viewing. This chapter then explores his retheorization of the paranormal mind through works like *Starfire, The Nostradamus Factor,* and *Natural ESP.* By delving into new literary areas like psychic manuals and intelligence reports, I illustrate the power of government, science fiction, and Cold War ideology in advancing paranormal science.

The last two body chapters in *Parascientific Revolutions* shift from Western science toward non-Western science in the ongoing development of the paranormal. In the process, I make the case for ethnic fiction as a critical frontier for understanding how contemporary culture now engages with the paranormal. Chapter 4 examines the recurring trope of the spirit medium in Asian American literature to see how non-Western scientific frameworks reformulate supernatural beings and cognition. In recent years, science studies and postcolonial studies have reconceived traditional knowledge systems (e.g., native science, ethnoscience, traditional medicine) as valid sciences distinct from the Western tradition. Drawing from the decolonial epistemologies of Sandra Harding, Bernd Reiter, Walter Mignolo, and Jee Loo Liu, I analyze two canonical works by Amy Tan and Nora Okja Keller to highlight the ways Eastern metaphysics reconceptualizes the ability to communicate with ghosts. While spirit mediums in Asian American fiction have often been criticized as Orientalist tropes, I reinterpret them as characters grounded in the alternative approaches of non-Western science. In Tan's *The Hundred Secret Senses*, I focus on controversial spirit medium Kwan Li and how her interpellation within Chinese metaphysics—and specifically her ability to see qi—authorizes otherworldly powers of remembering past lives and speaking with ghosts. Similarly, in Keller's *Comfort Woman*, I highlight Akiko Bradley as a medium whose immersion in Korean shamanism affords her access to spectral vision and communication. These counterreadings of ghosts qua ghosts (instead of ghosts as literary symbols) are couched in what Bernd Reiter calls "mosaic epistemology"—an approach to knowledge respectful of cultural frameworks beyond positivistic science.[115] This chapter positions Kwan and Akiko as female psychical researchers, investigators who are empowered in ways unavailable to many women in orthodox science. It also showcases how multicultural science and fiction can drive the ongoing evolution of the paranormal mind while expanding critical approaches to the supernatural in non-Western contexts.

Chapter 5 explores how speculative ethnic fiction deploys traditional knowledge to envision bizarre new modes of telepathy. Building from the global science studies approach of the previous chapter, I analyze Ruth Ozeki's *A Tale for the Time Being* and Sesshu Foster's *Atomik Aztex* to show how non-Western chronologies (sciences of time) renew mind-to-mind communication for the twenty-first century. In *Tale*, Zen Buddhism serves as an alternative scientific system allowing crazed manipulations of space-time, including moving backward, forward, and even sideways through time. As such, it naturalizes occult phenomena like telepathy, time travel, and even astral travel. In *Atomik Aztex*, Mesoamerican circular time becomes the scientific backdrop for the protagonist, Zenzontli, to communicate with himself across the multiverse. The result is a powerful new mode of telepathy never seen before in either paranormal fiction or nonfiction. This chapter affirms speculative ethnic fiction as an important space in paranormal literature and theory today.

Chapter 6 is a brief conclusion that examines Daryl Bem's controversial ESP experiments at Cornell University en route to analyzing the durability of paranormal thinking in modern science and culture. Despite its outsider status for the better part of a century, the paranormal mind shows few signs of fully disappearing. While the scientific history of the paranormal might seem dubious, its cultural impact is not. In fact, when looking across cinema, television, literature, and the internet, it is increasingly clear that paranormal discourse is here to stay. This chapter argues that recognizing the lively interrelations across science and art is crucial for understanding the longevity, flow, and power of heterodoxy in contemporary culture.

In *The Troubadour of Knowledge,* Michel Serres crafts a wonderful analogy when describing knowledge as an ellipse with two focal points. One focal point is the rational (modern science and knowledge) and the other is the irrational (nonscience and premodern knowledge). Post-Enlightenment culture has valorized the former while ignoring the latter, but Serres argues that the true center resides in the third space between the two points.[116] "Legend, myth, history, philosophy, and pure science have common borders over which a unitary schema builds bridges," he writes.[117] He goes on to say the third space surrounding the point of reason must be investigated thoroughly, and that we require an epistemology dedicated to understanding "obscure, dark, non-evident knowledge."[118] *Parascientific Revolutions* is aligned with Serres's thinking, a pivot toward that third space immediately beyond establishment science where so much vital cultural production is happening. At an historical moment when misinformation, disinformation, and conspiracy theories cloud our collective vision, we cannot pretend heterodox science does not exist. Nor can we pretend it will magically vanish. It is far better, I think, to open our eyes and clearly see the paths through which marginalized ideas have evolved. The paranormal mind is an oddity, to be sure, but an instructive one, and through it, we can build meaningful bridges across the chasms of rational and irrational thought.

1 TOWARD A (META)PHYSICS OF PRECOGNITION

The Science and Aesthetics of the Psitron

THE APOCALYPSE COULD HAVE BEEN AVOIDED. In Alan Moore's classic graphic novel *Watchmen,* an alien the size of a corporate tower teleports into midtown Manhattan on November 2, 1985, and unleashes a massive psychic shock wave. In a single stroke, half the city's population is annihilated. All around Madison Square Garden, the bodies of men, women, and children lie in heaps, brains liquidated, faces contorted in agony. Shattered glass and wrecked cars litter the streets. Blood runs in torrents down Seventh Avenue. Beyond those immediately killed in New York, millions more around the world must cope with the phantasmagoric nightmares unleashed by the telepathic blast. The cruelest twist of this Hiroshima-level event is that Dr. Manhattan should have seen it coming. As the most powerful superhero in the comic and the master of several paranormal powers, including precognition, surely he, of all people, could have foreseen and preempted this catastrophe. One immediate question that comes to both his and the audience's mind is, how did this happen?

The answer, curiously enough, is tachyons. In modern physics, the tachyon is a hypothetical particle that travels faster than light. Dr. Manhattan's ability to see the future is connected to this particle, and he quickly discovers that an excess of tachyons in Earth's atmosphere has clouded his second sight. The unfortunate result of this mental static is the psychical holocaust in the streets of New York.[1] While tachyons might seem like a lazy plot device for bypassing Dr. Manhattan's omniscience, they actually perform a far more significant function in the text—as well as in scientific and literary history. The very mention of tachyons in the comic steers Dr. Manhattan's extraordinary powers from the realm of magic toward science. More importantly, by identifying the tachyon as a physical particle implicated in precognition, Moore yokes his groundbreaking text with one of the strangest—and strangely enduring—phenomenological models of the late twentieth century: the psitron. In 1965, a SPR member named Adrian Dobbs proposed the existence of a faster-than-light "thought particle" that emanated from future events and could impart prescient information on striking the brain. Drawing from the latest developments in psychology, neurology, and physics, Dobbs contended that this hypothetical particle—the psitron—constituted nothing less than a fundamental unit of consciousness and was the master key for explaining precognition. In his writings, he called for the field of psychical research to establish a new experimental program dedicated to investigating the psitron, confirming its existence, and ascertaining its marvelous properties.

It is worth acknowledging from the outset that this ambitious research agenda never transpired as originally conceived. Dobbs died from heart failure shortly after the publication of his psitron theory. As far as I know, no other parapsychologist in the following decades ever discovered evidence for his particle, and it never found purchase as an ongoing concern within the larger arena of psychical research. In this regard, the psitron could be considered a dud—just another bad scientific idea destined for the proverbial dustbin of history. Yet this is hardly the case because the psitron continues to persist throughout scientific and literary discourse. As we shall soon see, the idea of faster-than-light particles as the basis for precognition somehow endures across the disparate worlds of speculative fiction, comics, and marginal science. It may vary depending on the writer or the researcher, but it always seems to linger: Particles from the future! Faster than anything else in the universe! For this reason, I do not view Dobbs's psitron as a failed concept. Even if its origins and history are largely forgotten today, it has remarkably emerged as one of the most durable pseudoscientific ideas in contemporary culture, one that continues to wield influence across technoscientific and humanistic imaginaries.

This chapter provides an intellectual history of the psitron, and through it the parascientification of precognition. This thought particle may or may not exist, but my aim is to follow its trajectory as a parascientific object par excellence to illustrate how heterodox science is not always vanquished by empirical approaches to facticity but can instead lead a robust second life as an ideational entity traversing scientific and cultural thought. The psitron is both archetypal and unique in this respect. Any number of paranormal phenomena can exist as parascientific objects, as other chapters in this book will attest. But the psitron is special because it is an actual particle; it is the fundamental unit of revelation. It is a singular and material instantiation of a parascientific object, and by tracking its winding path across the twentieth century, we can understand how new sciences and literatures can modify precognition from a spiritualistic archaism into bleeding-edge physics.

In the realm of science, the psitron amalgamates three concepts from vastly different fields: thought particles from early psychology (the psychon); mind–brain interactions from neurology (dualist-interactionism); and particles of imaginary mass from quantum physics (meta-particles and tachyons). Each of these models is analyzed in turn. Physics will play an outsize role across this narrative because it authorizes the immaterial nature of fundamental particles as well as the lengthy time spans required to verify their existence. The prestige of modern physics further lends the psitron a certain cachet missing from parapsychology alone. Most importantly, quantum physics offers the counterintuitive logic that our universe is fundamentally occult. According to modern physics, the everyday world we see, hear, and touch only represents a sliver of a larger reality constituted by unseen particles and exotic matter, all of which suggest that physical existence resides atop a deeper metaphysical foundation. The psitron's journey thus foreshadows what we shall observe time and again over the course of the twentieth century, namely that the decline of psychical research invites thinkers to use alternative sciences to explain how the paranormal mind *really* works.

Literature plays an equally significant role in this endeavor. The idea of faster-than-light particles was quickly taken up in science fiction, and the following pages look specifically at Philip K. Dick's *VALIS* and Alan Moore and David Gibbons's *Watchmen*. Both are noteworthy for their production of the psitron as a parascientific object—not only in its dissemination to broader publics but also its conceptual formation. Each text builds on Dobbs's research by further hypothesizing the psitron's scientific characteristics, mechanisms, and significance. They hence provide robust examples for the role fiction plays in parascientification. For objects that exceed the ideological and empirical bounds of science, fiction is often the only place where they can be further developed and observed. Put differently, science cannot experiment on hypothetical particles, but literature can. As such, fiction serves as both a theoretical and laboratory space for the intellectual advancement of the pseudoscientific. Roger Luckhurst claims that science fiction is integral to pseudoscience because it represents a "cultural record" of marginal science; it is a testimonial to weird and wrong ideas before they are erased by the forward march of science.[2] I agree but would go even further: speculative fiction plays a seminal role in *producing* the pseudoscientific, which is to say in powering the theorization, distribution, and popularization of parascientific objects.

Contra traditional scientific ideologies regarding pseudoscience as relics of the premodern past, the psitron emerges here as a hybrid of contemporary science and its engagement with literature. Parascientific objects metamorphose via their interactions with diffuse fields of knowledge, and the fluctuations of the psitron through parapsychology, neurology, physics, and science fiction demonstrate how precognition can be rationalized through modern science and culture. Despite its obvious shortcomings, Dobbs's work is by no means an intellectual dead end. Rather, I see something revolutionary in his thinking, for his radically original theory of consciousness would unleash one of the most provocative mental models of the twentieth century.

Precognition and Psychons

Precognition, or "knowledge of the future, extending beyond the scope of our ordinary inference," has a tenuous history, even within the field of psychical research.[3] Early psychical researchers focused their attention on telepathy, which made more sense to them because it involved one mind communicating with another mind within the same space-time frame. Retrocognition, or "knowledge of the past, extending back beyond the reach of our ordinary memory," was trickier, although still logical because the historical event had already occurred.[4] The same could not be said for precognition, which involved insight into events that had not yet transpired. In the penultimate chapter of *Human Personality and Its Survival of Bodily Death,* preeminent SPR theorist Frederic Myers argued that in a world where gravity exists but cannot be explained, humanity must remain open to vast mysteries in space-time cognition.[5] According to him, there were three possible mechanisms for seeing the future. The most straightforward was "a heightening of inner sensations to a point where the future history of the organisation can be guessed or divined with unusual distinctness";[6] the subliminal self is powerful enough to assemble the flood of present data so clearly as to ascertain the future. Second, he wrote, "It is conceivable that predictions of these and other types may be communicated by disembodied spirits."[7] The minds of the deceased exist in perpetuity, and such entities could transmit futural knowledge to the subliminal selves of the living. Third was the possibility that all humans are part of a larger world-soul, "thinking thus of the Universe as no more congeries of individual experiences, but as a plenum of infinite knowledge of which all souls form part."[8] Knowledge of the future is perceived as ordinary because it is part of a universal mind cognizant of all things.

While Myers embraced the metaphysical possibilities of precognition, other psychical researchers disagreed. As Gardner Murphy noted, "To make contact with that which does not exist yet is, for many, a contradiction in terms, a philosophical paradox, an outrage."[9] In "The Notion of Precognition," C. D. Broad rejected foresight on purely logical grounds. "*Until* an event, which will answer in the present experience in such a way as to make that experience count as a pre-perception of it, shall happen, *nothing can be caused by it*," he wrote. "Therefore that event *cannot* have contributed, either directly or through a causal chain of intermediate events, to cause the experience which is said to have been a pre-perception of it."[10] Modern society largely subscribes to Broad's logic, which explains why precognition has no place in any traditional scientific paradigm. Humankind simply cannot see things that have not happened yet. Period. This is not just a philosophical tenet but a physical one as well because effects cannot precede causes. Nevertheless, both philosophy and physics can be used to naturalize this unnatural phenomenon. The psitron is an excellent case study because it illuminates how a scientific anomaly can engage with new sciences and literatures to reemerge as a perfectly rational entity.

As the proposed fundamental unit of the paranormal mind, the psitron did not arise fully formed in the mind of Adrian Dobbs in 1965 but rather as the synergistic product of a scientific progression that began in the late nineteenth century. More specifically, it begins with thought particles: the search for "atoms" of cognition, the ideology behind them, and their theoretical promise. We must therefore start with Melissa Littlefield's insightful work on the psychon. The very notion of a fundamental unit of consciousness was a real-life novum that emerged rhizomatically across neuroscience, psychology, and science fiction in the early decades of the twentieth century as a response to scientific legitimization.[11] At the turn of the century, psychology was a fledging discipline wary of its philosophical origins and unsavory connections to psychical research. In his 1892 essay "A Plea for Psychology as a Natural Science," William James wrote, "We have here an immense opening on which a stable phenomenal science must someday appear. We needn't pretend that we have the science already, but we can cheer others on working for its future, and clear metaphysical entanglements from their path."[12] James stressed the elimination of "metaphysical entanglements" because the material practices of hard science imparted greater cultural authority. One of most attractive means for achieving epistemic legitimization was a *disciplinary unit of matter.* The psychon, also known as the *psychikon* and the *psychome,* would soon emerge as this fundamental particle: it described "the least psychic event or unit of matter related to thought—but not fully encompassed by—nerves, synapses, and reflect arcs."[13] Serving an analogous role to the photon in physics or the electron in chemistry, the psychon used the logical atomism of Bertrand Russell during the early twentieth century to become a disciplinary cornerstone on which psychology could erect itself as a thoroughly *material* science.[14]

The psychon's genealogy can be traced back to the 1895 essay "Project for a Scientific Psychology," where Sigmund Freud speculated on locating "specific material particles" to ground the myriad processes of the mind.[15] In doing so, he hoped to find a psychological analog to the recently discovered neuron, which had helped legitimize neurology as a new subfield within biology.[16] Although Freud eventually abandoned his search for thought particles in favor of developing psychoanalysis, other researchers would take up the task. As a concept situated more in theoretical inquiry than experimental praxis, the psychon remained fundamentally fluid. For many, the psychon served a symbolic purpose. For example, in "Hypnotism, or Suggestions and Psychotherapy" (1906), Auguste Forel described the "psychome" as a conceptual placeholder for a basic psychic unit.[17] In 1943, Warren McCulloch argued if the neuron was the material embodiment of thought, then the psychon should be considered its "semiotic" counterpart.[18]

Others viewed the psychon in more concrete terms. In *Emotions of Normal People* (1928), William Marston described the psychon as "the totality of energy generated within the junctional tissue between any two neurons," which is to say an electromagnetic presence capable of effecting change on matter.[19] In the 1935 short story "The Ideal," pulp science fiction writer Stanley G. Weinbaum introduced a machine, the Idealizer, that extracts psychons from the brain and projects them on a screen. "I will make your thoughts visible!" the brilliant Dr. van Manderpootz explains. "The psychons of the mind are the same as those of any other mind, just as electrons are identical whether for gold or iron."[20] In Paul Ernst's "From the Wells of the Brain" (1933), a fictional professor, Wheeler, similarly develops a machine that converts the "electrical energy" of the mind into thought particles that can be materialized in the actual world. As Littlefield notes, such fictional and nonfictional psychons capture the psychological ideology that mentation exists in an atomistic form ripe for study and even harvesting.[21]

The psychon quickly proved attractive to psychical researchers as a mechanism for explaining occult phenomena. In *Thought Transference* (1946), Whately Carington used *psychon* as a generic term denoting any constituent sense element of the mind.[22] For him, the mind was "the whole aggregate of all existing psychons," and moreover, autonomous subgroups of psychons linked through "forces of assertion" can exist independently of the brain itself.[23] Frederic Myers believed the subliminal self could survive the passing of the material body, and here Carington essentially layered twentieth-century atomism over Victorian spiritualism by reconceptualizing the immortal soul as a collection of mental atoms. "What is it that survives when the body perishes?" Carington rhetorically asks his readers. "It is the psychon system."[24]

The psychon also played a prominent role in the work of L. L. Vasiliev, Russia's foremost parapsychologist. While his groundbreaking text *Experiments in Mental Suggestion* was only published in English in 1963, the book details experiments going back to the 1920s and 1930s during the heyday of psychological atomism. Vasiliev's own investigations focused on the telepatheme, a highly charged particle of thought that traveled from a psychic agent to a percipient. The primary goal of Vasiliev's research was to determine if telepathy was a form of electromagnetic radiation (such as a microwave or an X-ray), as previously proposed by psychiatrist V. M. Bekhterev,[25] and his experiments involved sending telepathic messages over extraordinary distances (e.g., Leningrad to Sevastopol) while using Faraday cages designed to shield percipients from electrostatic influences. Within the parapsychology community, his research proved that the telepatheme did not follow the known rules of electromagnetism, and thus telepathy represented a novel form of energy transmission unbound from the classic laws of physics.

As this brief overview illustrates, theorization over a fundamental unit of consciousness was well underway across the psychological sciences by the time Dobbs developed his own idiosyncratic version in the 1960s. More importantly, it foreshadows how a parascientific object like precognition can gain legitimacy by drawing from the authority, frameworks, and suppositions of other sciences. As William James made clear, metaphysics was insufficient to fulfill psychology's destiny as a proper science in an increasingly positivistic culture; to achieve that lofty goal, a material practice (or particle) was deemed necessary. Likewise, as Carington and Vasiliev illustrate, the paranormal enhances its ideological appeal when couched in the theories and rhetoric of physics. If psychical researchers could someday unlock the psychon/telepatheme just as physicists had deciphered the properties of the electron, the great mystery of human consciousness could at last be unraveled. At the same time, the indeterminacy and uncertainty authorized under modern physics offered tremendous leeway. That psychologists never settled on the materiality or immateriality of the psychon is telling. Alternatively described as synaptic, linguistic, electrical, and semiotic, the psychon was a wondrously polyvalent concept leveraging the fluidity of space, time, and energy spreading across scientific discourse.

John Eccles and the Brain Antenna

In addition to psychology, Dobbs's psitron theory drew from contemporary neurology, specifically a model of mind developed by Nobel laureate John Eccles. Born in 1903 in Melbourne, Australia, Eccles is best known today for his work on the synapse in the peripheral nervous system. However, the animating force behind his life's work was mind–body interaction. This long-running Cartesian problem revolves around a fundamental question: how do metaphysical states of thought relate to the physical processes of the brain? Most modern scientists and philosophers are monists, who assume that the mind arises as the phenomenon, or epiphenomenon, of normal brain function. In contrast, Eccles was a dualist who viewed the mind and brain as distinct entities. Much of his professional labor was dedicated to understanding how the former controlled the latter, and in his 1953 text *The Neurophysiological Basis of Mind,* he developed an unorthodox theory called dualistic interactionism that laid the groundwork for Dobbs's precognitive research.

Eccles's model focuses on the physiology of the mind–brain interface where metaphysical thought directly engages with the physical mechanisms of the body. Dualist-interactionism hinges on the brain having "the function of a 'detector' that has the sensitivity of a different kind and order from that of any physical instrument."[26] Eccles attributes such radical responsiveness to dendrons, which are the branch-like projections located on every neuron that receive electrochemical stimuli from other cells. Modern neurologists agree that these structures play a vital role in the reception and transmission of neural information, but Eccles argued that dendrons not only detect electrochemical transmissions (material) but also particles of "mental will" (immaterial).[27] He is referring, of course, to psychons.[28] In *Basis of Mind,* he describes particles of mentation so small that their motion is governed by the probability physics of the Heisenberg uncertainty principle. Although the subatomic size of such quanta would normally invalidate their effect on matter, Eccles contends that dendrons are sensitive enough to receive information from them, instigate a neural cascade, and prompt a bodily reaction.[29] Under dualist-interactionism, the mind is an autonomous entity that can issue its will to the brain via psychons. Indeed, it is psychon–dendron interaction that allows us to "see colors, hear sounds, and experience the existence of the body."[30] In contradistinction from a monist model, where the physical brain produces the metaphysical mind, dualist-interactionism presupposes the latter exists independently and delivers commands to the former.

It is worth pausing to examine some of the extraordinary claims housed within Eccles's neurophysiological model. First is the argument that particles of mental will are so infinitesimal that they represent a category of matter existing "outside physics."[31] Second is the claim that the human brain can receive and act on subatomic stimuli no other physical technology can register, which is to say it is the most sensitive antennae on the planet.[32] The brain is undoubtedly a responsive instrument, but subatomic particles governed by uncertainty physics (roughly the size of Planck's constant, or 6.63×10^{-34} joule seconds) are so small that they typically pass straight through matter. Yet even these two postulates are tame compared to a third assertion: that dendrons can receive psychons originating from *outside* the individual brain.

In the last chapter of *Basis of Mind,* Eccles observes that dualist-interactionism aligns with much of the ongoing telepathic research of J. B. Rhine and R. H. Thouless by proffering a physiological mechanism explaining mind-to-mind communication. As a subatomic particle operating beyond physics, a psychon originating from the mind of one person could conceivably escape the confines of that specific brain, travel across geographic space, and interact with the dendrons of another brain. Eccles confirms this scenario: "It will be agreed with Rhine (1948) that, if the so-called psi-capacities (psycho-kinesis and extrasensory perception) exist, they provide evidence of slight and irregular effects which may be similar to the effects which have here been postulated for brain–mind liaison, where they would occur in highly developed form."[33] Though couched in the rhetoric of speculation, Eccles contends that individuals with highly developed brains, like mediums, could communicate with others via interpersonal psychons. Such persons could project psychons beyond their own brains; alternatively, they might possess powerful dendrons capable of capturing the faintest of escaped psychons. In either case, Eccles suggests that the same neurophysiological processes giving rise to consciousness and will within one brain holds true *between* two brains as well. Dualist-interactionism thus provides a theoretical and biological framework for explaining paranormal phenomena.

Eccles's cognitive model is notable for several reasons—including reprioritizing the mind over the body, rescaling the physics of thought, and embracing both atomistic psychology and contemporary parapsychology—but its greatest parascientific contribution is developing the logic for telepathic thought reception. By providing a concrete locus for physical/metaphysical exchange, the human brain emerges from *Basis of Mind* as a fantastical antenna where immaterial thought is captured and converted into material action. In this respect, Eccles's dendron emerges as a biological version of Dr. van Manderpootz's Idealizer: the ultimate physical/metaphysical mental conversion machine.

Tachyons and Other Impossibility Physics

The third and most important intellectual pillar of Dobbs's psitron model is the theorization and discovery of antimatter, exotic matter, and other counterintuitive particles during the middle of the twentieth century.[34] More than either psychology or neurology, quantum mechanics would naturalize the occultism of the natural world. By demonstrating that the subatomic particles constituting all things were fundamentally marvelous, physics conjured the perfect ideological conditions to authorize supernaturalism within the mind.

The cultural effect of quantum mechanics on the early decades of the twentieth century is well established. As Allen Thiher has argued, the triumph of Newtonian physics in explaining everything from pulleys to planetary motion had produced a widespread belief by the nineteenth century that the world could, and would, be fully known.[35] This positivistic certitude was punctured by the emergence of quantum mechanics, which describes physics at the smallest scales and highest velocities. Einstein's Annus Mirabilis papers equated mass and energy ($E = mc^2$), determined subjectivity over objectivity (special relativity), and described the particle–wave duality of light (photoelectric effect). Niels Bohr's work on complementarity placed fundamental limits on our ability to observe and measure the natural world, an insight Werner Heisenberg formalized with his uncertainty principle. Within a few decades, relativity, uncertainty, and fragmentation were understood as the *real* ontology of the universe, forcing society to perceive its world anew. Such a claim may seem hyperbolic, but historians and philosophers alike cite the quantum revolution as a critical turning point in modern art and culture, affecting everything from Cubist paintings to our cultural understanding of time to narrative itself.[36]

The effects of midcentury theoretical physics have been less scrutinized, however. While the theoretical advancements in 1930s to 1960s quantum mechanics remain obscure to the public, they are critical to the intellectual development of a field like psychical research. During this period, the positron, the antiproton, and the neutrino were discovered in rapid succession, and their quirky characteristics opened new worlds of paranormal possibility.

The positron, or antielectron, looms large in psychical history because it confirms the existence of a shadow world beyond our known reality. Paul Dirac first postulated the concept of antielectrons in 1928 when he noticed that a relativistic interpretation of the Schrödinger wave equation predicted electrons with both positive and negative energy. The latter particle, which has the same mass as an electron (−1e) but carries the opposite charge (+1e), had never been observed before. Incredibly, Carl David Anderson would discover the antielectron just one year later, when he found that cosmic rays passing through a cloud chamber and a lead plate yielded a new particle aligning perfectly with Dirac's hypothesis. The discovery of the positron was a momentous event in physical history because it verified the existence of a new class of matter known as antimatter. Today, antimatter is considered relatively common, with every known element having an antimatter doppelgänger with an opposite charge to its positive-energy counterpart. At the time of its discovery, though, antimatter was revolutionary because it implied that what we had long accepted to be the totality of material world, which is to say matter, only represented half of what actually existed. The universe, as we knew it, was *fundamentally occluded.* According to Alex Owen, what united many occult groups during the fin de siècle was the belief that behind everyday reality existed a more complex and fantastical ultimate reality.[37] By verifying the existence of antimatter, physicists confirmed the universe was in fact not quite what it appears. Arthur Eddington puts it best: "In the world of physics we watch a shadowgraph performance of real life. The shadow of my elbow rest on the shadow-table as the shadow-ink flows over the shadow paper. . . . The frank realization that physical science is concerned with a world of shadows is one of the most significant of recent advances."[38] That physicists dealt in the world of "shadows" is not insignificant, for now they served as necromancers calling forth dark energy from the ether—not as magic or spirits, but as physical truths.

The 1950s proved a banner decade for occult physics. The second antimatter particle, the antiproton, was discovered in 1955 by Emilio Segrè and Owen Chamberlain, twenty-two years after Dirac postulated its existence at his 1933 Nobel Prize lecture. One year later, Clyde Cowan and his collaborators (1956) announced in *Science* that they had discovered evidence of the neutrino, an electrically neutral particle with near-zero mass that had evaded detection for a quarter century.[39] In *The Roots of Coincidence,* journalist (and paranormal enthusiast) Arthur Koestler dubbed neutrinos "ghost" particles because, for all intents and purposes, they did not exist.[40] Quoting science writer Martin Gardner, he writes, "As you read this sentence, billions of neutrinos, coming from the sun and other stars, perhaps from other galaxies, are streaming through your skull and brain."[41] For Koestler, the discovery of the neutrino illustrated just how little we know of the universe. If neutrinos are all around us, passing through us at the speed of light, completely undetected, what other ghost particles might exist in this, our occult universe? What are the properties of unseen matter? What natural phenomena might they explain? If nothing else, the succession of major discoveries demonstrated ghost particles were both real and ubiquitous. The unknown world was clearly vaster and weirder than the known world, and this epistemological blindness implied an onto-logical blindness as well.

In the 1960s, quantum physicists began to hypothesize the strangest class of hypothetical matter yet: particles of imaginary mass. In their 1962 article "'Meta'-relativity," O. M. P. Bilaniuk, V. K. Deshpande, and C. G. Sudarshan introduced the possibility of "meta-particles," which conform to the energy equation $E = (mc^2)/\sqrt{(1 - (v^2/c^2))}$. Such particles are imaginary because the formula's denominator takes the form of the imaginary number i ($\sqrt{-1}$), which means its velocity must be faster than light. This superluminal property is significant. Ever since Einstein's work on special relativity, the speed of light (3×10^8 m/s, or c) has been widely accepted as the absolute upper speed limit of the universe. Nearly every other aspect of the universe—matter, energy, time, observer perspective—is relative, but c remains constant. Indeed, c dictates the very rate at which time unfolds because time consciousness is a phenomeno-logical product borne of photons traveling at c and striking the eye. As we are beginning to see, though, nothing is sacred when it comes to midcentury quan-tum physics. Bilaniuk's conjectures were bolstered in 1967, when Gerald Fein-berg published "Possibility of Faster-than-Light Particles" in the prestigious *Physical Review.* In this seminal paper, Feinberg argued that superluminal par-ticles did not inherently contradict special relativity and consequently legiti-mized their potential existence within mainstream physics. Feinberg called his faster-than-light particles tachyons (*tachy* in Greek means "rapid"), which immediately caught on and have remained popular within physics discourse ever since. The tachyon still remains undiscovered, and much debate over its formal characteristics persists, but what remains unquestioned is that super-luminal particles created new vistas of cultural and scientific possibility.[42]

While the theorization and discovery of positrons, neutrinos, and tachyons in the middle of the twentieth century might appear haphazard, a long view of these developments suggests a turning point for both physics and epis-temology. Physics has long held a privileged position as the queen of the sciences—not just because its findings reverberate through other disciplines like chemistry and engineering, but also because it is the natural science most authorized to dictate the principles of reality itself. In the mid-twentieth century, antimatter and exotic matter transformed common notions of the real. The world the public thought it knew so well was in fact only half (or even less than half) of the material universe. Particles that only existed as placeholders within mathematical formulas were being discovered decades after the fact. Physicists continually challenged former absolutes like the speed of light. If the first wave of the quantum revolution in the early 1900s broke humanity from the structures of classical physics, then the second wave in the 1950s and 1960s broke the idea of structure itself. Any notion of Newtonian wholeness was banished, with researchers populating quantum mechanics with increasingly bizarre particles and postulates. "Quantum mechanics is a science of the bizarre," writes David Kaiser, and it increasingly seemed to authorize the impossible.[43]

Such a cultural shift altered the epistemological conditions of possibility for parapsychology. For a science where investigations of clairvoyant dreams, ghosts, and mediums transpire as a matter of course, the legitimization of a paranormal universe opened new avenues for research. "The seemingly fantastic propositions of parapsychology appear less preposterous in the light of the truly fantastic concepts of modern physics," Koestler observes, for the metaphysical claims of the latter implicitly condone the supernatural tendencies of the former.[44] More importantly, Koestler recognized in physics a different methodological onus and telos that could serve parapsychology well. "Physicists are not shy of producing *ad hoc* hypotheses—or speculations—to accommodate newly discovered phenomena which do not fit into the existing framework," he writes.[45] In fact, it is completely legitimate for physicists working in the standard model to hypothesize the existence of new elementary particles (quarks, gluons, bosons, etc.) to explain observed phenomena and then locate them empirically at a later date.[46] This methodology is now orthodoxy in physics; what began with the positron continues today with the discovery of the tau neutrino (2000) and the Higgs boson (2012). According to Koestler, psychical research ought to adopt a similar methodology, for only through an increasingly "theoretical" parapsychology could the discipline make the same strides quantum mechanics had achieved in previous decades. This was not a condemnation of Rhine-style empiricism, which Koestler valued for its confirmation of paranormal phenomena. But a more conjectural approach was necessary to build on the physicists' occult universe. Koestler saw in modern physics' embrace of pure theory a model for a new kind of parapsychology. Neither anecdotal nor experimental programs could fulfill parapsychology's epistemological destiny because they were overly rooted in the real. The field required more speculation, more of the imaginary.

In addition, quantum physicists' successful track record of verifying ghost particles provided psychical research with the rationale for *long-term* study of the paranormal. The neutrino took twenty-six years to discover, but it unveiled the hidden mechanisms of matter and energy. What might the telepatheme reveal about human consciousness? Even if it took decades, hundreds of scientists working around the world, and millions of dollars, the data arising from such a dedicated effort would be worth the effort. At stake was what H. H. Price called psychical research's promise to "transform the whole intellectual outlook upon which our present civilization is based."[47] For over three-quarters of a century, psychical researchers had risked their reputations, careers, and fortunes for what they believed was a greater truth, and the midpoint of the twentieth century seemed to presage a moment when favorable epistemic conditions and theoretical advancements across science had synchronized, at long last, to make a great leap forward in their discipline.

So the parascientific environment was primed for a breakthrough. Psychology gave the mind a fundamental particle. Neurology located a site for capturing it. Physics offered both an occult ontology and superluminal logic for rationalizing it. The stage was now set for Adrian Dobbs and the psitron.

The Psitron in Theory and Practice

Adrian Dobbs is a minor figure in the history of psychical research, although an exemplary one for showcasing the parascientific process. Born in Waterford, Ireland, in 1914, Dobbs studied at Trinity College in Cambridge, England, before spending most of his career working for the British government in colonial administration. He joined the SPR in 1957, where he remained an active member until his death in 1970. The focus of his Cambridge studies was philosophy and physics, and both fields influenced his inquiry into paranormal consciousness.

Dobbs developed his psitron theory over two major essays, "Time and Extrasensory Perception" (1965) in the *Proceedings of the Society for Psychical Research,* and "The Feasibility of a Physical Theory of ESP" (1967) in the edited collection *Science and ESP.* His chief concern at the time was precognition. As noted earlier, many parapsychologists argued that precognition was a logical impossibility because effects cannot precede causes. Dobbs believed the opposite, and he tackled this issue by utilizing all the latest sciences at his disposal. Building from psychology's psychon, Dobbs contended that a fundamental unit of consciousness existed and that it could explain the paranormal. His updated version of the psychon, which he called the *psitron*, was not a symbolic concept in the manner of Auguste Forel's psychome, but a real particle producing real mental effects. However, this particle was not real in the same way one would consider an electron or a neutron real because Dobbs argued this elusive unit of matter possessed imaginary rest mass, which Bilaniuk had hypothesized three years earlier in "'Meta'-relativity."[48] That such particles had never been observed (and might never be observed) did not bother Dobbs. "This fact [undetectable nature] about virtual particles does not make them any less *physically* real or *material* than an electron or a neutrino," he writes. "They are still an essential feature in the physical world as envisioned by the physicist."[49] As this passage suggests, Dobbs was inspired by quantum mechanics, which had proven that mathematically possible particles could eventually be found. For him, theory rightfully preceded experimental proof in both physics and parapsychology.

The psitron's imaginary mass—and therefore its superluminal velocity—represents a critical advancement over the psychon because of its role in Dobbs's interpretation of time. He believed in a five-dimensional universe in which the three traditional spatial dimensions are joined by two temporal dimensions: objective time and subjective time. The former reflects time as it actually unfolds in the world, whereas the latter represents probabilities of its unfolding. One of his most crucial claims is that psitrons move through subjective time faster than photons move through objective time. "Because such particles travel faster than the velocity of light," Dobbs writes, "it implies a mode of the Progress of Time other than the objective mode."[50] Put differently, the discrepancy in velocities between photons and psitrons allows the latter to move backward in time.

This brings us to Dobbs's key assertion in "Time and Extrasensory Perception": "This kind of development [two dimensions of time] leads to the introduction of a new type of entity which has not hitherto to my knowledge been considered as a possible cause factor in paranormal phenomena: namely, particles of imaginary rest-mass which I call *psitrons*. These are conceived to be emitted and absorbed by particles of mathematically real rest mass as are the photons which quantum physics holds to be the constituents of an electromagnetic field surrounding a charged particle such as an electron."[51] In these two sentences, Dobbs amalgamates much of the scientific background traversed so far. He argues that in the same way photons are constantly emitted by matter of real mass (e.g., photons emerging from the filament of a light bulb), so too are psitrons emitted by future events. Like photons, they travel through space and time before being captured and processed by the brain. For photons, the human eye is the site of reception and the brain the site of analysis. For psitrons, the brain serves both roles, for its dendrons are "so delicately poised that they can be caused to trigger off a cascade or chain reaction or neurone discharge, in consequence of capturing particles of imaginary mass."[52] Drawing from Eccles's research, Dobbs views the brain as a uniquely sensitive detector, and he uses dualist-interactionism to posit that dendrons can capture particles of infinitesimal size as well as particles of imaginary mass. Hence, when Dobbs says his theory posits "a concrete and specific idea in place of Eccles' vague notion, of a 'small influence' acting from outside the brain but with decisive effects upon its behavior," he refers to psitrons unveiling prophetic information from the future.[53]

Dobbs goes on to argue that psitronic information emerges from the cerebral cortex as "precasts" in the subconscious mind, which can take any number of forms: "such human responses to objective probabilities, initially at the subconscious level, can manifest themselves behaviorally (as in the Soal-Shackleton experiments); or they can trigger off introspectively conscious states, such as visual or auditory images, or sensory hallucinations[54] found in 'spontaneous' cases."[55] In other instances, the precasts would manifest as a "hunches, which come before the mind with a peculiar conviction."[56] Regardless of its ultimate form, any precast provides subliminal data about a future event before that event had occurred in objective time. Their predictive power is inherently limited, though, because not everything in the subconscious rises to consciousness. The vibrant hunch or visual image in Dobbs's model remains unpredictable. In addition, because psitrons navigate probabilistic time rather than deterministic time, their veracity is not guaranteed: "these *objectively* highly probable precasts of events . . . are very *likely* but are not *certain* to happen in the future."[57] In other words, precasts are accurate . . . but not always.

At this point, a brief recap of Dobbs's theory of mind is in order. Psitrons (or tachyons) originating from all types of future events and objects are constantly traveling backward in time throughout the universe at superluminal velocities. When they strike the dendrons of the cerebral cortex, they catalyze a neural cascade wherein their information is translated into a precognitive hunch or image. Sometimes, especially in the case of talented individuals, these precasts reach the conscious mind, in which case that person experiences a genuine moment of precognition. For an unorthodox idea that a scientist might dismiss out of hand, Dobbs's psitron theory is unusually detailed, suffused throughout with cutting-edge physics and defensible logic. His work is no doubt speculative in nature, but it is not far removed from the trailblazing ideas of Paul Dirac before him. Like that giant of modern physics, Dobbs was hardly settling for the pedantries of ordinary science but for what Thomas Kuhn would call "extraordinary science"—the pathbreaking, paradigm-shifting work that alters intellectual history. He was, in effect, attempting to reconceptualize the human mind by remapping its connections to time, space, and matter. In addition, by liberally borrowing from quantum physics, neurology, and psychology, Dobbs imbued precognition with scientific legitimacy.

For Dobbs, one of the strongest aspects of psitron theory was its eminent testability. As a strong supporter of Karl Popper's falsifiability, Dobbs outlined in great detail an empirical research program to locate the psitron. As a particle with imaginary mass, the psitron cannot be easily detected; instead, an indirect visual method using an electroencephalogram (EEG) was necessary. Dobbs describes his experiment setup as follows:

> As soon as it is possible to get another subject like Shackleton, who can perform feats of ostensible precognition with an agent, tests should be set up to record their EEG patterns in an experimental situation in which the agent tosses a coin, or performs some suitable classical randomizer, and the percipient guesses the outcome. If a correlative phase change can be established in the coherent after discharges of their frontal lobes in the cases of responses scoring "hits" above chance my theory would receive positive experimental confirmation. If no such change could be established my theory would be refuted.[58]

Dobbs describes a Rhine-style laboratory card-guessing experiment involving a talented medium monitored via EEG. In instances when he guessed cards correctly, the medium would ostensibly display distinct phase shifts in the alpha, beta, tau, or delta brain waves of his EEG reading. The irregular phase shifts observed during "hits" would thus represent visible, albeit indirect, confirmation that psitrons had struck the dendrons of the brain and generated a precast. Dobbs was confident enough in his methods that he declared:

> This approach at once suggests an entirely new experimental program of ESP research with the particular object of elucidating its physical basis. The object would be to test the hypothesis that general ESP and telepathy in general is due to the stimulation of ordinary particles in the brain ("critically poised" neurons) by patterns of particles of imaginary rest mass, which should lead to correlated shifts in the phases of the relevant EEG waves of two persons en rapport.[59]

A dedicated research program was essential because psitron detection would be a laborious process. Finding a medium as talented as Basil Shackleton would be difficult, as would deploying EEG to visualize psitronic strikes on the brain. But as physics had shown all too well, the results justified the high cost.

"It took approximately thirty years, and huge strides in technology, to make possible the 'observation' of the neutrino; and even now it is only detectable by very indirect methods," Dobbs argues. "Yet every competent physicist today is convinced of the particle's existence."[60] Physics illustrated what was possible with a valiant experimental program: deep insights into matter, energy, and the underlying mechanisms of reality. With patience and effort, the psitron too might conceivably unveil the hidden secrets of the paranormal mind.

What makes the psitron such an unusual model of consciousness are three shifts from previous approaches to paranormality. When psychical researchers like Myers first developed the subliminal self, they were operating in a world coming to grips with electrons and X-rays. Dobbs's unit of consciousness operates on even smaller scales. His paranormal mind occupies a subatomic space of neutrinos and tachyons—orders of materiality where the known laws of the universe no longer apply. In an occult shadow world, ghost particles logically belong more than ever.

The psitron also signaled the potential of "theoretical" parapsychology. A new particle required new approaches to study it. The first wave of psychical research (1890s–1910s) was "natural-historical" and emphasized the organization of supernatural phenomena into larger taxonomies.[61] The second wave (1920s–1940s), exemplified by J. B. Rhine, was empiricist and focused on producing statistically relevant data in laboratory settings. Dobbs envisioned a third wave in which mathematics, physical theories, and interdisciplinary methods played a dominant role in exploring the human mind. Such a program did not depend on finding empirical evidence; that would come later. What mattered was the physicalist logic to explain the paranormal. Modern physics had proven that the theoretical could precede the empirical en route to extraordinary science, and parapsychology should similarly embrace hypotheses and imagination as powerful tools for psychical investigation.

In addition, the psitron bridged an ideological division within psychical research. As Arthur Hastings notes, the field was long divided into rival camps.[62] On one side there were materialists like Charles Richet, who believed psychical research was a branch of the existing mechanist sciences; its underlying principles "do not contradict any accepted scientific truths. They are new; they are unusual; they are difficult to classify but they do not demolish anything of what has been so laboriously built up in our classic edifice."[63] On the other side were immaterialists like Frederic Myers, who believed that new sciences would be required to explain the paranormal. The psitron blurred such distinctions. As a subatomic particle, the psitron was simultaneously material and immaterial. It was a "real" particle, but it did not have the properties of a "real" particle. It was justified by the laws of physics, but it operated beyond the bounds of physics. The psitron fit within the paradigm of quantum mechanics, but that paradigm was admittedly shadowy. The epistemological influence of modern physics in the development of this parascientific object is therefore clear. By naturalizing the occult quintessence of the universe, quantum mechanics offered an epistemic platform for rationalizing precognition anew.

Unfortunately, Dobbs did not live long enough to oversee the psychical revolution he sought. He died from heart failure only three years after the publication of "Feasibility." As far as I can tell, no other parapsychologists made locating the psitron a research priority; nor would other scientists ever find EEG evidence of its existence. Even worse, none of his SPR colleagues advanced psitron theory or even challenged its assumptions. Kuhn claims that scientific ideas that fail to produce supporters are doomed to oblivion; such is the natural selection of paradigmatic science. One might assume the psitron's end was nigh.

Surprisingly, the psitron has survived deep into the contemporary era. Though essentially defunct as a parapsychological concept, the psitron found new life in the 1970s and 1980s, but not in the speculative theories of psychical researchers. Rather, it found life in the speculative fictions of writers. Such a migration from one field to another is hardly unusual. From its beginning, science fiction has served as the intellectual breeding ground of the scientific imaginary, be it geographic exploration (e.g., *Journey to the Center of the Earth*) or alien contact (e.g., *War of the Worlds*). The ontological passage of the psitron from science into fiction is particularly noteworthy because it is a hallmark of the parascientific process in which heterodox ideas rejected from mainstream paradigms find a second life in alternate bodies of thought.

In the introduction, I argued that science and literature play vital roles in parascientification. While psychological, neurological, and physical models have taken center stage in rejuvenating precognition so far, fiction also wields its own set of logics. Literature plays a key role because it can serve as the primary medium for the evaluation and dissemination of anomalous science. This is not a case of literature merely reflecting scientific thought but fully engaging, advancing, and promulgating it. Establishment science may not take up heterodox subjects like the psitron, but literature can, and its effects are considerable. Starting in the 1970s, several authors steeped in modern science and occultism began investigating the psitron in their fictions, teasing out its mechanisms, secondary properties, and overall functionality. In so doing, they have effectively taken up Dobbs's esoteric research program and bolstered the particle's fantastical status as a simultaneously real and imaginary scientific object. As it turns out, two of the psitron's most prominent theorists are also the biggest names in science fiction and comics. I consider Philip K. Dick and Alan Moore to be literary experimenters because they use their fictions to advance psitron theory in particular and the paranormal mind in general. The sections that follow will examine their explorations via two of the oddest characters in modern science fiction: Horselover Fat in *VALIS* and Dr. Manhattan in *Watchmen*.

VALIS and the Psitron

Philip K. Dick was no stranger to the paranormal. Like many other science fiction writers, he was intrigued by parapsychological phenomena and populated many of his stories with telepaths and psis. "Minority Report" (1956), for example, envisions a world in which psychic mutants known as precogs can predict the future so accurately that the government imprisons future offenders before any crime is actually committed. *Ubik* (1969) features a world in which psis are so common that companies employ "anti-telepaths" to prevent corporate theft. Beginning in 1974, though, the walls separating Dick's fiction and reality would come crashing down, leading him into an arcane world of precognition, self-directed ESP experiments, quantum mechanics, ghost particles, and cosmic consciousness in which he was the unwitting protagonist. These experiences—and the psitron theories underlying them—would eventually work their way into *VALIS* (1981), a novel beautifully capturing the role fiction plays in the epistemological production of parascientific objects.

Dick's psychic journey began while recovering from oral surgery. After calling a pharmacy for painkillers, he was captivated by the young woman who delivered his prescription, and more specifically by the necklace she wore.[64] The fish-trinket necklace triggered what Dick believed was an awakening of ancient genetic memories, divine messages from the future, or something weirder altogether: "In that instant, as I stared at the gleaming fish sign and heard her words, I suddenly experienced what I later learned was anamnesis—a Greek word meaning, literally, loss of forgetfulness. I remembered who I was and where I was. In an instant, in the twinkling of an eye, it all came back to me."[65] Inspired by the mystic forces now arising within and communicating with him, he began consuming incredible amounts of vitamins to "re-attune" his brain into a giant antenna. As a longtime member of the 1960s California counterculture, Dick was no stranger to psychedelic drugs, which are known for enhancing the senses, modifying perceptions of space and time, and promoting internal and external awareness.[66] In this case, Dick behaved more like a neuroscientist than an addict. "Vitamins in megadoses can improve neural firing and produce vastly increased brain efficacy," he would later explain.[67] His goal was to "block out ordinary man-made fields, which we consider 'signal' and at the same time . . . directly transduce what we usually think of as 'noise' in particular weak natural electrical fields."[68] Like John Eccles, Dick had come to believe that the brain could receive external signals—supernatural information that surrounds us at all times but is usually ignored as static. Incredibly, his citizen science experiment worked. Over the next two months of what he would later call "2-3-74" (for February–March 1974), his antenna brain began receiving a dizzying array of images and colors. The following excerpt describes a single night of his so-called divine invasions:

> While lying in bed unable to sleep for the fifth night in a row overwhelmed with dread and melancholy, I suddenly began seeing whirling lights which moved away at such a fast speed—and were instantly replaced—that they forced me into total wakefulness. For almost eight hours I continued to see these frightening vortexes of light, if that's the word; they spun around and around, and moved away with incredible speed. What was most painful was the rapidity of my thoughts, which seem to synchronize with the lights, it was as if I were moving, and the lights standing still—I felt as if I were racing along at the speed of light, no longer lying beside my wife in our bed. My anxiety was unbelievable.[69]

In the following weeks, Dick saw abstract art resembling the paintings of Wassily Kandinsky and Paul Klee, ancient Rome, entire novel manuscripts, mystic scrolls, and miscellaneous religious imagery. Calling himself the "involuntary recipient of an ESP experiment," he claimed these images were no mere hallucinations, for embedded within them was information of a higher order.[70] In fact, one of the futural pieces of information he gleaned from 2-3-74 was knowledge of his infant son's inguinal hernia, an undetected birth defect that could have proven fatal had Dick not told the pediatrician to seek out and operate on it.[71]

Dick had several theories about the cause of 2-3-74. In fact, he spent the better part of eight years and eight thousand pages analyzing his experiences in the epic diary known as the Exegesis, where he hypothesized prophetic information coming from any number of sources, including three-eyed aliens and suppressed Christian sects. His otherworldly encounter has fascinated critics for years, and they have attributed its causes to schizophrenia, drug addiction, and genuine mystic revelation, among other explanations.[72] The interpretation I find most intriguing is Dick's own belief that his brain had become a telepathic receiver for psitrons. In a letter to Peter Fitting dated June 28, 1974, he credits the transmission of 2-3-74 to "the concept of tachyons, which are supposed to be particles of cosmic origin, (I am quoting Arthur Koestler) which fly faster than light and consequently in reverse time direction. 'They would thus,' Koestler says, 'carry information from the future into our present, as light and X-rays from distant galaxies carry information from the remote past of the universe into our here and now.'[73] Koestler is one of many intellectual figures Dick cited in the Exegesis, but he gains considerable significance in this context because he became the chief proponent of Adrian Dobbs's psitron theory. In *The Roots of Coincidence* and other articles for the mass market, Koestler discussed the psitron and how quantum physics revealed the underlying paranormality of nature. "In light of these developments [Feinberg's and Dobbs's research], we can no longer exclude on a priori grounds the theoretical possibility of precognitive phenomena," Koestler wrote in a 1974 *Harper's* article that would later enthrall Dick.[74] In tracing the psitron's trajectory from Dobbs to Koestler to Dick, we see how a parascientific object has not faded from view but instead solicited greater contemplation. Such theoretical parapsychology was not limited to Dick's journals. It actually gave rise to the fictional entity known as VALIS.

Dick investigated the meaning and nature of 2-3-74 in the VALIS trilogy, a triptych of religious-infused novels including *VALIS, The Divine Invasion,* and *The Transmigration of Timothy Archer.* Of these texts, *VALIS* is the most relevant from a parascientific perspective because it clearly builds on Dobbs's work while operating outside scientific paradigms. *VALIS* is an admittedly difficult book to summarize because it blurs the lines of metafiction, autobiography, science fiction, and postmodern literature. The novel is ostensibly narrated by a science fiction writer named Phil Dick, but its main protagonist is Horselover Fat, an imaginary alter ego whom Phil Dick has conjured to "gain much needed objectivity" from the conundrum that is 2-3-74.[75] Fat believes he has received prophetic information from a divine entity called VALIS, which stands for Vast Active Living Intelligence System. On one level, VALIS is "essentially an alien from outer space" as Laurence Rickels glibly puts it.[76] On another level, though, VALIS is an entity of living information corresponding to the notion of cosmic consciousness, the one true God flowing through all things and of which we are only part and particle. *VALIS* follows Fat's experience of 2-3-74, including the discovery that our material world is a false reality known as the Black Iron Prison, which exists to occlude mankind from the true Edenic reality of the Palm Tree Garden. He eventually finds others who have been communicating with VALIS, and his life's mission becomes locating the human incarnation of VALIS (i.e., Jesus), who will lead humanity into the Palm Tree Garden.

While previous scholars have addressed the Gnostic, postmodern, narratological, and psychoanalytic characteristics of *VALIS,* none has analyzed its scientific principles or the broader transformation of the paranormal mind Dick propels.[77] The psitron, a parascientific progression of precognition with nineteenth-century origins, in Dick's hands becomes the central novum undergirding a new system of mind. This hypothetical particle is too small and too esoteric to be taken up in mainstream science. Instead, its theorization and experimentation transpire via fiction. As such, *VALIS* becomes a literary laboratory where the psitron's elusive properties and characteristics can be articulated.

Though it is never identified as such, Fat's conception of VALIS corresponds directly with the psitron. In the novel, his experience with 2-3-74 is an autobiographical reference to Dick's actual 2-3-74 episode. Fat's "personal theory" should sound familiar: "Sites of his brain were being selectively stimulated by tight energy beams emanating from far off, perhaps millions of miles away. These selective brain-site stimulations generated in his head the *impression*— for him—that he was in fact seeing and hearing words, pictures, figures of people, printed pages, in short God and God's message, or as Fat liked to call it, the Logos."[78] Through the Exegesis, we know Dick learned of this theory from Arthur Koestler. All the features of the psitron are represented. The tight energy beams from far away signify psitrons from future events. The brain sites receiving the beams are dendrons of the cerebral cortex. The stimulations of words and images are precasts rising into consciousness. Such a reconstitution of a parascientific concept through fiction is noteworthy because it transforms Horselover Fat into a literary test subject. He becomes a Basil Shackleton figure whom we can analyze via literary analysis rather than EEG. It further suggests that the entire novel is an experimental space where Dobbs's psitron model can be theorized in ways impossible within scientific praxis.

To be clear, Fat's theory is not a passing reference to the psitron but a departure point for deeper conjecture. As any reader of his fiction knows, Dick is nothing if not a theorist of the mind. In *VALIS,* Dick probes the imaginary particle and makes three "discoveries" concerning psitron precasts, ubiquity, and toxicity. The first discovery, obviously based on his 2-3-74 episode, is that precasts manifest as striking images of erratic legibility. Throughout the novel, images generated by tachyonic "energy beams" are dazzling and clear: pink lights, printed pages, endless sequences of abstract art, and so on.[79] Yet the clarity of these images often fails to convey actual meaning. "At any moment a beam of pink light could strike us, blind us, and when we regained our sight (if we ever did) we could know everything or nothing," Fat laments.[80] In some instances, like his son's inguinal hernia, Fat's precasts are remarkably clear: "he knew, specifically, that his 5-year old son had an undiagnosed birth defect, and he knew what that birth defect consisted of down to the anatomical details."[81] In other instances, Dick acknowledges his precasts are complete nonsense:

> In my reception of tachyon bombardment (assuming this is what it is, of course), I frequently either fail to transduce properly (error at the receiving end) or else there is a lapse of accurate transmission (as if the teletype operator has his fingers on the wrong line of keys, etc.). When that happens, instead of seeing, in my dreams, the perfectly articulated English prose passages which would be the result of all components functioning correctly, I get gibberish like this: meaningless "names" and "words" and sequences of numbers which have no significance.[82]

It is noteworthy how drastically this differs from Dobbs's original interpretation of precasts, which appear in consciousness as vague ideations with strong conviction. For him, the psitron produced subtle phenomenological effects reinforced by confidence in their veracity. For Fat, it is the opposite. The psitron produces strong visuals; sometimes they mean something, but usually they do not. Throughout both the Exegesis and *VALIS*, Dick theorizes tachyons as a type of electromagnetic signal intertwined with "noise." Tachyons are invisible to most people, but even if one has the dendronic capacity to capture these particles, most messages remain garbled. In the nineteenth century, when mediumship was understood as "spiritual telegraph," separating signal from noise was a common problem that only the best sensitives could overcome.[83] A century later, Dick dealt with the same issues. "Considering the distance over which these packets of information travel and their velocity," Dick writes, "much contamination, signal-loss and other familiar invasion of the material contained must take place—cross-talk from other fields, so that when the tachyons at last impinge on us even if transduction is superb (as in the case of 'mystics' and 'saints') there would be something quite less than a perfect meaningful construct."[84] And therein lies the rub. Psitrons carry definitive information from the future, but their meaning is lost in transit and translation. Their data transfer is fundamentally uncertain.

Another psitron characteristic Dick theorizes in *VALIS* is the universality of precognition. In Dobbs's original framework, special individuals like mediums have sensitive brains that permit psitron capture. To possess second sight is the supernatural condition. Dick makes a contrarian argument: our natural condition is to see prophecy, and the problem is our psitron-processing mechanisms have been neutered. "We should be able to hear this information, or rather narrative, as a neutral voice inside us," Fat explains in *VALIS*. "But something has gone wrong. All creation is a language and nothing but a language, which for some inexplicable reason we can't read outside and can't hear inside. So I say, we have become idiots."[85] In his own life, Dick overcame his idiocy thanks to the fish necklace and a vitamin regimen, but most humans are blind to their paranormal birthright. The belief that humanity has regressed over time is a common occult narrative. In *The Secret Doctrine,* for example, Madam Blavatsky claims premodern man had wisdom, powers, and foresight now lost to the contemporary age.[86] *VALIS* hinges on this pastoral vision of paranormality because our knowledge of the Palm Tree Garden is occluded by the Black Iron Prison. Similarly, *The Divine Invasion* details the battle between the one true God Yah and the false god Belial, who clouds mankind's perception of things. For both Dick and Fat, man's *normal* state is precognitive. What is abnormal is the loss of second sight as a daily phenomenological experience.

The third wrinkle that *VALIS* adds to Dobbs's theory is psitron toxicity. At the end of the novel, Fat and his friends travel to Sonoma, California, where they meet several musicians who have been communicating with VALIS. However, we soon learn that electronic composer Brent Mini is dying from brain cancer due to his extended contact with VALIS's tachyons.[87] Among the novel's characters, Mini possesses the most extensive knowledge of VALIS—its purpose, logic, symbols, and meaning. Such insights have come at great cost. Mini explains that his multiple myeloma, which has left him partially deaf, wheelchair bound, and with less than two years to live, comes from his "proximity to VALIS, to its energy."[88] In the same way that excessive exposure to X-rays lead to skin burns, nausea, and cancer, so too can the psitron injure the brain. "The levels of radiation can sometimes be enormous," explains Mini. "Too much for us."[89] This is an intriguing theoretical development. Whereas Dobbs views the psitron as a neutrino-like particle barely interacting with matter, Dick returns to an older conception of energy in which the psitron represents a high-energy blast with deleterious effects on the body. Like gamma rays and microwaves, the psitron will literally cook the brain. Prophets pay a steep cost for seeing the future: to listen to VALIS is to hear whispers of your coming death.

The psitron's toxic effects are both physiological and psychological. All the characters in *VALIS* who receive psitrons are considered "mad" at some point or another. Musicians Eric and Linda Lampton speak of chakras and aliens from Albemuth. Mini ends up murdering Sophia, the human incarnation of VALIS. Even Fat, we must recall, is an imaginary projection of former mental patient Phil.[90] For Dick, to receive the psitron is "to go nuts"—to suffer from erratic visions and conspiracy theories with cryptic messages hidden everywhere.

Just as the prophets of yesteryear lingered in the fantastical space between genius and insanity, so too do the psitron receivers of *VALIS,* who see the world in paranoid, futural, and fatal ways.

VALIS underscores a key aspect of parascientification, namely how fiction plays an epistemic role in updating a failed scientific concept. Dick is not simply presenting Dobbs's view of precognition to the masses; he pushes it forward in terms of theory, logic, and personal experience, and in doing so, he contributes to its ideational evolution. Precognition used to be a mystery, but no longer. Dick posits it as a physicalist phenomenon powered by superluminal particles and the brain's unique ability to capture its data. The capacity to receive psitrons is man's natural condition, perhaps even his original state of being. Precasts are not subtle and compelling but rather concrete and confusing, a "language which we have lost the ability to read."[91] This is not to suggest Dick's psitrons are meaningless. Instead, they contain too much meaning and too much power, for they decimate the minds of those (un)fortunate enough to comprehend its meaning. In this parascientific process, precognition has transformed from a divine metaphysics to a dangerous hyperphysics.

While the major thrust of *VALIS* clearly articulates a neo-Christian vision of a Gnostic universe where the psitron plays a seminal role, a broader analysis of the psychological, neurological, and physical sciences informing Dick's reconceptualization of the paranormal mind reveals parascientification in action. Koestler and Dobbs introduced him to tachyons and ghost particles, but Dick's additional theorizations of the psitron through science fiction, religion, communication theory, and citizen science demonstrate how this parascientific object attains a new and hybridized form. The intellectual transformation of precognition is not pseudoscientific because it fails to uphold the standards of "real" science. Rather, it is parascientific because it blurs the difference between science and nonscience, representing a form of epistemological inquiry—and production—that alters what it means to see the future.

Dr. Manhattan, the Psitron Superman

If *VALIS* hypothesizes the confusing and debilitating effects of precognition, then *Watchmen* (1986–87) speculates on its infinite potential. Indeed, the comic subtly views the psitron as an exceedingly powerful unit of consciousness. Alan Moore's groundbreaking series is one of the most critically acclaimed comics in literary history. Interspersed with fictional newspaper clippings, book chapters, psychiatric evaluations, and even other comic books, the central plotline of this postmodern crime saga revolves around the Minutemen, a team of superheroes populating an alternative universe where Richard Nixon is still president in 1985 and America stands at the brink of nuclear war with Russia. While most of the Minutemen have long since retired, several have reassembled after a series of suspicious murders. The investigations of Rorschach, Nite Owl, Silk Spectre, and the Comedian eventually reveal that one of their former comrades, Adrian Veidt (aka Ozymandias), has hatched an insidious (or ingenious?) plot to kill millions of Americans in order to achieve a lasting global peace. Much has already been written about the novel's political, formal, and ethical concerns, but little has been said about its theorization of the paranormal.[92] *Watchmen* is significant in this respect because it serves as a literary laboratory where the full power of the psitron is revealed.

Much of the comic's supernaturalism focuses on the most powerful member of the Minutemen, Dr. Manhattan, the lone character in *Watchmen* who possesses actual superpowers. His origin story parodies the creation myths of other atomic-age superheroes like Spider-Man (bitten by a radioactive spider) and the Fantastic Four (exposure to cosmic rays): Jon Osterman is a nuclear physicist at the Gila Flats government test facility who is accidently trapped in an "intrinsic field experiment" chamber and disintegrated. Moore is a well-known occultist versed in the history of Theosophy, spiritualism, and magic, and in *Watchmen,* he extrapolates Myers's contention that the subliminal self can survive the material body.[93] Jon's mind persists after the accident (perhaps as a system of psychons?) and attempts to reconstitute its former body. Through sheer force of will, Jon successfully rebuilds himself and reaps vast new powers: "In the process of reconstructing its corporeal form, this new and wholly original entity achieved a complete mastery of all matter; able to shape reality by the manipulation of its basic building blocks."[94] The praxis of "thinking" his body back into existence results in Jon's supernatural ability to exert mental control over all matter in the universe. Renamed Dr. Manhattan by the government to brand his mastery over the atom under the guise of American political force, Jon possesses unfathomable telekinetic power. As shown in Figure 2, he disassembles a gun using only his mind. His neutral facial expression suggests this impossible task is as easy for him as breathing. In the right panel, he shoots a laser from his hand to destroy a tank. This act is, of course, a performative gesture for the global audiences watching him on TV (signaled through the CRT television-shaped panel); he just as easily could have transformed the tank into wine or a burning bush. Throughout the comic, Dr. Manhattan alters his shape, teleports across the galaxy, and conjures Martian fortresses, all of which is to say he is a god.

Figure 2. Dr. Manhattan's demonstrates his telekinetic mastery over atomic structure. *Watchmen* (1986), DC Comics.

While the presence of an Übermensch in a world of normal menches is politically and philosophically intriguing, what I find most fascinating is how Dr. Manhattan's mastery of atoms also makes him a psi par excellence. His collective powers of clairvoyance and telekinesis are without peer in the world of fiction. His precognition merits particular focus because it is made possible via Dobbs's psitron. Dr. Manhattan's precognitive powers are typically associated with Eternalism (also known as the block universe), a philosophical and ontological approach that views space-time as an unchanging four-dimensional block;[95] this is in contradistinction to the standard Western view of time as constantly unfolding toward the unknown future. In such a block, where past, present, and future are known quantities because they are static, a god like Dr. Manhattan can see things to come as clearly as things that are. "For Dr. Manhattan, time is out of joint," Christopher Drohan writes. "Past, present, and future seem to blend together."[96] This temporal simultaneity is masterfully captured across chapter 4 of *Watchmen*, where Dr. Manhattan's multitudinous story lines barrage readers as a bricolage of old, new, and futural memories. While Eternalism provides the superstructure of time in the comic, tachyons (and therefore psitrons) provide a physicalist mechanism explaining how Dr. Manhattan actually cognizes the future.

Let us follow the clues. Adrian Veidt successfully masterminds his global conspiracy without interference because he preemptively launched several satellites into Earth's orbit to create "tachyon interference"—a massive influx of artificially generated tachyons. As Dr. Manhattan notes, "There is some sort of static obscuring the future, preventing any clear impression."[97] That "static" is the hum of Veidt's manufactured tachyons, which have overloaded Dr. Manhattan's antenna brain from receiving real tachyons, consequently muddying his futural vision. We can thus infer Dr. Manhattan's precognition is *powered* by a constant flow of superluminal particles striking his brain and producing precasts; that is, his Eternalist ability to see the future follows the psitron model. Unfortunately, Veidt sabotages this system by overloading Dr. Manhattan's dendrons with artificial psitrons, which leaves him as blind to the future as any normal human. By the time Dr. Manhattan finally realizes what has happened, it is too late. In Figure 3, Dr. Manhattan inspects the wreckage of Times Square, unconcerned by the human carnage surrounding him. In fact, he revels in the "delights of uncertainty" that come with not knowing the future,[98] which reinforces our interpretation that, under normal circumstances, he experiences the ennui of psitronic certainty. In this regard, Dobbs's psitron is what makes Veidt's diabolical plan—and Dr. Manhattan's precognition— mechanistically possible. Dobbs had previously argued testing the psitron depended on finding talented psis like Basil Shackleton. In the laboratory that is *Watchmen,* Moore conjures such a subject via Dr. Manhattan.

Moore is not content to simply regurgitate the psitron model, though. Like Philip K. Dick, he uses his characters to conceptually advance the boundaries of the paranormal. More specifically, he theoretically elevates the psitron into the ultimate particle of prophecy. Unlike Dobbs's strong hunch or Dick's cryptic symbology, Dr. Manhattan's precasts offer clear and accurate knowledge of the future. The flow of information from future event to psitron to dendron to subconscious to conscious is free from data loss; it is high fidelity. In fact, Dr. Manhattan sees the future in the same way the rest of us see the present: "In 1959 I could hear you shouting, here, now, in 1963," he explains to his girlfriend, Janey Slater. "I can't prevent the future. To me it's already happening."[99] Most readers would read the sharp exchange between the two lovers in Figure 4 as an example of Dr. Manhattan's curious and omnipotent powers. I see it as indisputable evidence of psitron accuracy. For Dr. Manhattan, precasts of the future are indistinguishable from the past and the present because all three time dimensions reach his consciousness with undiluted clarity. As David Barnes notes, "Throughout the novel Manhattan seems increasingly to experience time as a singularity and to remember events, including future events,

Figure 3. Tachyon interference blocks Dr. Manhattan's precognitive abilities. *Watchmen* (1986), DC Comics.

as a well-crafted if sometimes disturbing simultaneity"—which speaks to the stunning lucidity of psitron-based foresight.[100] Dr. Manhattan's perfected prophecy—his complete knowledge of things to come—suggests a phenomenological breakthrough superseding anything either Dick or Dobbs ever imagined. Moore's psitron does not suffer from information loss while traveling through vast tracts of space and time, as Dick had postulated. Nor is it toxic, for Dr. Manhattan's health is unaffected by his ability to capture psitrons. Dobbs believed psitronic information was filtered through the subconscious and did not always reach consciousness, but Moore's particle conveys the future with perfect accuracy. Dr. Manhattan thus experiences a paranormal

Figure 4. Alan Moore's interpretation of the psitron delivers high-clarity precasts. *Watchmen* (1986), DC Comics.

mode of cognition in which the future is elevated from vague hunches into material truth. Whether Dr. Manhattan acts on this precognitive information is another issue altogether; much critical ink has been spilled analyzing his choice to act or not act because he always knows the future.[101] His decision-making is a separate issue from the physical ability to see the future. Ethics notwithstanding, Dr. Manhattan represents the theoretical apex of the psitron. It is an idealized vision of precognition, courtesy of a biophysical system so advanced that second sight and normal sight are indistinguishable. It is also an idealized vision of the prophetic medium. As we shall see in greater detail in chapter 4, mediums occupy a vexed position in psychical research for both their extraordinary gifts and guile. There is no deception with Dr. Manhattan, though. Tachyon satellites aside, his prescience is perfect.

VALIS and *Watchmen* thus illustrate two different theoretical possibilities of the psitron as a paranormal model of mind. In one, the psitron is a faster-than-light particle that proffers insight and insanity in equal doses; in the other, it makes its recipient a god. Despite these differences, each text serves the same epistemological function. Both Dick and Moore start from Dobbs's theory and use their fictions as experimental spaces for probing the psitron's elusive properties: its electromagnetic characteristics, physical and biological effects, cognitive outputs, ubiquity, and puzzling relationship with standard space-time. Instead of offering the ethical critique often associated with science fiction, these books perform epistemological work on ideas science has deemed beyond the pale, for anomalies existing outside current paradigms.

As a material object, the psitron has never been found and remains absent from establishment science. But as a literary object, it still exists in the minds of scientifically attuned writers, and it is as a literary object that our culture has continued to examine its esoteric properties. Such conjecture is parascientific precisely because it amalgamates the scientific and nonscientific realms to produce new conceptual forms.

In bypassing the strictures of experimental science yet rising above pure fictionality via its epistemological engagements, literary studies of the psitron perform crucial theoretical work. In fact, we might say Moore and Dick have been key participants in Dobbs's ambitious research agenda to study the psitron. Their methods may not involve EEG experimentation, but their fictions serve the same purposes: to cultivate our understanding of a cryptic particle and to recognize its significance to mankind.

The Future of Precognition

Watchmen and *VALIS* have played a role in not only the intellectual development of the psitron but also its dissemination across contemporary culture. Whether intentional or not, Dick and Moore have exponentially expanded the audience for the paranormal mind. Dick is one of the most popular and influential science fiction writers of the late twentieth century.[102] Moore's graphic novel is widely considered the greatest superhero comic ever produced.[103] Although legions of readers may never know the full history of the psitron, let alone the name Adrian Dobbs, their awareness of faster-than-light particles and the brain's native ability to capture them now constitute a part of modern speculative discourse. Thanks to Dick and Moore, tachyons and precognition are now common features in contemporary science fiction. In Gregory Benford's *Timescape,* for instance, scientists in a dystopian 1998 use tachyons to send messages back to 1962 to warn scientists about coming ecological disasters. In Robert Sawyer's *Flashforward,* a CERN experiment to discover the Higgs boson accidentally interacts with a passing squall of neutrinos and precipitates a global blackout wherein the entire human population sees twenty-one years into the future. While this MacGuffin is never fully explained, it obviously deploys psitron logic on a worldwide scale. Literature has always served a role of ideological amplification and signification. Through speculative fictions like *Watchmen,* the psitron gains a larger cultural presence than Dobbs could have predicted half a century ago.

The popularity of the psitron in science fiction delineates the untraditional path of a parascientific object. The epistemological conditions of possibility opened by psychology, neurology, and physics allowed nineteenth-century notions of precognition to transform into a real/imaginary particle now thriving outside paradigmatic science. The paranormal mind has always been a protean entity produced at the nexus of science, culture, religion, fiction, and myth. Through its hybrid existence between speculative science and art, the psitron has been theorized, retheorized, observed, analyzed, and disseminated through modern society. In this manner, it has successfully escaped the intellectual closure predicted by Thomas Kuhn, for parascience allows the psitron to exist beyond scientific paradigms in more fluid and rhizomatic forms.

Yet the psitron does not exist only in literature today. The very nature of parascientification allows its objects to appear and reappear in any number of disciplines. A case in point is neuroquantology, a fringe field dedicated to studying the interactions between physics and neuroscience. In the 2003 debut issue of *NeuroQuantology,* editor in chief Sultan Tarlaci writes, "The past decade has seen a rising tide in interest in between quantum physics and neural sciences, accompanied by a surge of publications, new journals, and scientific meetings."[104] The mission of his journal, and by extension the entire field, is "the interdisciplinary exploration of the nature of quantum physics and its relation to the nervous system," with those disciplines including physics, neuroscience, cognitive science, psychology, and artificial intelligence.[105] This academic boilerplate may sound traditional, but the list of problems the field addresses does not: mind–body interaction, history of consciousness, theory of consciousness, quantum entanglement, and so on. Interestingly, neuroquantology advances ideas like panpsychism and dualist-interactionist theory, which closely align with the theoretical frameworks devised by Eccles and Dobbs several decades before.[106]

One particular article is worth highlighting because it demonstrates the durability of parascientific objects. In the 2008 article entitled "Eccles's Psychons Could Be Zero-Energy Tachyons," Syamala Hari argues that psychons might be the tachyons proposed by Feinberg in 1967. "We believe we are the minority who believe that tachyons can be found if attempts are made in the right place," Hari writes. "We believe a living brain is a right place to look for them."[107] Moreover, she considers "a tachyon as a non-local object (but not a material particle) which can be associated with an imaginary-mass value and with waves which are not localized in space but localized in time."[108] We can observe here a recapitulation of the psitron as a real/imaginary particle that interacts with the brain to process consciousness. Ironically, Hari calls for a familiar mode of experimentation at the end of her article: because "EEG is the primary means by which electric and electromagnetic activities in the brain are measured and their features inferred, the well-known alpha, beta, theta, and delta rhythms offer [a] suitable area of investigation."[109] As far as I can tell, Hari did not realize Dobbs had already merged the psychon and the tachyon fifty years before. Nor is she aware that he proposed the same experimental EEG methodology in the *Proceedings of the SPR*. At face value, the coincidence is remarkable, but perhaps such an overlap should not surprise us. In the epistemic sprawl of parascience, the rich interplay of literature, myth, and unconventional science can give rise to similar, even identical, concepts time and again as marginalized ideas contact newer theoretical frameworks and find more outlets of possibility.

Such epistemic recycling is not new, of course. What it suggests more than anything is the indestructability of a parascientific object. Where Dobbs once used imaginary mass calculations, Hari now deploys Klein–Gordon equations and Schrödinger wave mechanics.[110] Parascientific objects rely on new narratives and sciences to subsist. In the late nineteenth century, Theosophy hypothesized the pineal gland as the physiological site of the third eye. In the early twentieth century, statistical analysis provided mathematical evidence for telepathy. Eccles and Dobbs similarly reformulated precognition into newer scientific terms better suited for the ideologies of the day. In *A Thousand Years of Nonlinear History*, Manuel DeLanda writes that history is not a linear progression but rather a series of accumulations and feedback loops, a "flow of stuff" in which past, present, and future perpetually coexist.[111] Such is the case for the psitron, in which quantum mechanics, particle psychology, and neurology layer fresh strata onto precognition in the ongoing development of the paranormal.

The psitron brings much-needed attention to the parascientific processes keeping marginal scientific ideas in circulation as well as to the acts of scientific and aesthetic renewal allowing heterodoxy to reach new audiences. For this reason, it has been a useful starting point, a symbol of parascience that is a real *and* imaginary particle shuttling between speculative science and speculative fiction. As such, it has propitiously escaped the intellectual closure predicted by paradigmatic science and continues to linger in our collective conscious.

2 GAIA, THE PARANORMAL PLANET

> We cannot understand what the Earth is and is meant to be without ques-
> tioning what we ourselves are and are meant to be.
>
> —JACOB NEEDLEMAN, *AN UNKNOWN WORLD*

IN THE PREFACE TO NEW AGE MANUAL *Gaia Speaks* (2005), author Pepper
Lewis makes the unusual point of deemphasizing her role in the book's creation because it was not written so much as channeled. During the text's production, she had regularly entered a trance state and allowed Earth to assume
control of her body and mind. With the third planet from the sun as the primary speaker and Lewis merely a scribe, ancient truths are revealed for all
humanity to behold. And so Lewis—or rather Earth—begins:

> I am the non-physical sentience of this planet. Simply put, I am the planet Earth.
> I am the body and soul of the planet you currently inhabit. My sentience is con-
> scious and aware of itself and its purpose . . . You call the physical planet "Earth"
> but she is also known by other names.[1] Today you are familiar with terms such
> as Mother Earth, Gaia's Voice, Mother Nature and Terra Firma; other civiliza-
> tions have had names of their own. My sentience guides and enlivens all that
> surrounds the planet as well as all that is upon and within her. My sentience
> animates the air you breathe, the energy you burn and the water you drink. All
> of the elements are under my care and direction, as are the seasons and the
> weather. I am a devoted companion and trusted friend to all life forms, including
> the animal, plant, and mineral kingdoms that share the planet with you. My sen-
> tience, the most aware, advanced, and attuned aspect of my being, is that which
> directs these words.[2]

In ventriloquizing the voice of our living planet, Lewis delivers more than
holistic advice to the spiritualistic members of her New Age audience. By ref-
erencing the higher consciousness of the planet as well as its direct agency

over aquatic, atmospheric, geological, and thermodynamic interrelations, she presents a recognizable, albeit hyperbolic, version of one of the twentieth century's most controversial scientific concepts: the Gaia hypothesis. Moreover, she illustrates the ways literary texts promulgate paranormal thinking through contemporary mass culture and vast networks of orthodox science, alternative science, theology, fiction, and philosophy.

First conceived in 1965 by British engineer James Lovelock and developed over the next decade with American microbiologist Lynn Margulis, the Gaia hypothesis asserts that Earth, through its interconnected systems of gases, minerals, water, energy, and biological life, is best construed as a single living entity, "a superorganism comprised of all life and tightly coupled with air, oceans, and surface rocks."[3] In contrast to the conventional idea that Earth is a nonliving object populated by living things, the Gaia hypothesis contends that our planet is a living system capable of self-regulation, action, reaction, and even cognition. Lovelock's concept, disseminated through a steady flow of scholarly articles and best-selling books, has captured the public imagination in a way few scientific ideas do. From the start, it attracted criticism from mainstream scientists, who assailed Lovelock's extraordinary claims as the pseudoscientific ravings of a "holy fool."[4] But it also drew fierce support from geologists, environmentalists, and philosophers as a promising means of rethinking everything from biomineralization to humanism. This debate, which has evolved drastically in the intervening decades, still rages today.

What I find salient in terms of parascience is how such an unorthodox theory captured the attention of scientists and writers, who have found in Gaia a platform for preternatural conceptions of worldwide consciousness or global mind actively guiding the planet's ecological, historic, and evolutionary processes. The esoteric speculations of Pepper Lewis may make critics scoff thanks to its synthesis of scientism and spiritualism, but Gaian literature is a revealing discourse because it demonstrates how marginalized and discarded sciences can persist in our cultural consciousness. When the public thinks of Gaia, it does not typically envision "cybernetic processes," "Daisyworld," or any of the scientific models researchers typically use when discussing Earth as a living system.[5] To be honest, the public thinks of *Gaia Speaks;* that is, it imagines a sentient being as the shepherd of all life and the beholder of past, present, and future, a fusion of contemporary science and premodern living matter generating a truly majestic figure. Furthermore, it sees in Gaia a planetary mind with miraculous powers like telekinetic agency over geological epochs and microcellular processes alike. For these reasons, speculative Gaian texts deserve our utmost attention. They constitute a parascientific medium through which marginalized ideas circulate through society while also acting as an ideational space where biology, philosophy, religion, and aesthetics can collectively reinvent outmoded concepts like telekinesis (the ability to affect physical objects with the mind in ways inexplicable to modern science).

The pages that follow focus on Gaia as an overlooked instantiation of the paranormal mind in late twentieth-century discourse as well as a key site for investigating the technics of parascientific production. In the same way the psitron in chapter 1 served as a useful object for understanding how quantum physics helped to reinvigorate precognition, this chapter uses Gaia to explore how the biological sciences recalibrated telekinesis for the late twentieth century. More specifically, it examines how the interplay of systems biology and microbiology inspired a range of scientists, philosophers, New Age theorists, and writers to reimagine planetary consciousness and its mechanisms of agency. Critics like Donna Haraway, Isabelle Stengers, Bruno Latour, and Bruce Clarke have all written about Gaia as a hub for exploring posthumanism, cybernetics, climate change, and the Anthropocene.[6] While this chapter touches on several of these areas, it nevertheless foregrounds the stranger mythologies and fictions bubbling *around* the margins of Gaian thought. While many scholars find the pseudoscientific conversations surrounding Gaia unhelpful, I believe their presence matters deeply. When it comes to the occult, the margins of science cannot be discounted. Tracking the interactions between burgeoning science and speculative literature is generative because it reveals the expansive nature of the parascientific process. The fields we will traverse, from classical philosophy to modern systems theory, depart dramatically from those covered in chapter 1, but the larger pattern remains steady: time and again, peculiar mental powers condemned by modern science circulate back into mainstream discourse through emerging scientific approaches, speculative imaginaries, and a public eager to understand a greater truth of mind. The plenitude of thinkers we locate across physics, psychology, philosophy, and the arts suggests that parascientification is a natural process, a normal outgrowth of scientific thought in service to heterodox ends. As such, the diversification of the paranormal mind in psitronic and Gaian directions over the past century is not an anomaly so much as an epiphenomenon of scientific progress.

Our paranormal investigation of Gaia begins by tracing the conceptual origins of the living Earth in classical philosophy and its decline in the post-Enlightenment era courtesy of Cartesian logic. We then turn to the rise of systems biology, which laid the epistemic groundwork for the Gaia hypothesis. Systems theory, an anti-Cartesian reframing of science away from individual units toward interconnected parts and processes, reintroduced metaphysics into establishment science through the so-called emergent properties it located within complex networks. This paradigm shift challenged classical notions of life and intelligence and made the dissident concept of planetary consciousness epistemologically feasible. After examining Lovelock and Margulis's development of the Gaia hypothesis, we follow several New Age writers who pushed their ideas in ever bolder directions, including the possibility of Gaia mind. The New Age remains a minor field in religious studies, but it emerges here as a compelling nexus where science, pseudoscience, and speculation fuse to produce popular knowledge forms. The texts of popular science writers like Fritjof Capra therefore deserve serious consideration because of the outsize role they play in parascientific production and dissemination. This chapter concludes with the science fictions of Nebula Award–winning author Greg Bear, who takes Gaia into truly esoteric territory in *Darwin's Radio* and *Vitals,* two novels that deploy virology and bacteriology to explain how planetary telepathy works. In these apocalyptic texts, Gaia has the psychical ability to manipulate physical reality—and to crush humanity.

For those who argue that paranormality has faded over the course of the twentieth century thanks to greater scientific scrutiny, Gaia offers an unequivocal rebuttal because it is only through new sciences like systems biology that Gaia exists today. For those who say psychical models carry little heft in the twenty-first century, Gaia illustrates an alternative mode of disciplinary development. In contrast with the subliminal self, the psitron, and other mental models rooted in the psychological sciences, Gaia pulls from an entirely fresh set of disciplines to establish a new type of paranormal consciousness grounded in macroscopic systems rather than microscopic particles. The movement from logical atomism to planetary interconnection signals how the paranormal mind has grown exponentially. Curious too is the decentering of the human, who only plays a bit part in the schema of things when mind itself is a global phenomenon. This oscillation from material particles to abstract systems, and from individual to planet, highlights the infinite variations and possibilities of the parascientific process.

The Rise and Fall of Living Matter

The idea that Earth is "a living creature" with an active and powerful mind has a long tradition in classical philosophy.[7] As early as 500 BCE, Greeks like Anaximander, Anaximines, and Pythagoras all believed the Earth was alive.[8] From the springs of freshwater burbling to the surface to the copious vegetation feeding fish and fowl, the Earth was undoubtedly a living thing to the ancients. Thales of Miletus, often considered the founder of natural philosophy, argued that everything in the universe was infused with life: "Earth is part of the universe, and hence it is ensouled, a being that is godlike."[9] Plato, of course, was the most influential Greek proponent of "living matter." In *Timaeus,* he speaks of the Demiurge's creation of the material world as a simulacrum of the eternal world of forms: "He constructed it [the Earth] as a Living Creature, one and visible, containing within itself all the living creatures which are by nature akin to itself."[10] (As a living thing constituted by other living things, Plato's Earth can be understood in modern terms as a superorganism, not unlike Lovelock's eventual conception of Gaia.) For many early Greeks, then, the claim that Earth is alive was not a metaphor but a physical truth: our planet, its inhabitants, and all of its geological features collectively form a living entity.

Greek theories of living matter, or hylozoism, would pass into Western culture via the Stoics and later the Romans, but it was effectively marginalized by the scientific revolution.[11] René Descartes played a significant role in disallowing Gaian discourse. According to his mechanistic theories, the human body (and the universe more broadly) was best conceived as an automated clockwork of *nonliving* matter guided by mathematical principles and physical forces. In contrast to hylozoism, mechanism dictated that the activity of any object was not attributable to the thing itself but to its internal parts or external influences on those parts. Accordingly, what we consider biological life is not a property of matter but simply the proper functioning of an organism's inner machinery. Furthermore, only by studying the constituent parts of a clockwork can the entire clock be understood. Cartesian mechanism has dominated scientific epistemology in the post-Enlightenment period. In modern physics, for example, the ever-shrinking scale of discovery from atoms to protons to quarks highlights the ideological presumption that finer granularity reveals greater truths. In biology, scientists have progressively studied cells, organelles, and finally DNA, at each step extracting deeper facts about nature. Even in chapter 1, locating thought *particles* trumped analyzing the brain.

Mechanism has clearly benefited humanity. But according to Mary Midgley, a moral philosopher who uses Gaia to question human–nature relations, mechanism has unfortunately evolved into an ethical and epistemological straitjacket. In terms of ethics, the migration of natural philosophy from living wholes to nonliving parts authorizes selfishness, even cruelty. "Descartes taught us to think of matter essentially as our resource," she writes. "Animals and plants were machines and were provided for us to build into more machines."[12] In terms of epistemology, the cultural hegemony of mechanism has narrowed our ability to perceive the natural world. Whereas Descartes believes scientific inquiry should focus on the individual and the microscopic, Midgley sees value in the interconnected and the macroscopic. She makes the analogy that nature is best interpreted as an aquarium which we can view through different epistemic "windows," which are biological, mathematical, historical, spiritual, Western, non-Western, and so on.[13] Mechanism unfortunately blocks all other philosophical windows. She writes: "When Galileo first expressed his [mechanistic] views about the world, not only the Pope but the scientists of his day found them largely incomprehensible. Yet those ideas, when developed by Descartes, Newton, Laplace, and the rest shaped the set of windows through which the whole Enlightenment looked into the vast aquarium which is our world. Many in our own age still want to see everything through this set of windows."[14] In its parochial focus on the cellular, the atomic, and the superstring, science has lost sight of the larger wholes defining reality. The visceral rejection of Gaia during the 1970s is testament to the epistemological walls scientific orthodoxy has built. "Our moral, psychological, and political ideas have all been armed against holism," Midgley concludes,[15] and the result is a scientific myopia akin to "looking into a vast, ill-lit aquarium through a single window."[16]

Living Systems

The banishment of hylozoic thinking is a matter of scientific history, but less understood is the role systems theory has played in reviving its premodern logic. My emphasis here is how systems theory (inadvertently) reauthorized neovitalist concepts of life—and mind—for circulation back into mass culture. While it may seem odd to view a widely respected framework like systems theory as inherently parascientific, one of my larger points is that scientific discourse can encode metaphysics and esoterica into itself just as easily as fiction. Bruno Latour argues that science and culture are co-constituted, and one of the positivistic assumptions we ought to abandon from the outset is the purity of scientific thought.[17]

   In the twenty-first century, systems theory has emerged as one of the most vibrant fields in the natural and social sciences.[18] It is broadly understood as the interdisciplinary study of systems, wherein a *system* is defined as any entity with interdependent parts, defined boundaries, and gestalt properties (greater than the sum of its individual components). The ambiguity of the system is one of its strengths: cells, organisms, social clusters, environments, populations, and even nations may all fall under its expansive purview. Systems theory consequently exists across multiple realms of study, with systems biology, systems ecology, and systems psychology only encompassing part of its growing domain. While academic scholarship tends to emphasize its transdisciplinary principles (law of exponential growth, feedback laws, etc.),[19] I focus on its antimechanistic assumptions and the parascientific logic it has vitalized.

At its core, systems theory rejects the Cartesian stress on individual units and instead concerns itself with the interrelationships between those units as it relates to the overall entity. This emphasis on the relations within and the properties arising from the system as a whole characterizes its main epistemological difference from mechanism. Crucially, built into systems thinking is the potential for new and unexplainable properties to arise that cannot be attributed to individual units. To understand this unexpected swerve into metaphysics—so integral for understanding Gaia—we must first turn to Ludwig von Bertalanffy.

Bertalanffy is widely known today as the chief architect of systems theory. While his fame rests on his work on systems biology, or what he called *organismic biology,* his training at the University of Vienna actually began with philosophy of science. His landmark text *General System Theory* (1965) is not only concerned with the principles of systems but the historical and ideological trajectory of modern science. In this text, based on his research from the 1920s onward, Bertalanffy sees "general system theory" as a paradigm shift in scientific method and purpose.[20] In fact, the initial chapters do not address system principles at all but the problems of Cartesianism. I would therefore argue that Bertalanffy's major contribution to science (and Gaia) is not biological but epistemological: his goal was nothing less than revolutionizing the implicit frameworks for doing science.

Bertalanffy chafed against science's mechanist mind-set, in which "the aimless play of atoms, governed by the inexorable laws of causality, produced all phenomena in the world, inanimate, living, and mental."[21] For him, neither biology nor nature more generally was reducible to "mere" laws of chemistry and physics.[22] For example, organisms make active decisions in everyday life; a deer may choose to drink from a stream or cross an interstate highway. Organismic intention is a crucial aspect of nature, yet perhaps because of its sheer complexity, decision-making, adaptability, and other higher-order activities have been largely bracketed from study despite existing everywhere in the biological and behavioral sciences.[23] The problem with mechanism as a scientific ideology is that "no room is left for any directiveness, order, or telos."[24] General system theory aims to correct this flaw; only by looking at wholes can the true properties of the object in question become scientifically visible. Hence, the stated goal of systems theory is to reach "a higher generality than that in the specialized sciences"—to locate the sophisticated properties emerging from system interactions rather than reducing everything to clockwork physics.[25]

Concepts like agency cannot be explained by mathematics, yet they are undefinable outputs of systems we conventionally accept. However, this tacit acknowledgment of unexpected and even unknown emergent properties opens a back door for metaphysical speculation. Bertalanffy admits as much. Under mechanism, "teleology and directiveness appeared to be outside the scope of science and to be the playground of mysterious, supernatural, or anthropomorphic agencies," he notes, but "these agencies exist, and you cannot conceive of a living organism, not to speak of behavior and human society, without taking into account of what variously and loosely is called adaptiveness, purposiveness, goal-seeking, and the like."[26] Bertalanffy was no vitalist, but he saw vitalism everywhere in nature. As Manfred Drack et al. argue, "When Bertalanffy outlines the difference between science and metaphysics, mythical thinking and mysticism, he argues that a dialogue between them is necessary."[27] This is the crux of Bertalanffy's paradigm shift: systems theory redirects the focus of scientific inquiry from individual units to holistic systems, and in those gestalts arise unforeseen and vaguely unscientific properties like will. Metaphysical concerns like intentionality did not belong to the traditional purview of biology, but Bertalanffy argues that they should be, and systems theory provided a means to scientificize it. At issue, as we shall see, is the potential snowball effect of scientifically authorizing metaphysics wherein "will" eventually leads to something approximating "mind."

In the decades after the publication of *General System Theory*, second-order systems theorists from a variety of other fields joined Bertalanffy's bandwagon.[28] Ilya Prigogine won the 1977 Nobel Prize in chemistry for his work on dissipative structures, open systems such as tornadoes, hurricanes, and convection currents characterized by the spontaneous appearance of order from disordered states.[29] In certain conditions, Prigogine argues, matter can interact with its environment in such complex ways that it "begins to be able to perceive, to 'take into account' in its way of functioning, differences in the external world."[30] In 1972, Chilean biologists Humberto Maturana and Francisco Varela coined the term *autopoiesis* to describe the dynamics of autonomy within living systems. "An autopoietic machine," they write, "is a machine organized (defined as a unity) as a network of processes of production (transformation and destruction) of components, which (i) through their interactions and transformations continuously regenerate and realize the network of processes (relations) that produced them; and (ii) constitute it (the machine) as a concrete entity in space in which they (the components) exist by specifying the topological domain of its realization as such a network."[31] Across their research, systems of weather and cells alike demonstrate remarkable properties like adaption, self-organization, regeneration, and even awareness. This groundswell in postmechanistic thinking suggests metaphysical qualities like cognizance can arise from physical interactions within and between complex systems, and moreover that "mystic" qualities are scientifically grounded in the intrinsic relationships between the material stuff (atoms, object, larger entities) of the universe. On the one hand, this work simply follows from recent advances in modern science. On the other hand, it showcases a reversion toward vitalist tendencies where matter is magically imbued with unaccountable characteristics. In recent years, Ray Kurzweil and other thinkers have popularized the notion of the Singularity, in which technology becomes so advanced it coevolves with human evolution to infuse matter with intelligence itself.[32] Systems theory authorizes a similar claim sans human technoculture; the basic systems of nature are sufficiently complex to produce awareness and possibly even consciousness. The increasingly ambiguous space between premodern and modern thought would prove fertile ground for Lovelock and Margulis's Gaia.

James Lovelock, Lynn Margulis, and the Metaphysics of Mind

In 1965, James Lovelock had a revelation. The brilliant engineer, who was consulting for NASA's Jet Propulsion Laboratory in Pasadena, California, had been tasked with developing instruments capable of detecting signs of life on Mars for the upcoming Voyager missions. While other members of the space biology unit had proposed searching for proteins or amino acids in the soil, Lovelock rejected these suggestions out of hand because they assumed Martian life imitated terrestrial life.[33] Instead, he began to ask fundamental questions about life informed by systems theory. Can something beyond Earth's prototypical biological systems (i.e., organisms) maintain and regulate itself? If so, what were its properties and how might we recognize it? Around the same time, Lovelock also had the early opportunity to see some of the first satellite pictures of Earth. He observed how the "cloud-speckled, ocean-blue sphere" of Earth stood in stark contrast to its neighbors, Venus and Mars; the former was a boiling maelstrom of deadly gases and the latter an icy rock.[34] Situated between these dead zones was a system teeming with self-generating rain forests, clouds, undersea vents, and countless organisms. It was at that precise juncture, when he was reformulating the very definition of life and bearing witness to Earth's unique place in the cosmos, that Lovelock experienced his eureka moment: *the Earth was alive.*[35]

From this epiphany and Lovelock's eventual collaboration with Lynn Margulis would soon emerge the Gaia hypothesis: Earth as a giant super-organism, "a complex entity involving the Earth's biosphere, atmosphere, oceans, and soil; the totality constituting a feedback or cybernetic system which seeks an optimal physical and chemical environment for life on this planet."[36] While the Gaia hypothesis has been criticized from its inception for its "unwelcome teleological, feminist, and animist connotations" (much of which derives from its mythological etymology),[37] what often goes unnoticed is the physical/metaphysical blurring arising from its system theory underpinnings. From the interactions of known quantities like forests and carbon cycles arise unknown characteristics like self-regulation and decision-making—which is to say that from the natural emerges the super-natural.

Lovelock believed the traditional characteristics of life we are forced to memorize in high school biology (reproduction, metabolism, movement, etc.) are too arbitrary for defining life in generalist terms. For him, the key to life resided in homeostasis, or a system's ability to continually regulate itself.[38] "One of the most characteristic properties of all living organisms," he writes, "from the smallest to the largest, is their capacity to develop, operate, and maintain systems which set a goal and strive to achieve it through the cybernetic process of trial and error."[39] Norbert Wiener defines cybernetics as "the branch of study which is concerned with self-regulating systems of communication and control in living organisms and machines,"[40] and it is a critical term because Lovelock views life as an emergent property arising from the self-maintaining interactions of a system rather than the epiphenomenon of organismic clockwork. Just as the system is a scalable concept, so too is life. This was Lovelock's major conceptual leap. As a system property, life operates across microscopic and macroscopic levels. A bacterium is a living system pulling energy from its external environment. So is a human. By the same logic, is not Earth also a living system, drawing solar energy from the sun to continually organize its constituent parts?

Lovelock interpreted planetary life as a unique cosmic outcome arising from Earth's various (gas, mineral, energy, etc.) subsystems. Carbon dioxide, for instance, is produced by bacteria and animals (biological systems), distributed through the air (atmospheric systems), and ultimately regulated by bodies of water (oceanic systems) at a rate perfectly adjusted for all its other biological systems to flourish.[41] Similarly, oceans act as a "global steam engine" to carry salt, energy, and gases around the planet to sustain organic life yet is itself a "thin soup of living and dead organisms."[42] This ceaselessly unfolding state of interlocked being is not the result of mechanistic chance but the teleological hand of Gaia herself creating the optimum conditions for life. For Lovelock, Earth is special in a cosmological sense because its biotic and abiotic systems maintain each other in the same way human nervous, muscle, and digestive systems all sustain the larger body. The logical end point of these systemic confluences is Gaia, "a single living entity capable of manipulating the Earth's atmosphere to suit its overall needs and endowed with faculties and powers far beyond those of its constituent parts."[43]

The bold conclusion Lovelock reached at a macroscopic scale resembles the one Lynn Margulis recognized at a microscopic scale. Around the same time Lovelock conceived of Gaia, Margulis published "On the Origin of Mitosing Cells" (1967) in the *Journal of Theoretical Biology*. In this landmark paper, the Boston University professor argued that mitochondria, photosynthetic plasmids, and the basal bodies of flagella commonly found in eukaryotic cells (the cells of multicellular organisms like animals and plants) first developed millions of years ago, when their progenitors successfully absorbed prokaryotic cells (the cells of bacteria) and subsequently developed a symbiotic relationship with them. "The eukaryotic cell is the result of the evolution of ancient symbioses," she controversially claimed at the time.[44] Margulis was not the first to advance this heterodox theory,[45] but she was its chief proponent when other researchers confirmed the bacterial origins of mitochondria and chloroplasts in the 1970s.

True symbiosis, like the cyanobacteria–fungi relationship in lichen,[46] was once considered rare, but Margulis argues that mutually beneficial networks were ubiquitous throughout nature as well as integral to the development of complex life. Symbiosis between bacteria led to unicellular eukaryotes and then to multicellular organisms. In other words, higher life is a system of systems. Moreover, when systems exchange resources and information with other systems, ever sophisticated life-forms can and will emerge. Margulis's thinking further dovetails with Lovelock's when she contends that life is not a unique phenomenon of biological organisms but instead the outgrowth of autopoietic systems.[47] "Soil," she writes, "is not unalive. It is a mixture of broken rock, pollen, fungal filaments, ciliate cysts, bacterial spores, nematodes, and other microscopic animals and parts."[48] Life consequently escapes the confines of traditional, individualizing structures like "the animal" to include all the inputs and outputs necessary for self-maintenance, including organism-to-organism connections and organism-to-environment relationships.

As it is for soil, so it goes for Earth, our largest supersystem. In both *What Is Life?* (1995) and *Symbiotic Planet* (1998), Margulis claims the complex inter-relationships between prokaryotes and eukaryotes model the system life of Gaia. "The Earth, in the biological sense, has a body sustained by complex physiological processes," she contends.[49] "It is the sum of these uncountable interactions that yields the highest level of life: the blue biosphere, in all the holarchic coherence and mysterious grandeur of its budding in and from the black cosmos."[50] Like Lovelock, Margulis perceived life on scales that defied the establishment science of the 1970s. Given the systems approach each of these rebel scientists brought to their respective fields, it is unsurprising they would join intellectual forces over the last few decades to increase public awareness of Gaia.[51]

It is worth noting that Lovelock and Margulis's Gaia hypothesis exemplifies an atavism that Ludwik Fleck calls a "thought style." In *The Genesis and Development of a Scientific Fact* (1935), Fleck observes that older scientific paradigms often leave behind residues of archaic or mystic belief.[52] Such thought styles cannot be eradicated. The hylozoism of Plato and Thales has never completely disappeared; in fact, it has been partially reconstituted in the twentieth century via systems theory, which provides the rationale for perceiving life in nonbiological entities and across expanding scales. Bruce Clarke says it best: "Gaia theory *is* systems theory."[53] Indeed, Gaia is the telos of several decades' worth of systems thinking wherein Bertalanffy's war on mechanism established the intellectual foundation for resuscitating classic theories of living matter. It would also lead to further metaphysical speculations about the mind.

One of the more taboo areas in Gaian discourse, especially among scientific supporters, is planetary consciousness. Tracing these pseudoscientific elements is necessary because, as Steven Shapin has argued, they are crucial sites for understanding the core processes of scientification.[54] I concur that the notion of Gaia mind is a paragon for understanding how literary and scientific texts can blur the difference between rational and irrational thinking, as well as how a stepwise series of logical intimations makes the implausible ever plausible. When Lovelock writes that Earth's atmosphere is "being *manipulated*" or that Gaia can "*adapt* the environment to *its* needs," he intimates Gaia has something approximating agency.[55] Sutured to the rhetoric of Gaian life is the more radical claim of Gaian will. Although Lovelock has long emphasized the metaphorical nature of his claims, the cultural reception of Gaia as an active, thinking planet cannot be ignored. Furthermore, if we can accept Bertalanffy's proposition that gestalt properties are intrinsic to complex systems, then the emergence of Gaia mind is at the very least scientifically permissible.

In the final chapter of *Gaia: A New Look at Life on Earth,* Lovelock directly addresses planetary intelligence. Similar to the concept of *life,* he notes we cannot easily define *intelligence* and at best only categorize it. "Intelligence is a property of living systems and is concerned with the ability to answer questions correctly," he argues, "especially questions about those responses to the environment which affect the system's survival, and the survival of the association of systems to which it belongs."[56] Intelligence is a system property. It is a higher characteristic not born of chemistry between individual parts but arising mysteriously from the system as a whole. For bacteria, the search for food and favorable conditions are automatic processes, yet "it must be recognized that some form of intelligence is required even within an automatic process to interpret correctly information received about the environment."[57] Every cybernetic system must have a basic level of cognition to successfully maintain itself. "If Gaia exists," Lovelock writes, "then she is without a doubt intelligent in this limited sense at least."[58]

This view corresponds with Margulis's interpretation of Gaia mind. In *Symbiotic Planet,* she argues that all living organisms, from humans down to bacteria, are *proprioceptic.*[59] Defined as the perception of movement and spatial orientation arising from stimuli within the body, *proprioception* essentially describes organismic self-awareness. For example, it is through internal proprioception that an amoeba knows it requires more nutrients or that a dandelion recognizes it has collected enough water. Similarly, Gaia has basic awareness and "has engaged a proprioceptive system for millennia, since long before humans evolved."[60] If we can distance ourselves from Cartesian ideology and think like system theorists, then what we call mind begins appearing in places far beyond the human, the animal, or even the organism. For Margulis, mind is everywhere, "an emergent multispecies phenomenon, a kind of evolutionary synesthesia rather than a uniquely human let alone divinely anthropocentric trait."[61] Without the need for a physical brain, Gaian self-awareness is an emergent property of the network rather than the output of any single node. This is an important distinction for comprehending Lovelock's final phenomenological speculations.

"To what extent is our [human] collective intelligence also a part of Gaia?" Lovelock asks. "Do we as a species constitute a Gaian nervous system and brain which can consciously anticipate environmental changes?"[62] In the closing chapters of *Gaia,* Lovelock wonders if mankind represents a new, higher stage of Gaia mind providing advanced cognition characterized by self-reflection, self-responsibility, technological ability, and foresight. If Gaian intelligence is an historically simplistic system property, then the emergence of *Homo sapiens* suggests a great leap forward into complexity. Through mankind, Gaia has attained marvelous proprioceptic insights into her ontological self:

> She is now through us awake and aware of herself. She has seen the reflection of her fair face through the eyes of astronauts and the television cameras of orbiting spacecraft. Our sensations of wonder and pleasure, our capacity for conscious thought and speculation, our restless curiosity and drive are hers to share. This new interrelationship with Gaia with man is by no means fully established; we are not yet a truly collective species, corralled and tamed as an integral part of the biosphere, as we are as individual creatures. It may be that the destiny of mankind is to become tamed, so that the fierce, destructive, and greedy forces of tribalism and nationalism are fused into a compulsive urging to belong to the commonwealth of all creatures which constitutes Gaia. It might seem to be a surrender, but I suspect that the rewards, in the form of an increased sense of well-being and fulfillment, in knowing ourselves to be dynamic part of a far greater entity, would be worth the loss of tribal freedom.[63]

The political utopianism is obvious here. Lovelock wants humanity to cast aside its petty capitalist and nationalist impulses so we can all live peacefully within the superorganism of Earth. But there is a phenomenological pivot as well, one in which humanity's arrival unlocks higher stages of Gaian cognition. Prehuman Gaia was marked by rudimentary processes of self-maintenance and self-organization. Posthuman Gaia is characterized by advance planning and technological ability. In the epilogue of *Gaia,* for instance, Lovelock argues that humanity could someday use its technological prowess to destroy an asteroid on a collision course with Earth;[64] we save ourselves, of course, but we also save the Gaian system! Through us, then, Gaia becomes a superorganism with enhanced insight, will, and proactive powers. In *Evolution and Consciousness* (1976), Erich Jantsch and Conrad Waddington contend that human consciousness has evolved upward through several stages over millennia and will eventually peak in "superconsciousness"—a transpersonal form of mind that transcends the individual and gains a *networked* awareness of history, biological evolution, and ontological existence.[65] The natural evolution of human consciousness is therefore akin to global mind. As such, Lovelock, Jantsch, and Waddington all see the geohuman amplification of mind as a significant, even inevitable, step in the rapidly evolving Earth system.

In his autobiography, *Homage to Gaia,* Lovelock writes, "To my astonishment, the main interest in Gaia came from the general public, from philosophers, and from the religious."[66] Such a reception does not surprise me, though. For philosophers like Mary Midgley and Jacob Needleman (whose quotation opened this chapter), Gaia connects science with broader questions of ethics, nature, and humanism. For the spiritual, Lovelock and Margulis's redefinition of mind draws from bleeding-edge developments in systems theory to resuscitate premodern beliefs in a world soul. In short, their progressive science provided a platform to substantiate marginalized concepts of living matter, distributed mind, and the collective unconscious. As Michael Ruse writes, "If you are true to the Gaia project—to see the world as an organism—you are committed to something deeply end-directed, deeply teleological."[67] Such metaphysics might give scientists pause, but it identifies the locus of Gaia's raw cultural power: its invocation of ancient myth as scientific truth. The Gaia hypothesis may be a natural extension of Bertalanffy's systems, but that progression is imbued with a latent supernaturalism pushing against the boundaries of mainstream science. The logical would further entangle with the illogical when Gaian science encountered the New Age.

Gaia and the New Age

The New Age is a widely misunderstood subculture often associated with "past lives, therapists, and crystal healers, earth goddesses and lost civilizations, mantras and gurus, Harmonic Convergence and shamanic voyages, Hollywood ghosts and California channelers, natural medicine and pagan rituals, Shirley MacLaine and Marilyn Ferguson."[68] Beyond this pejorative view, however, is a wide-ranging spiritual movement that emerged from 1960s countercultural and 1970s environmental movements.[69] According to religion scholar Sarah Pike, four principles united its members: (1) movement toward self-awareness, (2) disenchantment with traditional religion, (3) embrace of non-Western philosophy, and (4) interest in holistic healing practices.[70] The New Age is relevant because it serves as a parascientific hub bridging Gaian science, literature, and myth; in fact, several of its leading writers would push Gaia in paranormal directions.

Similar to other occult movements, the New Age exists at the crossroads of spiritualism, science, and mind. Its associations with physics are well known. In *How the Hippies Saved Physics,* David Kaiser documents how several of the most important physicists of the late twentieth century were members of the New Age, as well as how its "anything-goes counterculture frenzy" of psychedelic drugs and Eastern mysticism generated a productive space for reinventing quantum mechanics.[71] The Fundamental Fysiks Group (FFG) famously started as a theoretical physics discussion group at the University of California,

Berkeley, that later added LSD, séances, and mind reading into its activities. One of its founders, Elizabeth Rauscher, was an occultist who believed in remote healing and telekinesis.[72] Another FFG member, Jack Sarfatti, started a popular physics and consciousness workshop series (1976–88) at the Esalen Institute, the New Age incubator in Big Sur, California.[73] One of the most important physicists in the FFG network, David Bohm, had a long-running series of conversations with Indian spiritualist (and former Theosophist) Jiddu Krishnamurti on the nature of consciousness; both agreed the meditating mind, "having freed itself from the general and particular structure of the consciousness of mankind, from its limits, is now much greater."[74]

Systems theory also proved influential among New Age adherents because it provided a well-regarded framework for theorizing wholeness as a physical fact of nature. The gestalt logic of emergent properties coheres well with New Age ideas of higher forms of interconnected being. Gaia mind is the phenomenological extension of such being. Building from the work of Bertalanffy and Lovelock, several New Age writers began to explore the properties and possibilities of planetary mind in the 1980s. While I began this chapter with Pepper Lewis, authors Peter Russell and Fritjof Capra provide even more influential examples of parascientific inquiry in which science and spiritualist discourse blur into distinctive new forms. As we shall see, Russell's Gaiafield and Capra's global unconscious represent clear advancements in paranormal theory because they push systems logic further than orthodox scientists are willing to go. New Age texts are also effective tools of dissemination because they combine religious, mythic, and scientific sources for a public sphere that outnumbers scientific audiences. Even though Russell's and Capra's intellectual labors are more popular than scientific, their ability to extrapolate systems logic is analogous to the function of speculative fiction. Clearly, science is not the only way scientific ideas spread.

In *The Awakening Earth* (1982), Peter Russell advances the concept of Gaia mind by positing not merely intelligence but *global consciousness* as a natural by-product of Earth's systems, a new mental state hinging on the planet's most recent and powerful development: mankind. Russell was inspired by cyberneticist Gregory Bateson, who argued that "mind" was an inherent feature of any self-regulating system.[75] Along similar lines, Russell writes, "Any being which experiences has consciousness."[76] Humanity is nevertheless unique in Earth's history because of its self-reflective ability ("we not only experience the world around us and within, we are also conscious that we are conscious") and advanced communications (writing and telecommunications as linguistic technologies projected over time and space).[77] Russell further contends that humanity's expanding population and telecommunication capabilities represent a breakthrough in organizational sophistication—not just for our species, but for the larger Gaian system where we now play a leading role. "As worldwide communication capabilities become increasingly complex, society is beginning to look more and more like a planetary nervous system," he writes. "The global brain is being activated."[78] Earth may already have been conscious, but only the rise of mankind and its remarkable technologies have roused the planet's higher mental processes and holistic sense of being. Russell thus shares Lovelock's belief that humanity plays a special role in Gaia mind. Where Lovelock the scientist can only speak in metaphors, though, Russell the New Age theorist is remarkably frank. Humanity's ability to communicate instantaneously with all its members via radio, television, and telephone means the "global social super-organism" has been actualized, thereby mobilizing instantaneous, worldwide thought.[79]

Russell calls this exceptional development the "Gaiafield," a form of distributed consciousness arising from the human technocollective. Mankind makes this dynamic stage possible but can neither control nor comprehend it. "The Gaiafield will not be the property of individual human beings, any more than consciousness is the property of individual cells," he writes. "The Gaiafield will occur at the planetary level, emerging from the combined interactions of all the minds within the social superorganism."[80] Put differently, the Gaiafield transcends. It is constituted by the networked minds of all the other subsystems (human, bacterial, plant, atmospheric, aquatic, etc.) on the planet, and therefore its emergent properties are capable of "functioning at a new evolutionary level with faculties quite literally beyond our imagination."[81] This is where the mysticism of New Age thought crucially overlaps with the scientificity of systems theory. Bertalanffy insisted that unexpected properties arise within any complex system. Lovelock saw Earth as the most complex system of all. This allows Russell to posit a fringe theory of mind that joins hylozoism and technoutopianism into an all-powerful, all-thinking Earth.

Fritjof Capra adopts a less anthropocentric but no less powerful view of planetary mind in *The Turning Point* (1982). Capra is an FFG and Esalen stalwart most famous for writing *The Tao of Physics,* a New Age blockbuster connecting modern physics with ancient Chinese, Buddhist, and Hindu truths. Importantly, it claims quantum mechanics revealed the flaws of Cartesianism and advocates for more holistic, Eastern thinking. In *The Turning Point,* Capra turns his Bertalanffy-inflected lens onto Earth, which "functions not just *like* an organism but actually seems to *be* an organism—Gaia, a living planetary being. Her properties and activities cannot be predicted from the sum of her parts;

every one of her tissues is linked to every other tissue and all of them are mutually interdependent."[82] Like Russell, Capra is a writer who can take systems theory to places that Lovelock the engineer can only hint at. Capra's interpretation of Gaia mind is notable for two reasons. First, unlike the Gaiafield, Earth's current state of "mentation" is neither human centered nor human derived. When he writes "there are larger manifestations of mind which our individual minds are only subsystems," mankind exists on par with other organic and nonorganic systems.[83] In this respect, his view hews more closely to Margulis's posthuman Gaia than Lovelock's prohuman one. Second, he believes Gaian mentation is so complex its true dimensions are hidden from us. Capra argues: "Because the systems view of mind is not limited to individual organisms but can be extended to social and ecological systems, we may say that groups of people, societies, and cultures have a collective mind, and therefore also possess a collective consciousness. We may also follow Jung in the assumption that the collective mind, or collective psyche, also includes a collective unconscious. As individuals, we participate in these collective mental patterns, are influenced by them, and shape them in turn."[84] Capra's interpretation of Gaia mind includes both conscious and unconscious realms and drastically enhances the breadth of its mystery. More than Freud, Swiss psychoanalyst Carl Jung saw the unconscious as the true seat of the mind's power, an undiscovered continent housing our collective past, archetypal knowledge forms, and oldest myths. To the great unknown of Gaian consciousness, Capra proffers a Gaian unconsciousness further influencing our existence. If the human unconscious is powerful and mysterious, then Gaia's is infinitely more so. Once again, the logic of emergent properties point us toward divine, ethereal, and supernatural implications. The paradoxical power of New Age discourse is how deftly it weaves scientific possibilities into scientific impossibilities, and vice versa. The artful blend of science and esoterica suggests that contemporary science merely confirms what the ancients have always known: that humanity is not a collection of individuals but part of something deeper and wiser. Whether that something is called God or mind is irrelevant; what matters is that systems theory affirms this fundamental truth.

While some New Age theorists lean heavily on science to argue for the mystic unity of the Earth, others like Pepper Lewis rely on spiritualism. What both groups share is an ideological distaste for Cartesian reductionism, man–nature binaries, and positivistic ideology. In "Human Consciousness in Transformation," sociologist O. W. Markley describes a new mode of scientific inquiry emerging at the end of the twentieth century. This approach, which arises from scientific interest in meditation as well as Eastern philosophical perspectives on consciousness, represents a novel "knowledge paradigm" focused on systems and mind.[85] Its hallmarks include inclusivity (wisdom derived from prescientific cultures); methodological eclecticism ("extrasensory" modes of knowing are legitimate); systematization of subjective experience; holism (rejection of artificial binaries); and complementarity between material and spiritual systems.[86] New Age Gaianism fits this description well and, following Markley's lead, potentially represents a new form of inquiry "which denies none of the conclusions of science in its contemporary form, but rather expands its boundaries."[87] The works of Lewis, Russell, and Capra all participate in this expansion, extrapolating science in both archaic and avant-garde directions and subsequently redefining what counts as rational. In the process, mind becomes an increasingly powerful emergent property that is scientifically undeniable. In straddling the epistemes of science and nonscience, New Age texts are formidable instruments for reconfiguring the epistemological convictions of its readers as well as the boundaries of science. They are not the only literary mechanism for reconceiving the science of planetary mind in mass culture, though. For an equally powerful tool, let us turn at last toward fiction.

Greg Bear and Gaia Unleashed

Speculative fiction plays a special role in parascientification because its synthesis of scientific and nonscientific (mythic, philosophical, etc.) elements confers greater leeway—and therefore imaginative power—in reconceptualizing scientific heterodoxy. As we just saw in chapter 1, that imagination can retheorize the physical and neurological mechanisms for making the paranormal rather normal. In addition, the enduring popularity of the genre grants it wider cultural influence than purely scientific or theological sources. In Gaian discourse, speculative fiction has several functions. First, it has popularized the idea of a living planet as the ultimate posthuman entity. Second, it has frequently granted these godlike entities with unfathomable intelligence and will. Most importantly, it conjures an epistemological space for probing Gaia mind, its physical mechanisms, and its metaphysical possibilities. In extrapolating what Gaia mind is, how it works, and what it wants, fiction helps establish the ideological landscape for planetary consciousness. As such, it is more than just a tool for sociological reflection; it is also a source for epistemological production.

The idea of Gaia has a rich history in science fiction. Stanislaw Lem's 1961 novel *Solaris* features an ocean planet that "lived, thought, and acted."[88] In fact, it telepathically mines the memories of researchers sent to study it and manufactures people to screw with their sanity. In Isaac Asimov's *Foundation's Edge* (1982), the mysterious planet Gaia is a single being sharing collective consciousness with all its denizens: "The whole planet and everything on it is Gaia. We're all individuals—we're all separate organisms—but we all share an overall consciousness."[89] Other texts invoking Gaia include David Brin's *Earth*, Stan Constantine's *Hermetech*, and Piers Anthony's *Being a Green Mother*. Most recently, N. K. Jemisin's Broken Earth trilogy (2015–17) features Father Earth, a vindictive Gaian figure so enraged by humanity's environmental destruction that he rains down geological terror.

Greg Bear is unique in this collection of writer-theorists for his investment in systems theory and the biological logistics of Gaian telekinesis. Roger Luckhurst regards him as the most rhizomatic of science fiction writers: "As Bruno Latour has done for science studies, so Bear's fiction offers the opportunity to trace the networks that connect together wildly diverse hard and soft things: laboratories, parliaments, machine intelligence, galaxies, bedrooms, spaceships, survivalists, mitochondrial DNA, American presidents, geologists, viruses, high tech start-up capitalists, the undead, the posthuman, and the alien."[90] His interest in narrative networks extends to Gaia mind as a system phenomenon—how it arises, how it functions, its biophysical components and subsystem relations, and its connections to planetary being. Interestingly, he maps microbiology onto systems biology to explore the technics of Gaian will; for Bear, viruses and bacteria are agents of telekinesis in service to the planetary whole. Like Philip K. Dick and Alan Moore, Bear is a literary explorer of the paranormal whose fiction emerges as a key site for investigating parascientification.

Bear has published several "microbial" narratives since the 1980s, beginning with his nanotechnology viruses in "Blood Music" (1983), but *Darwin's Radio* (1999) and *Vitals* (2002) are noteworthy because systems theory assumes a dominant role in the development of Gaian consciousness. The plots of both texts feature a global entity that is not merely alive in a metaphorical sense but is an actual superorganism with extraordinary awareness, determination, and control over human affairs. In *Darwin's Radio,* the human genome is a Gaian entity that decides when and if *Homo sapiens sapiens* deserves a future on Earth. With its anthropocentric slant, the novel follows Lovelock's and Russell's view that humanity plays a special role in planetary mind. In contrast, *Vitals* posits bacteria as a global superorganism determining humanity's destiny and closely follows Margulis's and Capra's vision of a truly posthuman Gaia. A major theme in these works is the questioning of Cartesian individualism in the face of supernatural hylozoism. In Bear's fiction, systems of exchange take precedence over individuals, even if they are as simple as genetic or bacterial networks. If we define telekinesis as mind directly affecting matter, then viruses and bacteria operate in these novels as Gaia's telekinetic agents. Bear does not use fiction to simply reflect back existing scientific models but rather to think through new phenomenological models. Ironically, by using systems microbiology to explain the mechanics of Gaia mind and telepathy, he returns us to a premodern vitalism.

Darwin's Radio follows a worldwide public health crisis precipitated by the outbreak of a scattered human endogenous retrovirus code-named SHEVA. Government scientists initially believe SHEVA is the vector responsible for a global pandemic of miscarriages, but we soon learn SHEVA is not the "supreme destroyer," as its Hindu-esque name suggests, but a catalyst for the next evolutionary step forward for man: *Homo sapiens novus*. Miscarriage, it turns out, is just the first step in the multistage birthing of a new human subspecies with fifty-two chromosomes and incredible communication powers, including pheromone conversation and facial "flashing." The novel focuses on Kaye Lang, a Margulis-inspired microbiologist whose pioneering work reveals that human evolution depends on the symbiotic collaboration between viruses and DNA, and Mitch Rafelson, a rebel anthropologist who discovers that SHEVA outbreaks have catalyzed rapid evolution throughout human history. Positioned against these two outsider scientists are an array of government officials hoping to eliminate SHEVA and the mutant offspring it produces. Lurking behind all of them is a global superorganism advancing its own agenda.

While the narrative emphasis of *Darwin's Radio* is the fetus-killing SHEVA crisis, encoded in the backstory is the emerging intelligence of a global system invisibly guiding human evolution: our genome. One of Lang's major discoveries is that SHEVA is not a novel virus at all but rather an ancient genetic messenger located in the so-called junk DNA (noncoding regions of DNA once considered unimportant) of all humans. Once activated, SHEVA initiates a series of rapid mutations producing *Homo sapiens novus* in a single generation. One of the more radical propositions in the novel, besides its hyperbolic depiction of evolutionary cladogenesis,[91] is that viruses constitute a simultaneously primitive *and* advanced form of communication among human, nonhuman, and ecological systems. In *The Awakening Earth,* Russell claims that the Gaiafield is only possible because of radio and television. Bear argues that viruses already represent a form of worldwide telecommunication. Viruses jump from human to human, species to species, and continent to continent, so they have therefore served as information carriers since time immemorial. SHEVA thus acts as a "neurotransmitter of sorts" for a DNA neural net spanning the entire planet, one in which genetic development, environmental feedback loops, and adaptive mutations are secretly processed by the collective genome of the human species.[92] Individual people do not control the will of this meta-biological process; this is the role of a higher intelligence, an "evolutionary processor" existing in the distributed mind of the *human genomic system*.[93] As Lang explains, "There's a master biological computer in each species, a processor of some sort that tots up possible beneficial mutations. It makes decisions about what, where, and when something will change."[94] In *Darwin's Radio*, Gaia mind arises from the global system of DNA, and as such, it is both a physical entity existing within every human on Earth and a metaphysical force making operational decisions for the species.

Luckhurst has argued that *Darwin's Radio* explores the "genetic singularity," in which the human genome has reached such an advanced state it attains a higher consciousness capable of engineering itself.[95] What I find more fascinating from a parascientific perspective is how Bear explores the metaphysics and mechanics of the paranormal mind via systems theory and microbiology, respectively. Systems explain how the Earth is alive, and viruses explain how it telepathically enforces its will. The novel's system ideology is apparent from its decentering of individual bodies and valorization of emergent properties like agency and intelligence. The migration of life and consciousness away from biological Cartesianism toward one grounded in systems obviously derives from Lovelock. As Lang explains, mind is "the result of complex inter-actions of a network, with emergent thoughtlike properties," and the one we see in the novel is a worldwide transgenic system trading environmental and biophysiological data to make decisions.[96] The anthropocentrism of this Gaia also follows the logic of Russell's Gaiafield. The planetary consciousness in *Darwin's Radio* is "humanist" insofar it arises from and is dominated by the human genome. Distributed across time and space and instantaneously com-municating through viruses, humanist Gaia can interpret environmental stress-ors, strategize optimal mutations, and implement them on a massive scale as part of its autopoietic operations. There is little that anthropocentric Gaia cannot do.

Darwin's Radio excels at pitting Cartesian individualism against systemic needs. Individuals, like the misguided Centers for Disease Control official/villain Mark Augustine, want to preserve mankind's status quo; life has been great for hundreds of thousands of years, so why change it? But the evolution-ary processor chooses to eliminate *Homo sapiens sapiens* in favor of *Homo sapiens novus*. One explanation for this pivot is the health of Gaia. As Lang's mentor, Judith Kushner, argues, "For years I've been waiting for nature to react to our environmental bullshit, tell us to stop overpopulating and depleting resources, to shut up and stop messing around and just *die.*[97] Modern man has been a plague on Earth; hence, the Anthropocene must end as soon as possible to benefit the entire planet. SHEVA is an extermination event for Gaia's benefit: a "self-maintaining" correction for the larger system. Lang adopts a less pessimistic view and argues that the evolutionary processor is attempting a "system upgrade" to meet the stressors of twentieth-century life. Overcrowding, intense competition in all fields of education and employment, radiation, and war have all taken a toll on humanity, she argues.[98] For her, SHEVA is a repair mechanism enacted by a genomic Gaiafield, "an organized network that responds to its environment and issues judgments about what its individual nodes should look like."[99] The shift from *sapiens* to *novus* speaks to system concerns, which is why small-minded humans like Augustine cannot see the bigger picture. He is limited by his mechanistic individualism. Only system theorists like Kushner and Lang, and Bertalanffy before them, correctly recognize why this paradigm shift, this revolution, is so necessary.

Bear's use of virology to explain the biomechanics of planetary mind is noteworthy. "These pathways and methods of regulations [between organisms and viruses] point toward a massive neural network capable of exchanging information," he describes in a lecture to the American Philosophical Society. "Though cross-species exchange of genes, an ecosystem becomes a network in its own right."[100] Bear recognizes DNA is an ancient form of information that exists practically everywhere. If all that information is interwoven and interconnected (thanks to viruses), could we not say that DNA itself is a global system that has been regulating and maintaining itself for millennia? Would it be a stretch to say genes have adaptiveness and intelligence, as defined by Gaian theorists? "What if DNA can store up changes for punctuated equilibrium and edit it as needed?" Bear proposes.[101] As we can see, the extrapolations made possible by systems theory enable all kinds of (pseudo)scientific conjectures.

The same goes for viruses as psychical agents. Bear is correct to identify viruses as traveling bits of DNA and RNA; that is exactly what they are. If mind exists in the human genome, then viruses become the physical agents for enacting mental will; they effect change in the material world. This insight helps us reinterpret SHEVA. On one level, it is the diabolical source of a pandemic, but on a second level, it is the telekinetic means by which Gaia mind controls matter. The genetic processor initiates a subspeciation event by releasing viruses that invisibly reach out, manipulate, and alter human bodies around the world. This is an intriguing development because Bear essentially reinvents the scientific principles of telekinesis. In the nineteenth century, psychical researchers believed those with powerful enough minds could levitate objects and rap against tables using unseen forces. In *Darwin's Radio,* the world mind initiates miscarriages and cladogenesis through another kind of unseen force: the virus.

Bear also introduces clairvoyance in *Darwin's Radio* as another system property abetting planetary telekinesis. Much of the book's plot hinges on Mitch Rafelson's discovery of a mummified Neanderthal family in an Alpine cave. Incredibly, a Neanderthal couple had given birth to a fully formed *modern* human baby—incontrovertible evidence of cladogenesis. Even stranger is the fact that Rafelson receives uncanny visions of the Neanderthal couple's final days on Earth. In some dreams, he can view their physiological features, like broad noses and low foreheads, through the planet's external perspective: "He saw them first in profile, as if in some rotating display, and amused himself for a while viewing them from different angles."[102] In other instances, the Neanderthals' phenomenological experience enters his mind, such that their anxieties becomes his own: "They [other tribal members] were kicking us off the lake, out of the village."[103] Rafelson thus learns the family was murdered by other Neanderthals who feared the "mutant" child *(Homo sapiens sapiens)* growing in the woman's womb—a grim history threatening to repeat itself with the emergence of *Homo sapiens novus.* Here, as before, systems theory explains the paranormal. In *Darwin's Radio,* the human genome has a mind as well as memories. Everything that has ever happened to any one individual is stored within the larger system. The dream sequences can be interpreted as the delivery of phylogenetic memories from the DNA collective to Rafelson, a bestowing of the ancient past to a single node in the network. In other words, he gains direct access to the Gaiafield's vast archive.

I bring up this story line because Rafelson the clairvoyant is also an agent of Gaian telekinesis. Over the course of the novel, he emerges as a New Age prophet for *Homo sapiens novus.* Our divine genomic system has revealed that the SHEVA crisis is part of a world-historical evolution, that this transformation is necessary and good, and that it must transpire as decreed. Before anyone else in the novel, Rafelson understands that *Homo sapiens novus* babies deserve protection, not genocide. In fact, his mission by the novel's end is to spread a message of love and acceptance to the public. Like the SHEVA virus itself, Rafelson becomes a disseminator of Gaian information and an active enforcer of its supreme will.

If *Darwin's Radio* presents an anthropocentric Gaia inspired by Lovelock and Russell, then *Vitals* provides a posthuman version refracted through the work of Margulis and Capra. *Vitals* is not a sequel to *Darwin's Children*, but it is a powerful expansion of systems theory, the microbial mechanisms of Gaian telepathy, and the apocalyptic powers of a global mind. The plot revolves around yet another Margulis-inspired renegade microbiologist, Hal Cousins, who is obsessed with unlocking the secrets of human immortality. He discovers human aging occurs when cellular mitochondria decrease their energy output; however, if they could be tricked into continually maintaining energy flow, mankind could enjoy eternal life. Because mitochondria derive from bacteria (theorized by Margulis in 1967), Cousins's research program involves tilting bacteria–human power relations toward the latter. "We're big kids now," Cousins explains to an angel investor. "We made fire. We made antibiotics. Did the bacteria give us permission to go to the moon? We're ready to take charge and be responsible for our own destiny. Screw the old ways."[104] Cousins hopes to flip the existing biological power structure so humanity, not bacteria, controls the cellular mechanisms of longevity.

However, as death and disasters pile up around Cousins's immortality research, we soon learn of two secret global conspiracies. First, Cousins encounters Silk, a secret Russian organization that has deployed bacteria for half a century as a form of mind control to influence the highest levels of society. Led by Maxim Golokov, the "most brilliant biologist of the twentieth century" and the "Svengali of germs," Silk has infiltrated both the Soviet and U.S. governments using bacteria-derived neurochemicals to further its nefarious goals.[105] But Bear executes the ultimate posthuman bait and switch when we learn in the final chapters that Silk merely occludes a larger conspiracy: bacteria are the true masterminds of the planet. Silk has not been using bacteria; bacteria have been using Silk! Through Silk, prokaryotes have stealthily acted as puppet masters controlling the social, political, and biological history of man. Humans have long assumed they were Earth's dominant organisms, but *Vitals* argues that bacteria are the planet's first life-forms, as well as its most powerful. They dictate not only human existence but all of organismic life. After centuries of anthropocentric hubris and exploitation, the worldwide bacteria network—bacterial Gaia—has chosen to wage war against humanity.

Much of the bacterial primacy of *Vitals* stems from the growing recognition of the microbiome. In recent decades, the biological sciences has experienced a paradigm shift stemming from the realization that the notion of the human cannot be conceived as an individual when it is actually a superorganism consisting of both human and bacterial components. From digestive system to immune system to epigenetic development, human growth and physiological maintenance are increasingly understood as symbiotic processes between human and bacteria cells in our bodies. Scott Gilbert and Jan Sapp argue the shift from individual organism to group existence demands a new way of doing science because "our bodies must be understood as holobionts whose anatomical, physiological, immunological, and developmental functions evolved in shared relationships of different species."[106] Accordingly, we cannot think of bacteria as pathogenic but integral to life itself. In *The Origins of Sociable Life*, Myra Hird argues that society must adopt a "microontological" perspective taking microbial coexistence and ethics seriously.[107] Ninety percent of the cells in the human body are prokaryotic, she notes, whereas only 10 percent are eukaryotic (which themselves have prokaryotic roots, as Margulis showed). When Hird proclaims we are "symbionts all the way down," she means that "we are, ancestrally, made up of bacteria."[108] To see ourselves as disconnected from bacterial life is philosophically parochial and scientifically misguided. She therefore encourages an ontological shift from humanist individualism toward a "nonhumanism" that refuses to privilege human concerns over microbial ones.[109] Like Mary Midgley, Hird adapts a holistic ideology stressing the universal connection of human systems to other forms of being. Bacteria have a distinguished place here because "every living thing that exists now, or has ever lived, is a bacterium."[110]

Vitals starts from this microontological premise and expands it by an order of magnitude: not only is the human a bacteria-dominated superorganism, but so is our planet. Bacteria are located in the upper troposphere, the deepest ocean trenches, and everywhere in between. Bear posits that if such a thing as Gaia exists, it stands to reason that her body—and mind—would be bacterial in origin. This is actually one of Margulis's claims. "One legitimate answer to the question 'What is life?' is 'bacteria,'" she notes. "Any organism, if not itself a live bacterium, is then a descendent—one way or another—of a bacterium, or, more likely, mergers of several kinds of bacterium. Bacteria initially populated the planet and have never relinquished their hold."[111] There is, of course, the issue of complexity. Historically speaking, humans and other eukaryotes are called higher organisms because our multicellular bodies are physically larger with communication networks (e.g., nerves) built into those bodies; bacteria are lower organisms because they lack such features. That said, the scale of bacterial systems in nature easily dwarfs the human. And whereas Russell's Gaiafield depends on human telecommunications to achieve global coverage, *Vitals* suggests that the perpetual swapping of genetic material, plasmids, and chemicals among bacteria—which cover most every surface on the planet— creates a planetary network larger and more sophisticated than anything humanity has created. Moreover, the network constitutes "the oldest mind on Earth"—an emergent property of the biggest system in history.[112] Like the evolutionary processor in *Darwin's Radio,* this Gaia mind is a distributed phenomenon with a ubiquitous presence and panoptic vision. It includes all bacteria around the world, and because they are constantly talking to each other biochemically, they are everywhere and nowhere, with limitless reach. The terrible realization for Cousins, and for us, is that a bacterial Gaia does not require mankind to survive. In fact, our very existence depends on her pleasure. To paraphrase Michel Foucault, Gaia has the hypersovereign power to take life or let live on an unprecedented scale.

This is best demonstrated when Cousins attempts to reach the vendobionts, ancient bacteria on the ocean floor that hold the secret to eternal life. Cousins successfully retrieves a vendobiont sample for his research but is almost killed when the pilot of his deep-sea submersible suddenly flies into a murderous rage: "[The pilot's] eyes went unfocused, wild, like an animal caught in a cage. . . . He yanked the control stick out of its socket and swung it around his head. Before I could raise my hands again, he crashed the stick hard against my temple."[113] This is only the first of many episodes throughout the novel in which friends, strangers, and even lovers turn into psychotic killers. Such bacterial brainwashing constitutes another extrapolation related to the mechanisms of telekinesis. In *Darwin's Radio,* genomic Gaia gains telekinetic control over human bodies through viruses; in *Vitals,* bacterial Gaia does so through neurotoxins. Silk has dominated global politics by spreading bacteria—and their vast array of behavioral neurotoxins—to manipulate their targets' minds. This is made clear when Cousins locates his dead brother's diary (he was killed investigating Silk) and discovers their odious plans: "G. [Golokhov] studied parasitic control of hosts. Parasitized ant climbs to grass tip, eaten by bird, parasite's next stage is in bird. Rats with toxoplasmosis have cysts in brain, not afraid of cats, get eaten, cats carry toxo. *Wolbachia,* a widespread bacterium, actually controls reproduction of host insects and other arthropods. G. then moved on to studying mind-altering substances produced by parasites and compared them with bacterial products. Many gut bacteria talk to intestinal cells. They too alter host bacteria, he found."[114] This is essentially bacterial telepathy. As the passage reveals, many bacteria produce chemicals allowing them to invisibly control animal behavior, reproduction, and mobility. By getting specific targets (e.g., presidents, senators, KGB officials) to ingest bacterial spores or neurotoxins, Silk—and their bacterial overlords—directly manipulates human thoughts and actions. Throughout the novel, Cousins is by turns attacked, seduced, and tricked by people under the influence of microbes. In this way, bacteriology provides a biochemical mechanism explaining how the world mind influences material objects from afar.

One of the late revelations in *Vitals* is that Golokhov is not bent on world domination per se. He has backstabbed a series of world governments to ensure Silk's continued existence, and he has given himself unusually long life (he is a robust 105 years old), but his agenda is ultimately not political. Indeed, Golokhov considers himself more of a "linguist" whose true goal is deciphering the messages of the bacterial world mind. In *The Turning Point,* Capra argues that Gaia is so complex that its full scope and cogitations are inevitably occluded from us. Such blindness to bacterial Gaia's intentions, machinations, and awesome knowledge drives Golokhov mad with envy.

Near the end of the novel, Cousins chances on a colony of mutated humans submerged in water tanks overgrown with primitive bacterial structures known as stromatolites. Golohkov has enlisted these genetically engineered people as Listeners to receive and translate the Voice of the Little Mothers ("Little Mothers" refers to bacteria as both the tiny creators of earthly life as well as the "little motherfuckers" he cannot control). Although he is a microbiologist, Golokhov also assumes the role of a psychical researcher here because his aim is to decode the hidden principles of communication underwriting the world mind. Like talented mediums awaiting psitron signals in Adrian Dobb's proposed EEG experiments, the Listeners lie in microbial baths full of plasmids, chemicals, and other genetic-based signals to tap into Gaia's system intelligence. The Listeners' literal immersion in a biochemical soup illustrates the way twentieth-century telepathy has been parascientifically reconceived through biology. Where mental control of matter was once the domain of invisible forces, in *Vitals* it emerges as the microscopic exchange of genetic bits influencing material life. This scene also emphasizes the superiority of bacterial systems over human ones thanks to the former's age, size, and complexity. Golokhov, one of the great human minds of the twentieth century, is reduced to eavesdropping on our bacterial overseers. "Larger and older minds live inside our bodies and all around us, speaking in languages I have worked all my life to interpret," Golokhov explains to Cousins.[115] He goes on:

> Do you know what the message is? What little I have intercepted and translated over seven decades, the sum of all my good work on this Earth, in this forsaken country ... All the Little Mothers whispering in our bowels and in the forests and jungles and in the oceans we are working so hard to destroy. They are not happy. They are not happy with us at all. We are a bitter disappointment to them. They will wage all-out war against us now. It's a judgment we cannot withstand.[116]

Gaia is a living system requiring healthy air, jungles, and oceans. Humanity has provided poor stewardship of the system: clear-cutting forests, mountaintop removal mining, extermination of plant and animal life, toxic waste, oil spills, nuclear blasts, Superfund sites, ozone depletion, global climate change—and the list goes on. In *Darwin's Radio*, Dr. Kushner suggests that Earth is fed up with humanity's "bullshit," and this same logic runs through *Vitals*. In contrast to the wise, nurturing Earth Mother found in New Age writings like *Gaia Speaks*, Bear draws on Margulis's famous claim that Gaia is one tough bitch.[117] She can be a loving caretaker, but she has her limits, and in *Vitals* she makes a self-maintaining decision to annihilate humanity for the benefit of the overall system. Bacterial Gaia does not play favorites; she is a posthuman bitch ready and willing to destroy her "highest" creation.

At the close of the novel, we find Hal Cousins locked in a California apartment, living on pasteurized foods, staring out the window while waiting for the world to end. This final, paranoid image is rich with irony. In Bear's dark vision of the paranormal mind, man learns he is part of a larger planetary consciousness but paradoxically gets locked out of it, marginalized at the behest of a supernatural unity he has always sought within himself.

Ever since the end of the nineteenth century, psychical researchers have wondered if the paranormal mind exists and how it performs unexplainable feats like telekinesis. In *Darwin's Radio*, Bear views the human genome as a global system guiding mankind with viruses acting as a telekinetic force. In *Vitals*, bacteria claim center stage as a global mind deploying spores, neurochemicals, and plasmids as its invisible agents. Bear's reimagination of the paranormal obviously differs from previous iterations we have seen, especially those derived from Victorian psychical research. Much of this emerges from twentieth-century systems thinking and New Age science. Yet his iteration of paranormality nevertheless perpetuates a long-standing belief that the human mind is part of something greater—an anima mundi possessing collective agency and otherworldly powers, surging through us via unseen pathways of lost memory, deep time, and end-directed evolution. Like a necromancer raising long-buried ideas through the magic of systems theory, Bear conjures a new kind of Gaia that does what the public has often demanded of Mother Nature: wisdom, nurture, and, when push comes to shove, discipline.

Mainstream science may find little value in studying the characteristics of Gaian consciousness, but fiction takes this barrier as a departure point for fresh modes of theorization and knowledge seeking. Science fiction editor John W. Campbell considered science fiction "a sort of thought experiment in which the author carefully creates a set of hypotheses regarding fictional events and lets the story grow out of those hypotheses."[118] Bear's stories follow this schema. His texts are literary laboratories for reconsidering Gaia mind through modern biology, and through his steady output, he informs (and misinforms) the public on the infinite possibilities of complex systems. Like Lewis, Lovelock, Capra, and others, he alters existing perceptions of planetary being, of our scientific and ideological acceptance of it, and of Gaia herself.

The Systems Logic and Illogic of Gaia

The Gaia hypothesis is increasingly accepted across scientific and popular culture. While it was initially rejected by traditionalists as environmental pseudoscience, Lovelock and his supporters have successfully scrubbed Gaia of its "teleological excess and other offensive features" such that it now enjoys support from a wide number of scientific fields.[119] Perhaps the best example of its mainstreaming is the 2001 Amsterdam Declaration on Earth System Science, where four major programs (the International Geosphere–Biosphere Program, the International Human Dimensions program on Global Environmental Change, the World Climate Research Program, and DIVERSITAS) issued a joint statement identifying the need for interdisciplinary global environmental science and an ethical framework for global stewardship. The first resolution of the declaration reads as follows:

> The Earth System behaves as a single, self-regulating system comprised of physical, chemical, biological and human components. The interactions and feedbacks between the component parts are complex and exhibit multi-scale temporal and spatial variability. The understanding of the natural dynamics of the Earth System has advanced greatly in recent years and provides a sound basis for evaluating the effects and consequences of human-driven change.[120]

This is a remarkable turnaround. In less than forty years, Lovelock's unorthodox revelation of a living Earth has emerged as a theoretical and ethical foundation for international environmental science. This is not to suggest Gaia lacks for critics. But the ground has shifted since 1965, and what was once beyond the bounds of acceptable science has emerged as the basis for interdisciplinary scientific collaboration. As a second-order paradigm shift of Bertalanffy's system theory, the Gaia hypothesis has expanded the epistemic conditions of possibility from the mechanistic philosophy of the nineteenth century toward holistic, organismic, and emergent modes of thought in the contemporary era.

Obscured in the cultural acceptance of Gaia, however, is the power of systems to open up new configurations of metaphysical inquiry. Bertalanffy was fully aware that systems theory, in all its potent ambiguity, could authorize occult elements, but he saw them as a small price to pay for the epistemic revolution we needed.[121] In this chapter, I have explored the parascientific side of systems theory that traditional science largely ignores. In the New Age, Gaia throws off its metaphors to become a fully fledged planetary being. In Bear's novels, that entity returns to hard science to become a vengeful god. From this most dynamic and complex of supersystems emerges the most improbable of properties: bacterial consciousness, genetic agency, phylogenetic memory, evolutionary teleology, telekinesis. We can only imagine how Plato would react to what *Timaeus* has wrought.

On its face, these extrapolations of Gaia might seem a far cry from the nineteenth-century ether physics or twentieth-century ESP statistics typically associated with the paranormal mind. But in blurring science, technology, myth, and belief, their underlying principles are the same. In a circuitous way, systems theory authorizes ancient notions of matter and mind. Genetics, virology, and bacteriology allow us to reimagine the mechanics of telekinesis. Orthodox science is one product of the scientific process, but so is the emergence of the divine, the illogical, and the arcane.

3 REMOTE VIEWING

Ingo Swann, Project Stargate, and Weaponizing the Paranormal

IN SEASON 1, EPISODE 5 ("The Flea and the Acrobat"), of the blockbuster Netflix series *Stranger Things,* the enigmatic character known as Eleven experiences a flashback. As part of her psychical training for a clandestine Department of Energy program, Eleven is given the photograph of an unidentified Russian and asked to spy on him using her unique clairvoyant gifts. Floating in a pitch-black chamber of water, she focuses her mind. All the surrounding voices and her own inner thoughts quickly disappear. The world falls away. When she opens her eyes, she finds herself in a new plane of reality. Whether it is a metaphysical or metaphorical space remains unclear, but it is silent, black, and void. Only one other person is present: the Russian. As Eleven approaches, we hear him issuing orders in his native tongue. He is a spy of some sort, and he speaks of disinformation, of rumors, and of exposing agents on American soil. Back in Hawkins, Indiana, scientists and government representatives record every word, awestruck at the conversation they are eavesdropping on from thousands of miles away. They are rightfully astonished, for in Eleven they have found the ultimate spy. While this scene lasts barely a minute and mainly serves to elucidate the U.S. government's nefarious agenda, it also highlights the incredible potential of the paranormal mind as a weapon of espionage. Any psychic who could use her skills to collect top-secret information from halfway around the world would be an invaluable political weapon. With minimal technology, expense, and labor, the government could obtain accurate, real-time intelligence on any person or organization from anywhere across the globe. No scientific discovery, cabinet decision, military maneuver, assassination attempt, diplomatic proceeding, or classified document would remain secret for long. Without question, the telepathic spy would mark the greatest achievement in the history of espionage, and any nation-state possessing one would tip the global scales of power.

While most viewers of *Stranger Things* might discount telepathic surveil-lance as just another fantastical element from a television show featuring monsters and parallel dimensions, its foray into psychic spy craft is actually based on real-life events and scientific developments. From the 1970s to the 1990s, the United States' military and intelligence agencies covertly employed mediums to spy on the Soviet Union and other countries as part of their larger Cold War geopolitical strategy. Variously known over the years as Project Grill Flame, Project Gondola Wish, and most famously Project Stargate, these clas-sified government programs trained personnel in a novel form of clairvoyance known as remote viewing. By combining the supernatural ability to perceive distant objects with the use of geographical coordinates, remote viewing sup-posedly allowed skilled psychics to spy on military bases, research facilities, and other sites of interest thousands of miles away from the safety of U.S. soil.

The very concept of remote viewing is odd enough, but even stranger is that its intellectual architect was a prolific writer of parapsychological fiction and nonfiction named Ingo Swann. Within the psi community, Swann is an estab-lished figure for his groundbreaking work with Project Stargate. In literary scholarship, however, he remains virtually unknown. This is an unfortunate oversight because Swann is one of the most significant figures in twentieth-century psychical discourse. In addition to single-handedly creating a new form of clairvoyance, he also developed a radically original scientific frame-work for explaining its mechanisms and generating methods for deploying it at scale. Moreover, as a founding member of Project Stargate, he helped usher an outsider (pseudo)science to the highest levels of government. When his partnership with the government ended, he dedicated himself to helping humanity achieve its psychical potential. By writing numerous novels, mem-oirs, how-to manuals, and ESP guides over the years, Swann participated in a long-running countermovement seeking to liberate psychical research from the laboratory and return it to the people. Yet this populism must not be con-fused with pacificism. One of the most important transformations Swann trig-gered was turning psychical research from a science into a technology. For over half a century, psychical researchers had sought to comprehend the under-lying principles for explaining phenomena like telepathy, precognition, and telekinesis. For Swann, such concerns were secondary to how we could deploy these incredible powers. By developing remote viewing and Project Stargate, he unveiled the potential applications of psi to the U.S. government and later the masses. In doing so, he altered the paranormal mind from a scientific object into a technological one—from something to study into something to use.

After Swann, the conscious and the unconscious were no longer curiosities to probe but instruments of power to wield on the international stage. At the same time, Swann's technological and militaristic proclivities reflect the historical context of the late twentieth century. For much of its existence, psychical research had been cloistered in obscure journals and laboratories. The Cold War shifted national priorities and government funding, and with it the trajectory of the field. The threat of nuclear annihilation brought disciplines like psychology and sociology into tight proximity with the U.S. war machine.[1] Despite its status as a marginalized pseudoscience, psychical research offered an irresistible value proposition for intelligence agencies. It was cheap and safe, and if it worked, psychic surveillance could be a potent tool for national security. Swann's own preference for practical psi over academic fussiness aligned well with the political moment and would eventually lead to a reconceptualization of psychical research still prevalent in twenty-first-century culture.

This chapter focuses on Ingo Swann and the ways his personal experiences, theories, and values changed our perceptions of the paranormal. In addition to exploring his major fictional and nonfictional texts, it touches on the broader geopolitical climate influencing the parascientific warping of psychical research. The result is a comprehensive profile of a figure overlooked in both literary history and science studies. His texts remain largely unknown to scholars, but Swann's government work, phenomenological theories, and literary influence make him one of the most consequential characters in paranormal history. I begin with a brief biography of Swann, including his childhood experiences with precognition and burgeoning interest in parapsychology. I then turn to his time spent at Stanford Research Institute, the California think tank where he first developed remote viewing and began training a new generation of psychic soldiers for Project Stargate. The Cold War logic underlying such unconventional programming is explored via declassified CIA files, which document the imminent threat of Soviet-era psychic weapons. Implicit with such perils is the need for new kinds of global surveillance (which I interpret through Foucauldian governmentality) to assure the nation's security.

This curious political history is bolstered by (pseudo)scientific history. Using Swann's training manuals, personal notes, memoir drafts, book proposals, drawings, and correspondence, I piece together the theoretical frameworks underlying remote viewing and argue that it represents a novel intellectual and methodological approach to the paranormal. By combining occult techniques (out-of-body travel, automatic drawing, etc.) with modern cartography and masculine military fantasy, Swann delineated an alternative model for explaining concepts like clairvoyance and the unconscious. Indeed, many of his ideas are so inventive that they rival the foundational theories of F. W. H. Myers himself. In this regard, we must consider Swann to be a psychical trailblazer of the highest order.

I then analyze Swann's novel *Star Fire* (1977) and his nonfiction texts *The Nostradamus Factor* (1983) and *Natural ESP* (1987). He was a prolific writer, and these three texts capture well his larger psi agenda. For Swann, books did more than entertain readers; they also spread paranormal knowledge, and in so doing strengthened the collective unconscious of humanity and maximized its psychic potential. Like many others, he believed the minds of all people were linked, and he hoped to turbocharge their cumulative power. Literature was hence a psychical "technology" that could catalyze a new stage of universal consciousness. While the actual impact of his paranormal writing remains debatable, Swann's intentions are clear: he hoped to lead humanity toward a "psychic renaissance" where the untold powers of the human mind would at long last be revealed. If these goals sound grandiose, even quixotic, it is only because Swann's ambitions equaled his belief. While mainstream parapsychology had curtailed its research agenda over the course of the twentieth century, Swann maintained a sweeping vision of the human mind till the very end. At the close of the chapter, I examine Swann's lofty conception of paranormal consciousness against the longer history of psychical research and make the case for viewing parapsychological fiction, ESP manuals, and psi memoirs as a hidden resource for critics to investigate scientific epistemology and culture.

My chief argument throughout is that Swann transformed the paranormal mind from an object of science into an instrument of technology. This is not to suggest Swann was uninterested in science. He certainly was; indeed, his theories of remote viewing constitute some of the most compelling writings in the history of psychical research. But it would be disingenuous to imply that science alone motivated Swann. He was openly contemptuous of "traditional" parapsychologists, whom he regarded as elitist, pedantic, and beholden to the past. Swann saw himself as a psychical rebel whose calling was to jailbreak the paranormal mind from the laboratory and to return it to the masses where it belonged. In this regard, he was a throwback to both the nineteenth-century occultists who birthed the field of psychical research and the twentieth-century spiritualists who railed against empiricism, for he understood the human mind as a vital object situated *in* the world instead of cloistered from it. At the same time, Swann was a Cold War ideologue who hated the Soviet Union as much as he loved America. If the paranormal mind was first and foremost a technology, then what better use existed than to advance the political interests of the state?

Similar to the psitron and Gaia from chapters 1 and 2, respectively, remote viewing is a testament to the parascientific process. By blending clairvoyance with cartography, remote viewing reformulates an outmoded scientific concept for a new audience and a new era. Parascientification remains a critical through line here because it illustrates how the unlikeliest of ideas can be reimagined for the cultural moment. Who could have predicted at the SPR's founding in 1892 that a sitting U.S. president would consult psychics about helicopters in Iran, or that generals would ask mediums for updates about nuclear tests in China? It is a history that reads like science fiction when it is actually estranged by fact.

The Strange Beginnings of Ingo Swann

When most people think of parapsychology at the end of the twentieth century, they likely picture a diminished field characterized by strip mall psychics, Uri Geller bending spoons on television,[2] and the harebrained scientists of *Ghostbusters*. However, the most significant development in psychical research during this period was its rising import in global affairs. Instead of a scientific dead end, the paranormal mind had, oddly enough, become a Cold War battlefront, emerging more politically and intellectually vibrant than ever before. The man at the center of this psychical renaissance was Ingo Swann.

Ingo Douglas Swann was born in 1933 in Telluride, Colorado. According to the memoir *To Kiss the Earth Goodbye*, relatives on both sides of his family possessed paranormal abilities. His paternal grandmother, Maria, once communicated with the ghost of her first husband, who advised her to remarry after his death to avoid loneliness.[3] His maternal grandmother, Anna, possessed the so-called evil eye and could divine the future. Swann experienced this extraordinary power firsthand as a child: he was driving with Anna on a mountain pass when she suddenly uttered, "I don't think we should go any further up the road."[4] Minutes later, an avalanche wiped out the path ahead.

With hereditary psi abilities, it is perhaps natural (or rather supernatural) that Swann demonstrated formidable psychic prowess as well. When he was six years old, Swann writes, he had a vision of flames bursting from his kitchen— a scene so terrifying he demanded his family spend the night at his grandmother's house. Later that evening, "our little house erupted into a towering inferno, and before the fire department, a mere three blocks away, could arrive, the kitchen, bedrooms, and dining room were gone."[5] Swann's prophetic gifts also saved his life while he served in the Korean War. While stationed near the demilitarized zone in Kaesong, South Korea, he was tasked with taking notes in the officers' tent. In the middle of one meeting, Swann heard a voice commanding him, "Get up, get up now!"[6] Swann leapt up, startling those nearby. Moments later a shot rang out: "A bullet ripped through the prefab wall and hit the back of the chair where I had been sitting, striking right were my heart would have been if I had remained seated."[7]

Whether or not Swann's recollections are factual is difficult to determine. His multiple memoirs showcase a natural storyteller who loves to embellish. What is certain, though, is that by the 1960s, he was increasingly drawn to psychical research as a means of understanding his own gifts. After his discharge from the U.S. Army, he worked at the United Nations while trying to launch a career in painting, where he specialized in flowers, cosmic spacescapes, and auras inspired by Kirlian photography.[8] His paranormal interests drew widely from family mythology, Scientology, and New Age beliefs.

In June 1972, Swann began volunteering for parapsychological experiments at the New York City offices of the ASPR. The ASPR had a long history of working with citizen scientists. "Ordinary people contributed their raw experience in the hope of building up a new field of scientific knowledge where none had existed," writes Alicia Puglionesi, and Swann was one of these "ordinary people."[9] One area where the ASPR hoped to gather data was out-of-body experience. The term was coined by G. N. M. Tyrrell in 1943 as a more scientific alternative to *astral projection,* and it described the ability to perceive the world from a location outside one's physical body.[10] For instance, a medium who could willfully relocate her consciousness or soul to an adjacent room would be said to have an out-of-body experience. For his ASPR experiment, Swann was asked to "float" out of his body while strapped to a chair and identify several objects placed on a tray suspended over his head (and therefore out of his line of sight).[11] Many of his guesses were off the mark, but he eventually showed remarkable accuracy in correctly identifying a black leather case containing a letter opener as a "long and black" thing and the concentric circles of a target as "a bullseye."[12] The experimenters were impressed enough to bring him back for additional sessions.

What Swann did not reveal to the researchers was that he was already projecting his consciousness outside the building while waiting for the experiment to begin. Swann claimed that ever since he was a child, he could move his mind beyond the confines of the physical body. At the very moment he was tied to a chair, he was also floating through Manhattan: "He noticed a brilliance flowing from Central Park, where the light was apparently bouncing on trees and openness, emanating into the street somewhat like a mist of silent molecular bubbles. His awareness flashed momentarily up through the overcast and smog. There he perceived the sun, brilliant, already sinking in the west."[13] Swann did not disclose the full extent of his powers because he thought traditional parapsychologists could not comprehend, let alone test, his abilities. The dominant paradigms of psychical research were incommensurate for gauging the true powers of the human mind. ASPR experimenters were so focused on trivialities like leather cases they could not imagine psychics like him projecting their minds into Central Park with full command and cognition. "The construct of consciousness that would explain this might be a long time coming, might even turn out to be not telepathy or clairvoyance at all but some unknown, unimagined human potential," Swann writes. "With all due respect to the noble researchers long in the field, the field was by no means played out. There was yet much to discover about man, about consciousness, about human potential."[14]

Swann occupies a bivalent role in the history of psychical research. On the one hand, he was highly indebted to the field's major theorists. Many of his own books reference the intellectual contributions of men like F. W. H. Myers, S. G. Soal, and René Warcollier. On the other hand, he was highly critical, even dismissive, of the current state of the discipline, its practitioners, and their methodologies. Swann believed that parapsychology had devolved into lethargy and irrelevance. Its researchers—rarely psychics themselves—obsessed over insignificant research questions and pointless experiments. What did statistical analyses of coin flipping have to do with "the endless mind of man"?[15] In some respects, Swann's perspective reflected the tensions between spiritualists and scientists that fractured the ASPR in the 1920s, with the former openly hostile to statistics and laboratory testing.[16] Swann was not opposed to science in general, just the conservatism of mainstream parapsychology. As such, he joined other citizen scientists who "viewed psychical research as a democratic science to which they contributed and in which they had a say."[17] In addition, Swann argued modern parapsychology consisted of three types of people: (1) frauds seeking attention or money, (2) researchers doing minor work, and (3) the select few who could actually change man's paranormal potential.[18] Swann saw himself in the third category. To change humanity's psychic destiny, parapsychology required fresh thinkers, new methods, updated rhetoric, and a truly ambitious agenda to win back a public uninterested in the field's scientificity. Fortunately, there were others who shared Swann's passion for pushing parapsychology in bold new directions. These influential players in the military-political complex similarly believed the paranormal mind was wasted inside the laboratory when it could be deployed on a worldwide scale.

The New Governmentality

In a 1978 lecture at the Collège de France, historian and philosopher Michel Foucault defined a new term, *governmentality,* as the "art of government."[19] In contrast with earlier means of wielding political power that were based on sovereignty and punishment over individuals, governmentality referred to the expanding range of administrative technologies used to manage entire populations: schools, prisons, hospitals, public health systems, and other familiar forms of biopolitical control. Governmentality is ultimately rooted in the security of the state, and this complex form of power requires a vast "knowledge of things."[20] Foucault writes:

> The things with which in this sense government is to be concerned are in fact men, but men in their relations, their links, their imbrication with those other things which are wealth, resources, means of subsistence, the territory with its specific qualities, climate, irrigation, fertility, etc.; men in their relations to that other kind of things, customs, habits, ways of acting and thinking, etc.; lastly men in the relation to that other kind of things, accidents and misfortunes such as famines, epidemics, death, etc.[21]

At roughly the same time as Foucault's lecture, the knowledge of things a government needed was becoming increasingly global in scope. By the middle of the Cold War, governmentality now included what *other* countries, especially enemy countries, were developing, funding, stockpiling, engineering, farming, building, supporting, and believing. For the United States, knowing what adversaries like the Soviet Union were doing at every possible moment, whether testing nuclear weapons or assembling troops, took on greater significance to ensure the safety of the nation. The Cold War was a "distinctive moment in the history of the American human sciences": the government consulted political scientists (Albert Wohlstetter), game theorists (Martin Shubik), psychologists (Anatol Rapoport, Martin Deutsch), and economists (Thomas Shelling, Daniel Ellsberg) to better understand the communist mind.[22] The president and the Pentagon craved insights into an enemy other that could kill millions with the press of a button. Consequently, any surveillance technology that could provide an informational edge possessed enormous strategic value. As it turned out, parapsychology offered a powerful new kind of mentality that could drastically augment the existing mechanisms of governmentality.

The United States was not the first nation interested in psychical research as a tool of espionage and population control. In fact, it was the Soviet Union's investment in psychic statecraft that spurred the United States' own commitment to remote viewing. According to historian Wladimir Velminski, as early as the 1920s, the U.S.S.R. hoped to modify human behavior via telepathic and cybernetic technologies. Russian electrical engineer Bernard Kazhinsky believed the human body was a living "radio station" in which thoughts were electromagnetic waves and the central nervous system was "a storage site for subtle apparatuses of biological radio communication."[23] Developing technologies to transmit information directly into human brains was thus a perfectly rational scientific endeavor. Similarly, neurologist Vladimir Mikhailovich Bekhterev spent much of his career developing technologies to transmit ideas and psychophysical states into test subjects.[24] "The mental effect of one individual on another is possible at a distance through some kind of living matter," Bekhterev writes. "Here, too, we are dealing with phenomena of electromagnetic energy, most likely with Hertz waves."[25]

The belief that the human mind is fundamentally electromagnetic—and therefore technologically controllable—gave rise to the Soviet psychoenergetic program. While the term *psychoenergetics* (mind energy) is less popular in America than either *psychical research* or *parapsychology,* its meaning is essentially the same insofar as the field prioritizes telepathic and telekinetic communication beyond the known limits of science. For a communist government well versed in state-run propaganda, psychoenergetics provided a dynamic means of winning the hearts and minds of an entire population. According to Velminski, leading engineers, radio technicians, and physiologists conducted experiments for years to master "electromagnetic emissions of thought."[26] The apex of psychoenergetics occurred in October 1989, when the Soviet Union attempted to telepathically calm its citizens from political upheavals in East Germany, Austria, and Hungary via six televised séances. "You can leave your eyes open for a while," psychotherapist Anatoly Mikhailovich Kashpirovsky announced at the first nationwide broadcast. "Have a look at your surroundings. There should be no pointed objects and no fire. Your posture should be stable. If anyone is seriously ill, for example, suffering from epilepsy—please do not participate on our séance; simply turn off the television."[27]

U.S. intelligence agencies were fully aware of Soviet psychoenergetics as both a form of governmentality and a threat to their national security interests. In the recently declassified Department of Defense (DOD) document "Controlled Offensive Behavior—USSR" (1972), the agency correctly determined that Russia aspired to harness mind-control technologies like telepathy. "The Soviets are attempting to apply ESP to both police and military use," the report notes.[28] This represented a serious threat because an advanced Russian telepath could "know the contents of top secret U.S. documents," "mold the thoughts of key U.S. military and civilian leaders," and even "cause the instant death of any U.S. official, at a distance."[29] Such proclamations may seem ridiculous today, but long-distance psi fit squarely within the range of psychical possibilities. Time and again, parapsychology had shown telepathy and clairvoyance transpiring between people separated by hundreds, even thousands, of miles.[30] The retooling of psychical research as a Soviet threat is important because it essentially transforms the human mind into a Cold War battlefront. "The discovery of the energy underlying telepathic communication will be equivalent to the discovery of atomic energy," Dr. Leonid Vasiliev writes gravely.[31] What had once been merely a space race and an arms race now appeared to be a *mind* race. Hence, when the DOD report concludes with a rhetorical question—"Is ESP a weapon of war?"—the implicit answer is an unequivocal yes.[32]

The underlying message of "Controlled Offensive Behavior—USSR" is clear: the United States had fallen woefully behind in the realm of paranormal technology. Any country that mastered psychical research would own a tremendous advantage gathering data about its citizens, understanding the plans of rival nations, and even manipulating the behaviors of specific targets. In short, the paranormal mind offered a bold new vision of governmentality based on supernatural surveillance. Cold War ideology was ostensibly concerned with winning hearts and minds, but the psychical spy craft of the 1970s and 1980s upped the ante to *controlling* hearts and minds. The United States and the U.S.S.R. were superpowers insofar they both wielded immense political, military, and economic might. But the nation-state that could fully weaponize the human mind would possess a superpower in a far more mythic sense: a phenomenological gift granting geopolitical control over all other "normal" countries. Fortunately, the United States had just the right man for the job.

Project Stargate and the Science of Remote Viewing

The United States' own psi governmentality program began in earnest at Stanford Research Institute (SRI), a think tank in Menlo Park, California, with strong military connections. Two of its laser engineers, Hal Puthoff and Russell Targ,[33] were intrigued by psychical research, and in early 1972, they received $50,000 from the CIA to develop a program, method, or technique demonstrating "repeatable psi phenomena."[34] For U.S. intelligence, this was a low-cost undertaking to see if there was actual substance to Soviet psychoenergetics, and if so, to pull even. For Puthoff and Targ, this was a chance to advance both parapsychological research and their nation's interests. For Swann, it was the opportunity of a lifetime.

Puthoff had heard of Swann's telepathic exploits from mutual acquaintances in the psi community, and he flew the renegade psychic out to California hoping he might develop something viable for the CIA. Swann immediately proved his bona fides. At his arrival on the Stanford campus, Puthoff asked Swann to use his abilities to "look inside" a magnetometer (a device used for measuring magnetic fields), which was buried five feet underground and impervious to outside influence. Not only did Swann perceive its inner mechanisms ("the cavity was filled with liquid helium, dielectric gold-titanium-coated glass and a few balls of niobium used in the quark process") but he also used his mind to affect the magnetometer's supposedly untouchable readings.[35] "I have become convinced that the seat of consciousness does not emanate from or exist solely as a result of the physical body," Swann explains in his memoir. "It can move in space and time."[36] In this case, he projected his consciousness with sufficient dexterity to earn a contract at SRI.

To be clear, Swann's invitation to join SRI was not about exploring the fundamental principles of ESP so much as "the establishment of new trial hypotheses upon which future work might take place."[37] This "future work" meant the application of the paranormal mind for government purposes. SRI did not pursue knowledge for knowledge's sake; purely scientific inquiry into the paranormal was beside the point. The CIA wanted an applied science of the paranormal; they desired a *technology* with strategic and practical use against Cold War targets. This suited the empirically adverse Swann just fine.

The phenomenon called remote viewing was born in April 1973 in a Menlo Park swimming pool. Swann had grown frustrated with Russell and Targ's tedious parapsychology experiments, such as guessing which of two identical cannisters contained a radioactive element.[38] Swann desired bolder tests that were based on his specialty: out-of-body experience. As he relaxed in the pool one evening, Swann apparently heard a voice in his head saying, "Try coordinates."[39] These were fateful words. Indeed, they crystalized an idea gestating in his mind: using geographic coordinates of latitude and longitude to focalize the clairvoyant eye. "I wondered if simple geodetic coordinates might not serve the psychic entity well in locating him in terms of the Earth's surface," he explains in *To Kiss the Earth Goodbye*. "The abstraction of the coordinate sufficed to provide orientation for the psychic probe—whether it be called clairvoyance or out-of-body perception (it is now called 'remote viewing')—to locate by transcendental means the place to which the coordinate referred and thence to inspect it and describe what was there."[40] Once his mind's eye was fixed on the target, whether a military base or a government office, he could share a visual description of the site with his colleagues. At a fundamental level,

Swann is describing deeply esoteric concepts like clairvoyance (perceiving things beyond the normal senses) and astral travel (projecting the soul/mind out of the physical body), but securing them to the precision and pragmatism of modern geodesy. The end result is a parascientific blend of occultism and modern science that deploys the paranormal to spy on any location in the world. It is worth acknowledging from the outset that Swann himself did not know how remote viewing actually worked when the idea first came to him. As we shall see, he continually refined his thinking over the years, and in doing so, the rationale and terminology evolved from the quasi-Victorian language used here into a more elaborate, idiosyncratic system of his own creation. But that would come later. In spring 1973, Swann postulated an innovative method of clairvoyance with obvious benefits for U.S. spy craft . . . if it worked.

Puthoff and Targ were initially skeptical of Swann's "coordinate clairvoyance" scheme.[41] After all, coordinates are arbitrary human conventions absent any inherent or natural meaning; projecting one's consciousness to that "location" thus failed to make any sense to them. But Swann insisted they give his technique a try. SRI consequently created Project Scanate (a portmanteau of *scan* and *coordinate*) to explore Swann's brainchild.[42]

The basic structure of a remote viewing session quickly evolved out of Swann's Project Scanate work. Each session involves three main components: a target, a viewer, and a monitor. A researcher or government official not directly involved in the session would provide the target, specifically the coordinates of latitude and longitude for some location of strategic interest: a military base, an embassy, a research facility, and so on. The viewer was a psychic, "the person who employs his mental faculties to perceive and obtain information to which he has no other access and of which he has no previous knowledge."[43] The monitor was an assistant who "provides the coordinate, observes the viewer to help insure he stays in the proper structure, records relevant session information, [and] provides appropriate feedback when required."[44] Below is Swann's description of a standard remote viewing session:

> In conducting a coordinate remote viewing session, a remote viewer and a moni-
> tor begin by seating themselves at the opposite ends of a table in a special remote
> viewing room equipped with paper and pens, a tape recorder, and a TV camera
> which allows either recording for documentation, or monitoring by individuals
> outside the room. The room is homogenously-colored, acoustic-tiled, and feature-
> less, with light controlled by a dimmer, so that environmental distractions can
> be minimized. The session begins when the monitor provides cuing or prompt-
> ing information (geographic coordinates in this case) to the remote viewer. The
> remote viewer is given no additional identifying information, and at this point

has no conscious knowledge of the actual site. For training purposes, the <u>monitor</u> is allowed to know enough about the site to enable him to determine when accurate versus inaccurate information is being provided. The session then proceeds with the monitor repeating the prompting information at appropriate intervals and providing necessary feedback. The remote viewer generates verbal responses and sketches, until a coherent response to the overall task requirement emerges.[45]

Clearly, coordinate clairvoyance is not a high-tech affair. Besides paper, pens, and basic recording technology, little is needed in the way of equipment. The remote viewer is the primary instrument. Keeping the remote viewer's mind clear is of paramount importance—hence the homogenous and darkened rooms for minimizing distractions. In addition, by keeping target information secret, the viewer's mind is left as blank as possible. This is crucial because it is the viewer's unconscious that will generate the geographic visuals that constitute a successful session. Several elements of Swann's methodology should sound familiar to readers of Mary Craig and Upton Sinclair's psychical classic *Mental Radio* (1930), which described experiments involving drawing images telepathically sent across their house. From relaxing in a dark room to sketching pictures suddenly "appearing" to the mind, both Swann and Mary Craig follow a similar template to access the impossible.[46]

Any doubts that Puthoff and Targ had about Swann's unusual methodology faded as Project Scanate began collecting experimental data. Incredibly, Swann and other psychics recruited by SRI demonstrated unusual accuracy as they began visualizing sites around the world. In an early test session, Swann envisioned a chunk of land in his mind where geographic coordinates indicated there should only be water. SRI researchers were stunned that Swann described a peninsula off Ukerewe Island in Africa's Lake Victoria—a plot of land so small it did not appear on most maps.[47] In another session, Swann described a military installation in West Virginia where there should only have been a cabin in the woods: "a strange place, somewhat like the lawns that one would find around a military base, but I get the impression that there are . . . some old bunkers around."[48] Fellow psi Patrick Price perceived something similar: "Looks like a former missile site—bases for launchers still there, but area now houses record storage area, microfilm, file cabinets."[49] When the CIA sent an agent to double-check the location, he discovered an undisclosed cryptography facility.[50] Psychic Hella Hammid accurately described a covered bicycle rack at a target site as "an open barnlike structure with a pitched roof."[51] To be sure, many other viewing sessions were complete failures. But the uncanny hits of viewers like Swann, Price, and Hammid more than made up for any misses. Their successes could be neither explained nor ignored. In a 1974 article in

Nature describing their Project Scanate experiments, Puthoff and Targ write, "A channel exists whereby information about a remote location can be obtained by means of an *as yet unidentified modality.*"[52] Similarly, in a 1976 article published in the journal of the Institute of Electrical and Electronics Engineers, Puthoff and Targ acknowledge, "The precise nature of the information channel coupling remote events and human perception is not yet understood."[53] SRI had discovered and verified a phenomenon unexplained by any known scientific principle; ergo, the human mind could see and know things that should be impossible to see and know. The science behind such an extraordinary breakthrough could wait, though. As an application of government surveillance, coordinate clairvoyance was a boon to the nation.

In their academic articles, Puthoff and Targ describe remote viewing as "the ability of certain individuals to access and describe, by means of mental processes, information sources blocked from ordinary perception."[54] Swann called remote viewing "a method of psychoenergetic perception" and "the acquisition and description, by mental means, of information blocked from ordinary perception by distance, shielding, or time."[55] In short, they speak of clairvoyance. Puthoff and Targ created the neologism *remote viewing* as "a neutral descriptive term free from prior associations of bias as to mechanisms" associated with occult terms like *traveling clairvoyance, out-of-body experience,* and *astral projection.*[56] For acceptance by scholars and government officials, replacing esoteric words with newer ones that sounded more scientific was a proven rhetorical strategy; the scientification of *psychical research* into *parapsychology* is perhaps the most famous example. There is, of course, nothing neutral about a term like *remote viewing.* By turning *clairvoyance* into *remote viewing,* SRI deliberately pivoted from the discourse of nineteenth-century occultism into the discourse of twentieth-century science and sanitized a marginalized concept for modern audiences. The fact that remote viewing operates via an "unknown mechanism" is similarly effective because it breaks from theosophical theories based on the third eye or the pineal gland. The uncertainty surrounding remote viewing hence emerges as a linguistic asset because clairvoyance becomes a blank slate for SRI to decorate anew.

While remote viewing is indebted to existing concepts in parapsychology, it is different enough that we should consider it a wholly original approach within psychical research. It is a type of clairvoyance unlike any other practiced before. For starters, the geospatial coordinate system used in the most common form of remote viewing[57]—coordinate remote viewing—is ostensibly superior to traditional clairvoyance because the mind is directed by modern geodesy; the supposed geographical exactitude of the newer discipline makes the older one more rigorous. Political context also matters. The psychical was historically a personal phenomenon involving communicating with deceased family members or predicting one's future. Remote viewing was ruthlessly geopolitical; it served the interests of national security, intelligence agencies, and the military. Remote viewing must be understood first and foremost as a technology designed to advance the interests of the state—that is, to enhance governmentality. Trainability was yet another area of distinction. For many, mediumship was a God-given ability. In contrast, remote viewing was a skill that could be nurtured. "It may be that remote perceptual ability is widely distributed in the general population," write Russell and Targ, "but because the perception is generally below an individual's level of awareness, it is repressed or not noticed."[58] Swann's basic premise was that practically anyone could perceive the impossible with proper training and practice—an egalitarian ethic he shared with Madam Blavatsky, the Sinclairs, and other occultists.

As the inventor, intellectual architect, and primary psychic behind remote viewing, Swann assumed a leading role at the SRI program eventually known as Project Stargate.[59] His primary task was developing a training regimen to help others learn remote viewing techniques and best practices—in other words, to create psychic soldiers. Swann believed remote viewing was a universal talent one could fully harness and weaponize through proper education. Consequently, the training manuals he developed for SRI are integral for understanding the practice and theory of remote viewing. They are also useful for observing the evolution of his ideas. In his memoirs, Swann initially explained his clairvoyance through the preexisting framework of out-of-body experience. In contrast, his SRI manuals use a brand-new vocabulary and grammar for articulating psi phenomena, descending from an intellectual genealogy that was born with F. W. H. Myers's psychical research, grew into J. B. Rhine's parapsychology, and matured into a metaphysical, militaristic technoscience. Swann's training manuals represent some of the most important works in twentieth-century psychical research because they fundamentally rethink the theoretical principles undergirding the paranormal mind.

The SRI manual *Coordinate Remote Viewing*, dated May 1, 1986, is an excellent example of Swann's scientific reconceptualizations. As he explains in the opening pages, the document is "a comprehensive explanation of the theory and mechanics of coordinate remote viewing (CRV) and a guide for future training programs."[60] In its dual scientific and technical writing context, this manual provides the theoretical background for understanding—and instrumentalizing—psychic governmentality via several new and rejiggered paranormal concepts: the Matrix, the signal line, and analytical overlay. In doing so, Swann comes into his own as a psychical researcher, transitioning from older esoterica to approaches distilled from his own burgeoning expertise.

Perhaps the most significant reworking of the paranormal mind found in *Coordinate Remote Viewing* is the metaphysical space Swann calls "the Matrix." In the report, he claims remote viewing is made possible by a "huge, non-material, highly structured, mentally accessible 'framework' of information containing all data pertaining to everything in both the physical and non-physical universe."[61] While modern readers likely associated the term *matrix* with a shared digital hallucination from William Gibson's *Neuromancer* (1984) or the Wachowskis' *The Matrix* (1999), Swann's version pulls from older sources.[62] "In the same vein as Jung's cosmic unconscious," he writes, "the Matrix is open to and comprises all conscious entities as well as information relating to everything else living or non-living by accepted human definition."[63] Jung's collective unconscious refers to the shared unconscious structures of the human species (e.g., instincts, archetypes, dream symbols), and Swann builds on the concept of species-wide, networked knowledge. In doing so, he joins an esteemed lineage of thought. Myers's subliminal self is, at its core, a "leaky" unconscious that merges with others. William James argued that "there is a continuum of cosmic consciousness, against which our individuality builds but accidental fences, and into which our several minds plunge as into a mother-sea or reservoir."[64] According to his "mother-sea" theory, our suppos-edly individual minds are linked just as islands are connected to the ocean's bottom, and this deep network acts as a pan-psychic archive "in which many of earth's memories must in some way be stored."[65] Psychical researcher Gardner Murphy similarly proposed an interpersonal field theory: "It is the present hypothesis that these various [paranormal] powers are not solely derived from the psychological make-up of individuals; that they depend in some degree upon interpersonal relations—indeed, that such inter personal powers are much richer and more complex than any possessed by the individual when isolated from his fellows. When a group of individuals functioning as an inter-personal entity is involved, they may have really extraordinary capacity to make contact with phases of reality which transcend time and space."[66] As I note in chapter 2, Ludwig von Bertalanffy argued that complex systems produce un-explainable gestalts; Murphy claims the same here, suggesting that paranor-mality springs from emergent properties of the human mental collective. Swann runs with this idea. For him, the Matrix is a metaphysical structure containing all information in the physical and nonphysical worlds—global geographical data, weather patterns, the location of inanimate things, the memories of all living entities, and so on. The human system shares not only archetypal myths but also real-time information about the planet itself. Moreover, this cornucopia of infinite data does not reside *in* one's unconscious but is rather accessed *through* it.

As Swann explains, "The Matrix can be envisioned as a vast, three dimensional geometric arrangement of dots, each dot representing a discrete information bit. Each geographic location on earth has a corresponding segment of the Matrix corresponding exactly to the nature of the physical location."[67] The totality of the world's geographic data thus exists within a metaphysical globe accessible to the human mind, and knowledge of a specific dot on the globe coincides with knowledge of the actual site on Earth. If we step back from the imperfect metaphors of frameworks and geometric arrangements, we can see that Swann posits the existence of a psychical world perfectly mirroring the material world. All information in the latter also exists in the former. The obvious benefit of the psychical world is that its data can be instantaneously retrieved by our minds. Much like the fantastical map in Jorge Luis Borges's classic story "On Exactitude in Science" (1946), the Matrix is a perfect simulacrum encoding within it the entire planet's store of information, from the largest geophysical landscapes to the tiniest pebbles on a distant beach— an infinite map, then, of infinite data. Unlike Borges's map, though, the Matrix is not useless because of its scale but highly practical because of its ethereal ubiquity.

To access the torrents of Matrix data, a remote viewer must develop a signal line, an informational bridge connecting the psychic to a specific location in the Matrix. "In a manner roughly analogous to standard radio propagation theory, this signal line is a carrier wave which is inductively modulated by its intercourse with information, and may be detected and decoded by a remote viewer," Swann writes.[68] His entire remote viewing program was designed to achieve a strong signal between physical and metaphysical realms; psychics who succeeded could obtain accurate, real-time data about airplane hangars, munition depots, and other locales from the pan-psychic archive. The signal line was difficult to master, though. Swann likened it to finding a clear radio station against an enormous background of static. Another common problem was data inundation during brief connections with the signal line. "When the remote viewer first detects the signal line," he writes, "it manifests itself as a sharp, rapid influx of signal energy—representing large gestalts of information."[69] Such "narrow aperture" only offers brief access to the data treasure trove of the Matrix. (As we saw in chapter 1, Philip K. Dick ran into similar issues with his vitamin-fueled brain antenna.) With formal training, however, a remote viewer could achieve "wide aperture" and "larger, slower, more enduring waves" of information yielding clearer visual data of the site.[70] In sum, the signal line is a direct link to unlimited geophysical knowledge, and with extensive training, one could establish a metaphysical connection robust enough to retrieve strategically significant intelligence.

In addition to receiving Matrix data, a remote viewer must block out any mental noise degrading that data. Swann's term for *noise* was *analytical overlay* (or AOL), the "conscious subjective interpretation of signal line data, which may or may not be relevant to the site."[71] Remote viewing is based on a psychical model of consciousness assuming Matrix-based information is pure as long as it remains in the unconscious; unfortunately, the conscious can taint this information with personal memories and imagination. According to Swann,

> It is apparently this part [the unconscious] of the individual's psyche that first detects and receives the signal line. From here, it is passed to the autonomic nervous system [ANS]. When the signal line impinges on the ANS, the information is converted into a reflexive nervous response conducted through muscular channels controlled by the ANS. If so allowed, this response will manifest itself as an ideogram. At the same time, the signal is passed up through the subconscious, across the limen, and into the lower fringes of the consciousness. This is the highest state of consciousness from the standpoint of human material awareness. However, the normal waking consciousness poses certain problems for remote viewing, occasioned largely because of the linear, analytic thought processes which are societally enhanced and ingrained from our earliest stages of cognitive development.[72]

The problem is that the conscious mind must interpret signal line data, but in doing so, it corrupts it. In his training manual, Swann provides the example of a steel girder bridge, which might appear to the remote viewer as images of angled metal and riveted girders. A viewer's consciousness might assemble these visual fragments into a sports stadium instead of a bridge: "instead of allowing wholistic 'right-brain' processes (through which the signal line apparently manifests itself) to assemble a complete and accurate concept, untrained 'left brain'-based analytic processes seize upon whatever bit of information seems most familiar and forms an AOL construct based on it."[73] The rational left-side brain can consequently misinterpret the accurate signal line data emerging from the creative right-side brain. Analytical overlay would be disastrous for military intelligence, so much of Swann's training was designed to minimize this effect. One of his solutions, for instance, involved drawing Matrix images as ideograms ("the reflexive mark made on the paper as a result of the impingement of the signal . . . to the arm and hand muscles") rather than verbalizing them, because language is a higher-order cognitive process.[74]

Puglionesi notes that "drawing gradually gained acceptance in mainstream psychology as a reliable way to read the mind via the hand while bypassing the conscious subject," with its apparent simplicity circumventing the misrepresentations of speech.[75] In time, adjacent fields like neuropsychology, psychiatry, and psychotherapy all accepted drawing as "a technology of direct inscription, able to record hidden mental processes."[76] A common theme in psychical research is that instinctual low-order brain activity is superior to learned high-order behaviors when it comes to interpreting paranormal phenomena, and Swann similarly thought that unconsciously doodling a naval shipyard trumped a verbal description.

According to Swann, two years of training would help talented remote viewers proceed through six stages of clairvoyant vision. At each stage, signal line access would strengthen and widen to enhance visual accuracy. In stage 1, a remote viewer would perceive large landmasses like islands, mountains, and deserts. In stage 2, smaller features like glaciers could be observed. In stage 3, sites possessing "specific dimensional characteristics," like bridges and airfields, become visible.[77] By stage 4, a viewer would form "qualitative mental percepts" and could determine if sites were civilian or military in nature.[78] At stage 5, a viewer could describe increasingly specific details like aircraft tracking radar, biomedical research facilities, and production plants.[79] Last, a stage 6 viewer could perform direct 3-D assessments of site elements like "airplanes inside one of three camouflaged hangars" or "underground weapons storage areas."[80] These stages are fascinating for two reasons. First, they evince a scientific precision uncommon to previous approaches to psychical research; the abstract visions prevalent to Victorian occultism have been replaced by architectural forms and concrete objects. Second, the martial theme of the targets (radar equipment, camouflage hangers, etc.) reveal the Cold War ideology guiding—and funding—this psychical enterprise.

The novelty of Swann's remote viewing theory is worth analyzing against a broader scientific context. Older psychical theories based on the subliminal self assert the human mind is unbound from the physical body.[81] Astral travel and out-of-body experiences are based on the same principle that the paranormal mind can separate from the body and wander geographic space. This is the very logic Swann previously used in his memoirs to explain his ASPR exploits. But in his training manuals, coordinate clairvoyance does not involve projecting the mind to a real physical location so much as gaining access to a simulacrum (the Matrix) of that location. Geographic coordinates provide precise targets for viewers, who then use their training to conjure a signal line to the Matrix while avoiding any analytical overlay that would degrade data accuracy. The geographical information retrieved then moves from the unconscious to either the hand (ideogram) or the conscious mind (verbal description) for U.S. intelligence to act on. This radical shift from the physical space of out-of-body experience to the metaphysical realm of remote viewing marks an intriguing evolution in Swann's thought. What was once the movement of the soul through the material world has been reconfigured as the movement of the mind's eye through the immaterial landscape of the Matrix. And what was once spiritualist is now parascientific and paramilitary; implicit in the technoscientific ability to see the world is the capacity for militaristic action upon it. The science of remote viewing signals Swann's growth as a researcher in particular and the transformation of twentieth-century psychical research in general. With remote viewing, Swann advanced the scholarship of James and Murphy while also altering the direction of psychical research, which no longer merely studied the human mind. By the end of the 1970s, the CIA, DOD, National Security Agency, and Army had a vested interest in harnessing the powers of telepathy, clairvoyance, and telekinesis for the world stage. Communist nations were actively researching psi powers, and the United States had to advance its own technological capacities in this unorthodox field.

Swann reveled in his transformative role: "I do not consider CRV developments to be related to any aspects of parapsychology," he wrote in a letter to Dr. Dennis Edmondson. "The entire paradigm and experimentation to be derived from it are completely different. The most strategic difference is that a parapsychological theory is not being tested, but that I took real experiential phenomena and linked them together in a way that is natural to the inherent capabilities. To my knowledge, parapsychologists have never approached their problems this way."[82] While Swann exaggerates in his letter, he is correct in highlighting the novelty of this theory, methods, and purpose. In its scientific principles, cartographic-based targeting, and geopolitical function, remote viewing draws from previous psi ideas and amalgamates them into something outrageously different. No parapsychologist before Swann united second sight with the precision of geodesy. No citizen scientist had sutured Jungian collective consciousness with radio propagation principles. No researcher (at least in America) recognized the strategic value of clairvoyance to national security quite like him. Therein lies Swann's seminal role in the history of psi: he converted the paranormal mind from an object of study into an instrument to deploy. By updating clairvoyance from a science into a technology, he shifted the onus of psychical research away from exploration of basic principles in the laboratory toward leveraging—and maximizing—its capabilities in the field. In so doing, he also layered the paranormal with fresh veneers of political, ideological, economic, and military meaning. The nineteenth-century outsider science had become a top-secret weapon for Cold War supremacy.

Over time, SRI's Project Scanate assumed several alternative names, including its most famous iteration: Project Stargate. Operating out of Fort Meade, Maryland, and SRI in Menlo Park, Project Stargate remote viewers conducted psychic surveillance throughout the 1970s and 1980s following the protocols outlined by Swann. According to science historian Jim Schnabel, the psychics viewed their work in patriotic terms, for "they held out hope that their achievements would not be entangled in some obscure parapsychology journal but instead would be used ultimately in support of the nation's highest goals."[83] These goals were real; military personnel, intelligence officers, and members of Congress were all capable of using remote viewing for any number of Cold War activities.

Just as strange as having an occult unit embedded in the U.S. military was the purported number of successes achieved by Project Stargate over the years. In 1978, for example, remote viewer Frances Bryan located a downed Soviet plane in Zaire and drew a sketch of the crash site with sufficient geographic detail that the CIA reached the plane wreckage before the Russians.[84] In 1979, two independent remote viewers spying on a Chinese nuclear test site observed an explosion, although nothing approximating the scale of a successful nuclear detonation. These visions proved correct; China's nuclear test at Lap Nor had failed.[85] In 1980, the National Security Agency asked for help to explain why classified information kept leaking out of a U.S. consulate in the Mediterranean. Remote viewer Joe McMoneagle not only found a listening device hidden in the consulate but also located the apartment across the street where a Soviet listening post had been stationed.[86] During the Iran hostage crisis, President Jimmy Carter and the Pentagon used remote viewers "literally hundreds of times" to learn what was happening inside the U.S. embassy in Tehran.[87]

Not everyone believed Project Stargate was a success, though. A 1995 report commissioned by the CIA ("An Evaluation of Remote Viewing") heavily criticized remote viewing data as "vague and general in nature," "not consistent across independent viewings," and plagued "by a large amount of irrelevant, often erroneous information," all of which rendered it largely useless for operational end users.[88] Furthermore, the report claimed that "the viewings were never used as a primary source of evidence in making decisions."[89] We thus see a disjunct in the evidence for remote viewing that echoes the larger ideological tensions around the will to believe. To the paranormal disciple, the anecdotes above offer incontrovertible proof of remote viewing's legitimacy. To the skeptical reader, the CIA report suggests the program was a failure. Regardless of belief, the fact is the remote viewers' baffling string of successes secured Project Stargate enough credibility and funding to last over two decades.

We might wonder why psychical research caught a foothold in the government when the field as a whole was disappearing from mainstream science. Several reasons exist. As noted earlier, the paranormal mind conveyed extraordinary military advantages that made remote viewing intriguing, no matter how unorthodox it seemed. The fact that Russia was already investing in the paranormal provided a further rationale for America to keep pace in the psychic arms race. Remote viewing was also cheap. In contrast with other forms of surveillance like satellites, aircraft, and agents on the ground, using mediums to spy on other countries from the confines of U.S. military bases was a tremendous bargain. "It seems to me that it [remote viewing] would be a hell of a cheap radar system," congressional representative Charlie Rose argued. "And if the Russians have it and we don't, we are in serious trouble."[90] The classified nature of government programs further allowed theories and experiments beyond establishment science to receive ongoing support. The U.S. government has a long history of conducting projects with dubious scientific and ethical standards (e.g., Project MKUltra) that were nevertheless deemed advantageous.[91] In the case of remote viewing, experimental data remained mostly sheltered within the project itself and thus beyond the scrutiny of outside peer review. Finally, the hierarchical nature of the military meant powerful supporters of paranormal research could keep a project funded for years, if not decades. Swann, it turned out, was the right person at the right place at the right time to bring the paranormal out of the laboratory and into the world.

Literature as Psychical Technology

While the operational history of remote viewing has been addressed through texts like Hal Puthoff and Russell Targ's *Mind-Reach*, Jim Schnabel's *Remote Viewers,* and Joe McMoneagle's *The Stargate Chronicles,* its literary history has received less attention while arguably wielding greater cultural influence. Swann's pioneering work has transformed modern conceptions of the unconscious, psychic warfare, and their relation to governmentality while also inspiring popular books and movies. The number of fiction and nonfiction books, how-to manuals, and personal memoirs that have emerged from Project Stargate is truly astounding, and few, if any, studies have analyzed how this emerging discourse has altered cultural interpretations of the paranormal. Although it is easy to dismiss such writings as scientistic balderdash, they constitute a seminal part of twentieth-century psychical research and parascientific culture.

Moreover, Swann's psychical texts in particular engage with paranormal history while also performing important epistemological and phenomenological labor. His works were purposely designed, as I discuss below, to disseminate paranormal knowledge to mainstream audiences, and in doing so augmented the collective psychic potential of humanity. For Swann, all minds are connected, and if the entire human species increases its understanding of psi, then individual psychics, remote viewers, and telepaths would become enhanced as well. In a manner of speaking, the rising psychical tide would lift all boats. If we consider remote viewing a geopolitical tool, then literature about remote viewing was a cultural tool for maximizing the world mind. In the following pages, we shall examine several of Swann's major works, including his science fiction novel *Star Fire* (1977); his memoirs/psychic manuals *Natural ESP* (1987), *Your Nostradamus Factor* (1993), and *Psychic Literacy and the Coming Psychic Renaissance* (2018); and his unpublished proposal for the International School of Applied Psychic Studies. Through these texts, we can see how Swann's writings operate (or attempt to operate) as catalysts for phenomenological evolution.

Star Fire is an excellent point of departure for exploring the cultural and epistemological power of psychical literature. This obscure work of science fiction has attracted little notice, but it is significant for capturing much of the hidden scientific, methodological, political, and ideological history of remote viewing as well as late twentieth-century parapsychological discourse more broadly. The narrative centers on Dan Merriweather, a rock musician who is also the world's first true "super psychic."[92] While using his tremendous mental powers to survey the world, Merriweather makes a terrible discovery: both the United States and the U.S.S.R. have covertly developed psychic weaponry that could annihilate the world's population. After threatening to reveal these technologies to the public, Merriweather unwittingly sets into motion a series of events wherein the two nations begin arming these weapons in anticipation of World War III. While waging psychic battle on two fronts, he is also pursued by U.S. general Judd Harrah and psychical researcher Dr. Elizabeth Coogan. With his life on the line and geopolitical tensions reaching a boil, the fate of the planet depends on the actions of a lone super psychic. This novel is compelling not only for its outlandish plot but also the myriad psychical histories encoded within it.

One of the novel's most prominent features is how it captures Swann's evolving thoughts on remote viewing theory and methodology. Written in 1977, *Star Fire* sits squarely between older spiritualist frameworks of clairvoyance and the newer Matrix-based ones he would eventually develop in his 1980s training manuals. Although the term *remote viewing* is never explicitly used in the text, Merriweather's ability to monitor the world from afar is clearly based on methods pioneered at SRI. In an early flashback, Merriweather describes a visit to the North California Biocybernetics Center—an obvious stand-in for SRI—where parapsychologists asked him to perform coordinate clairvoyance: "They would give him a set of map coordinates which meant nothing to his conscious mind, and he would draw whatever he 'saw' there, mountains, buildings, trees; sometimes he could see clearly enough to draw floor plans. Where they could be checked they seemed to be pretty accurate."[93] This is undoubtedly the first reference to remote viewing in fiction because its inventor is also the book's author. Importantly, Merriweather's clairvoyance is based on pre-Matrix concepts of soul/mind movement: "He experienced floating in space above Earth, moving about the planet, perceiving scenes in distant places which he knew were actual and not hallucinations."[94] In describing Merriweather's consciousness moving through geophysical space, the passage draws from occult notions of astral travel and parapsychological notions of out-of-body experience. This makes sense, of course, as Swann had not yet fleshed out the Matrix or the signal line by the time of the book's publication. His theories changed over time, but here we witness some of his earliest cogitations on clairvoyant mechanics.

Swann frequently incorporates his personal psi experiences into the novel. We first meet Merriweather in the middle of a global surveillance session: "Dan Merriweather relaxed in a state resembling a light reverie. His metasensory fields breathed, expanding and contracting, vast psychic fields reaching further and further from his centered self, coming finally, with practice, to encompass the entirety of planet earth. Probing! He could barely stand it, this awesome tide of consciousness that engulfed him."[95] The description recalls Swann's early experiences with surveying the world as a child: "There were the familiar swirls of clouds and the emerald and blue glints of oceans. [The Earth] was not truly round as he had already been taught, but somewhat oblate. From its poles radiated faint, almost invisible bands of electric swirls, resembling enormous, nearly invisible tornadoes reaching into space, arching around, entered the earth again at the opposite pole."[96] In this scene, viewing the Earth's curvature and magnetic poles is literally child's play for the ultratalented Ingo Swann. More importantly, there is a clear correspondence between two super psychics—one from fiction and one from reality—probing the magnificence of the planet. Nevertheless, this extraordinary power is inflected through Project Stargate, for Merriweather's clairvoyance is predicated on targeting specific locations (he "leafed through a large atlas of the world, his finger almost automatically falling on the new site") and visualizing enemy military installations ("the tall metal towers that had not yet been dismantled . . . looked lonely, denuded of the buildings that had only a few days ago clustered like bees around their bases").[97] Remote viewing is a military endeavor, and its martial focus underscores the clairvoyant politics of *Star Fire*. In addition, the way the brain processes psychic information in *Star Fire* foreshadows Swann's SRI manual descriptions. Elaborating on Merriweather's mental methods, Swann writes, "From the cerebral hemispheres, where the tenuous psychic-trace-perceptions were neatly filed as real memory, much as in a computer, he could activate his higher mental processes and reclaim those traces, making them dance in his psychic peripheries like holographic movies."[98] The passage essentially describes *images* of distant locations captured by the unconscious and then interpreted by the conscious mind. This recalls the exact process by which clairvoyant information flows from the Matrix to the unconscious and then to the fringes of consciousness as a mental picture. While Swann does not have his technical terms in place yet, the scientific ideas behind remote viewing already power his fiction.

Star Fire also provides insights into the ideograms created during a remote viewing session. Early in the novel, Merriweather locates an underground psychic weapons facility. To blackmail the U.S. government into shutting it down, he draws a sketch of the secret facility and mails it to the Army. The accuracy of Merriweather's drawings unsettles General Harrah: "The drawing was a little distorted, but only a little. The depth underground was accurate. The distance from surrounding towns was accurate. The list of microwave equipment housed in the secret installation contained all items he recalled . . . plus some that were new to him. That made it all the more likely that the list was accurate."[99] As another military officer notes, "Whoever it is, his data is very good. He's even shown the location of the latrines."[100] Swann's SRI manuals focus on collecting geographical and structural data about strategically important sites like military installations. Given that Merriweather can see microwave technology and latrines in high detail, we can infer he is a stage 6 viewer of the highest caliber. Figure 5 illustrates the details a talented remote viewer could perceive. This drawing was created during one of Swann's remote viewing sessions at SRI, and its geographic specificity (a light tower, ten rectangular buildings, a series of pipes, water channel, a refinery) reveals its military value. Such a picture is the equivalent of satellite imagery but offers added benefits like real-time accuracy and subterranean infrastructure. The scene thus demonstrates the enormous potential of remote viewing to military intelligence as well as the anxieties generated if such a power were possessed by an enemy nation.

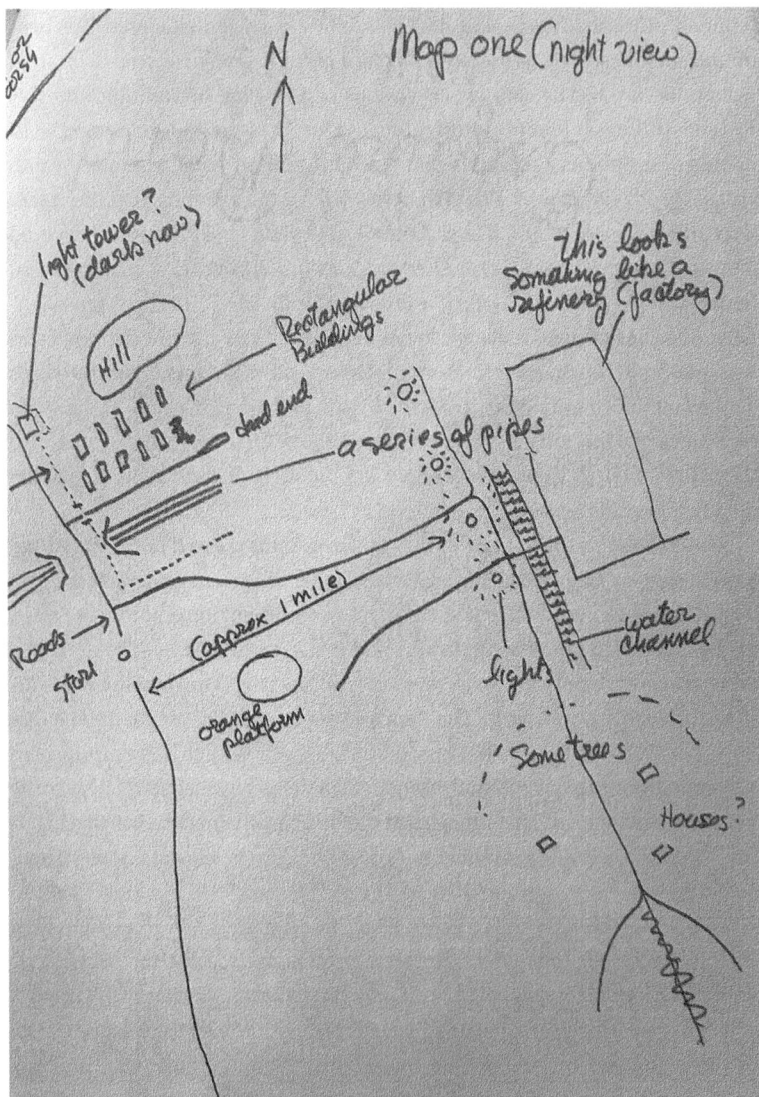

Figure 5. A map produced during an SRI remote viewing session. Box 96, Ingo Swann Papers, University of West Georgia Special Collections, University of West Georgia Ingram Library, Carollton, Ga.

Beyond the psychical methods depicted in the novel, *Star Fire* captures the political ideology underlying Project Stargate. It is revealing that Merriweather's primary function in the novel is surveilling the planet and safeguarding its security. A telepath of his ability could prognosticate the future, learn the scientific workings of the human mind, or take control of other people's bodies. That he never even considers such actions is worth acknowledging. Instead, Merriweather psychically gathers geographic data to supervise the state and the commons. In other words, his duty is governmentality. This is the militaristic ideology of Project Stargate, and it is the implicit rationale of *Star Fire* as well. Merriweather's obsession with eliminating "psychic weapons" of mass destruction is akin to destroying nuclear stockpiles, and his disarmament of the enemy ensures maintenance of the political status quo. As the novel makes clear, remote viewing does not serve a scientific or spiritual purpose so much as a nationalistic agenda. Compare this with James Tiptree Jr.'s *Up the Walls of the World* (1978), another Cold War science fiction featuring military-sponsored psychics. In this antiwar novel, a group of psychics are assembled to assist with deep-sea submarine communication. After they prove their skills, once-skeptical Major Drew Fearing immediately orders their murder because he could never control such powerful individuals. While Tiptree mistrusts the synthesis of nationalism and parapsychology, Swann has zero doubts: parapsychology is an obvious boon to the well-being of the state.

Star Fire further reflects the geopolitical battle lines established during the Cold War by exaggerating developments within the U.S.S.R.'s psychoenergetics program. In the novel, Russia's top-secret "Project Tolkien" features satellites with "experimental radio and microwave enforcement of mental telepathy by biological tissue wave transformers."[101] Merriweather discovers that the Soviets have taken brain slices from powerful mediums and installed them in satellites, which can consequently broadcast electromagnetic psi radiation over entire cities. According to Dr. Coogan, "the human body, or the body of any living creature for that matter, is potentially capable of receiving radiomic impulses through the central nervous system" and can thus receive telepathic instructions from the state.[102] In short, the Russians have developed the perfect mind-control device using psychic radio wave technology. While this satellite technology is hyperbolized, it actually aligns with U.S. intelligence reports on Russian psychoenergetics. For example, the 1972 DOD report notes "super-high frequency electromagnetic oscillators (SHF) may have potential use as a technique for altering human behavior," and "in espionage, one could telepathically hypnotize an individual with post-hypnotic suggestion to steal classified documents or detonate important military equipment."[103] As Velminski has shown, government scientists like Bekhterev and Kazhinsky viewed the brain as a biological device for receiving/transmitting electrical signals and telepathy as a hidden mode of electromagnetic communication. The key for mind control of an entire population, then, depended on creating the right kind of amplifier. In *Star Fire*, this hypothetical amplifier takes the form of the Tolkien satellite. If unleashed on an American city, its electromagnetic technology could brainwash millions of innocent U.S. citizens.

Ironically, the United States' own psychic weapon—Project Tonopah—also plays on fears of Soviet psychoenergetics, namely satellites using microwaves to physically damage human bodies. Project Tonopah involves "weaponry in the area of sonic control of the human mind and microwave demolition of the human nerve system."[104] To be specific, the satellite deploys electromagnetic waves to create "torpor, unconsciousness, or even biological destruction."[105] This technology once again derives from actual fears of Russian psychoenergetics, as the DOD report identified light and sound transmissions as ways of affecting human behavior.[106] As we can see, American fears of Soviet psychic weaponry are literalized in Tolkien and Tonopah technologies. While the former leads to mind control, the latter leads to loss of consciousness or death. In each case, a Cold War superpower has militarized the paranormal mind for malicious purposes. The concept of a psi satellite is hence a generative symbol of the mind's reconfiguration within paranormal governmentality. In both *Star Fire* and Cold War governance, the human mind becomes a weapon for controlling entire populations; the satellites are consequently not stand-alone objects but technologies belonging to larger networks involving media, military, governmental, corporate, and political institutions. This is an enlightening metaphor. Psychical research, the ultimate heterodox science, is not marginalized in the text. Instead, it is the ultimate technology of power, floating high above the Earth and deeply enmeshed in the power structures guiding twentieth-century life.

Dan Merriweather's role as critic and revolutionary of psychical research is another integral plot point. Merriweather is a thinly veiled avatar of Swann and an obvious figure of wish fulfillment. Both Swann (painter) and Merriweather (musician) are artists, though only the latter has reached the pinnacle of his field, with "money, wealth, publicity, [and] the raucous acclaim of an international following."[107] Like Swann, Merriweather is an outsider to the staid world of parapsychology, a rebel whose creativity and abilities augur a paradigm shift in the discipline. When Merriweather visits the North California Biocybernetics Center, he discovers the experimenters "were interested only in going over and over well-worn ground, following procedures they and others had laid down long ago, which to him seemed to have little to no bearing on the problems they were supposed to be dealing with."[108] Like the spiritualists of the ASPR who never understood the point of predicting dice rolls, Swann saw little real-world value in such experimentation. Likewise, his alter ego, Merriweather, vents frustration at such scientific pedantries. "Many of the tests seemed to him foolish, and he proposed new ones," Merriweather recalls, "but again the dead hand of the administration had come down hard with the dictum that research would follow the lines already laid down."[109] The "dead hand of the administration" captures Swann's view of parapsychology as a zombie field mindlessly going through the motions when it really ought to explore man's full psychic potential. When Merriweather demonstrates his psychic mastery at the lab by mentally entering a computer terminal and generating the digital image of a flower, a researcher is so unnerved he literally flees from the laboratory.[110] This scene not only alludes to Swann's flower paintings and magnetometer episode at Stanford but also reflects how the true might of the paranormal mind would send mainstream parapsychology running. For much of the twentieth century, psychical research was dedicated to the laboratory, a fact historian Brian Inglis believes disconnected the field from both the masses and the natural world. In contrast, Swann and Merriweather evoke a spiritualistic return to the marvelous along with a progression toward a practical science that makes things happen in the world.

Indeed, Merriweather's ability to monitor the globe and take action stands in stark contrast to parapsychology's decades-long indolence. When he locates the Tonopah and Tolkien sites, Merriweather immediately threatens the U.S. and U.S.S.R. governments. When they build alternative sites a few hundred miles away, he finds those too. When he discovers Dr. Cooper possesses files that could reveal his identity, Merriweather vaporizes them by destabilizing their molecules: "the region in which he existed seethed and roared with energy—magnetic, gravific, radiant energies surging, rending atoms and molecules, paced to and controlling the dance of electrons, neutrons, particles and antiparticles."[111] The telekinetic ability to alter the chemical structure of objects at a distance requires world-class psychic powers far exceeding the frivolous counting exercises transpiring in America's parapsychology labs. Merriweather is thus an antisymbol of midcentury psychical research, which is to say everything the field is not. By the 1970s, parapsychology had little scientific credibility and few practitioners performing work of consequence. Compare this diminished state with Merriweather: a rich, powerful telepath who cows heads of state, perceives all, controls matter, and dictates the future of the planet. Psychologist (and LSD enthusiast) Timothy Leary uses the term *psychonaut* to describe "the mind explorer as a new kind of heroic voyager."[112] In *Star Fire*, we witness the escapades of the ultimate psychonaut.

The thematic of psychical wish fulfillment reaches an apex when Merriweather transforms into pure consciousness at the end of the novel. "The very existence of psychic phenomena," Dr. Coogan explains, "suggests very strongly the existence of a condition of mind not bound by the laws of the physical universe."[113] In permanently detaching his mind from his body, Merriweather merges with the psychical unconscious connecting humanity. If we recall, Swann believed in a mode of collective unconscious linking the minds of all people on Earth. This is a popular notion shared by Jungians, psychical researchers, and the New Age Gaian theorists described in chapter 2. In *Star Fire*, Merriweather dissolves into this world mind. He becomes the reservoir of James's mother-sea and the field of Murphy's field theory, which is to say he universalizes himself: the psychic advances he can achieve as pure mind will become the achievements of all mankind. If the goal of psychical research at the end of the nineteenth century was to understand the paranormal mind and thereby unlock its full potential, then Merriweather achieves that mission by novel's end. His complete immersion into "the people" brilliantly captures Swann's populist impulses. Moreover, it foreshadows his future works of nonfiction.

While *Star Fire* provides keen insights into the historical and scientific context behind remote viewing, Swann's other literary and educational projects are also important to interpret as technologies of epistemological change. After Swann left Project Stargate in the late 1980s, his publishing endeavors focused on enhancing psi worldwide. The narrow scientificity of parapsychology had alienated the public from the paranormal, leading to a decline in the global collective unconscious. Swann thought if more people believed, studied, and practiced psi in their everyday lives, the psychic potential of the entire human system would strengthen. In this respect, Swann's various memoirs and ESP guides are not just primers for the individual reader but populist, phenomenological tools for advancing the paranormal mind on a global scale.

One of the best examples of literature as psychical amplifier is *Natural ESP* (1987).[114] *Natural ESP* is a layman's guide blending the history of psychical research with Swann's own clairvoyant methods to help readers master their own minds. One of its major claims is that ESP is fundamentally misunderstood because modern scientific attempts (laboratory based, elitist, etc.) to investigate the paranormal have led us away from its true potential. Only a practical, artistic, citizen science–style approach to ESP can truly unlock human consciousness.

The central conceit of *Natural ESP* is that ESP is a naturally occurring phenomenon intrinsic to all humans that anyone can master.[115] Just as in *Star Fire,* Swann blames establishment parapsychology for hindering humankind's ability to recognize this self-evident truth. "ESP is like the enormous, shifting dunes of sand in a desert, always changing, soft, and fluid," he writes. "Perhaps it will be seen that parapsychology has been trying to negotiate these dunes with horses, whose sharp hooves sink into the sand, when all along they should have been riding camels, whose big soft pads are perfect for desert travel."[116] The tools of science (e.g., statistical analysis) are too precise for the dynamic abstraction that is the paranormal. In addition, scientists are less capable of psychic exploration than mediums, or even members of the general population, and should therefore loosen their control over its discourse. The sad fact is that psychic exploration has been taken away from practitioners and the public and handed over to those least capable of understanding it: scientists. In their 1919 tract *Occult Chemistry,* Annie Besant and Charles Leadbeater claim that occultists like themselves must pave the way for establishment scientists. Madam Curie may have "discovered" that radium beta particles decay into helium, they argue, but alchemists have understood element transfiguration for centuries.[117] Swann similarly believes modern science can learn a thing or two about the paranormal from practitioners like him.

Natural ESP is an attempt to reconfigure the power relations and study of psi. Swann argues that mastering psi requires self-exploration and self-knowledge: "Only by becoming aware of *the extrasensory you* can you start to locate the processes that are taking place in your ESP core," he writes.[118] Instead of deploying older parapsychology terms, Swann develops a new vocabulary and grammar to empower the lay medium. For instance, the unconscious is relabeled the *ESP core,* "a hidden extrasensory perceptual system" that follows its own rules and logic.[119] Containing "the psychic nucleus, the preconscious processes, the subliminal barrier, and the area labeled 'past conscious experience,'" the ESP core represents stages of mind beneath conscious processing and culturally acquired knowledge.[120] Broadly speaking, it is the same thing as Myers's subliminal self—everything beyond the known conscious. However, Swann makes the ESP core his own because he claims it can access "the second reality" (Figure 6), a psychical space characterized by infinite information and the absence of space-time.[121] Here Swann reinscribes the Matrix in fresh terms and confirms the role of the unconscious in accessing it.

Similar to *Star Fire* and the *Coordinate Remote Viewing* manual, *Natural ESP* argues that the best way to access one's ESP core is by sitting in a dark room with a helper and drawing what one sees. In a chapter entitled "Taking

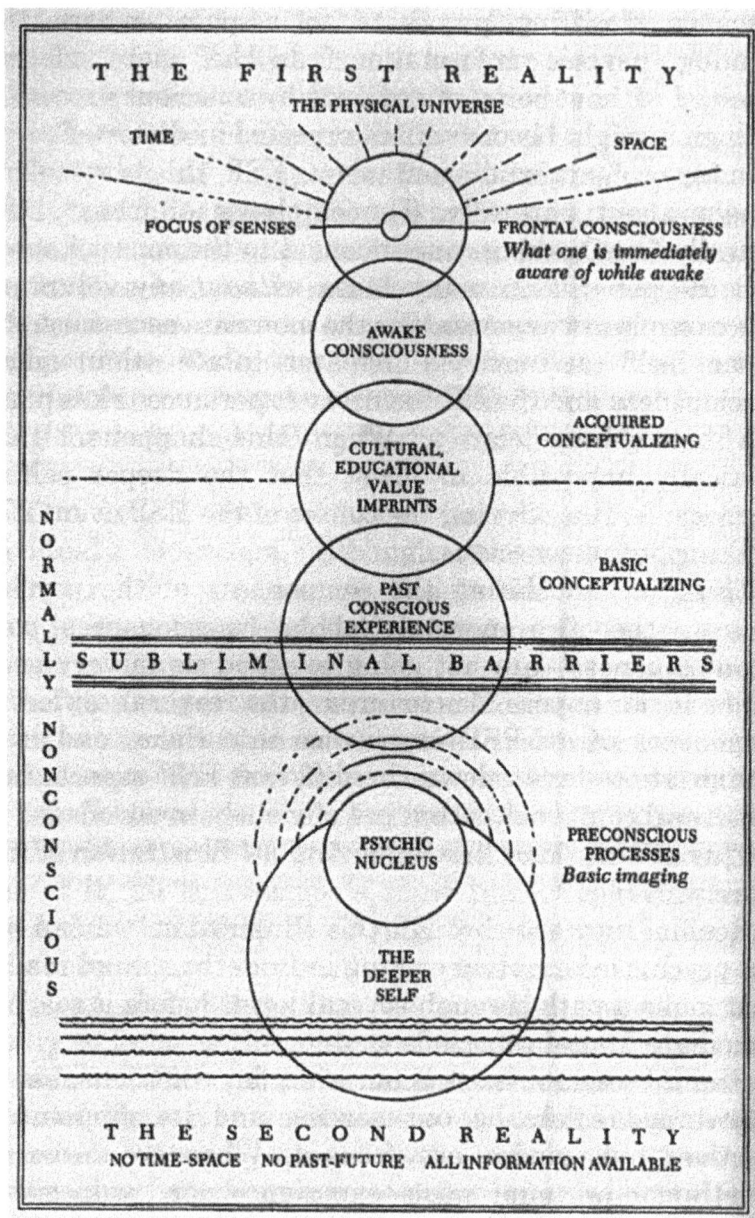

Figure 6. A diagram of human consciousness and planes of reality. *Everybody's Guide to Natural ESP* (2017), Swann-Ryder Productions.

the Plunge into Core ESP," Swann outlines a method of clairvoyance reminiscent of a remote viewing session. He recommends finding a meditative space without distractions: "You will need a quiet atmosphere. Take the phone off the hook and turn off the radio or stereo."[122] He suggests pairing with a friend "who is genuinely interested in the experiment, and who is not antagonistic to the idea of ESP."[123] And using pen and paper is crucial because drawing is the most direct way of accessing the unconscious. Throughout an ESP session, Swann advocates against consciously seeking information because this will only lead to analytical overlay: "If you find yourself thinking about what the target might be, take a break, and start over again."[124] The main idea is to let the ESP core, and by extension the Matrix/second reality, wash over you: "Bear in mind that the ESP information is partly gut feeling, partly intuition, and partly a sort of automatic response that does not actively engage your conscious mental processes."[125] As these passages illustrate, Swann applies the same techniques from remote viewing and applies them in a self-reflective practice stripped of its original militaristic context. Going back even further, we can see echoes of the relaxation and drawing techniques seen in *Mental Radio*.

Swann's how-to manual for visualizing distant objects has individual and collective goals. For the former, the aim is to provide individual readers techniques for retrieving second-reality data and experiencing the true power of the mind. For the latter, the aim is to marshal a new epoch of psychical ability. As he states in the introduction, "Only those individuals who have made some effort to locate and understand ESP as a personal experience will be in a position to comprehend more fully the potential of ESP. They are the advanced thinkers who will redefine the boundaries of consciousness and provide the foundation for the new age of applied ESP."[126] This last phrase—"the new age of applied ESP"—encapsulates Swann's desired telos. For too long the paranormal mind has been sequestered in the hands of know-nothing scientists neutering the field with banal experiments. The time has arrived for a broad-based coalition of mediums and hobbyists to conduct their own investigations into psi. Only this populist wave can help humankind fully comprehend the paranormal. In his book proposal for *Natural ESP*, Swann argues that true knowledge of ESP is only possible through widespread, practical application. "Its fundamentals need to be better understood and experienced by larger numbers of people," he writes. "The fundamentals of extrasensory perception need more in-depth exposure, an exposure the average reader can comprehend and in which he can participate."[127] Through texts like *Natural ESP*, Swann hopes to mobilize a popular movement building toward a phenomenological shift: psi as a universal ability powered by everyday practice.

Your Nostradamus Factor: Accessing Your Innate Ability to See into the Future (1983) is another important text designed to change the broader psychical culture. While *Natural ESP* emphasizes clairvoyance, *Nostradamus Factor* focuses on precognition as a natural talent anyone can deploy. This book operates less as a how-to manual than a general treatise on prophecy, or what Swann calls "future-seeing." By mixing personal anecdotes, predictions of twenty-first-century life, and "fundamental laws" of future-seeing, Swann establishes a quasi-scientific discipline around precognition.[128] Despite surface differences from *Star Fire* and *Natural ESP, Nostradamus Factor* participates in Swann's populist vision of applied psi, a widely practiced citizen science piercing through scientific skepticism and cultural ideologies to recognize prophecy for what it is: one of mankind's greatest tools.

Swann defines the Nostradamus Factor as "a special mind-dynamic presence beneath the intellect, which foresees the future and spontaneously (though only occasionally) alerts the individual through dreams, working visions, and autonomic urges."[129] Like the ESP core, it is another of Swann's neologisms for the subliminal self. Myers's subliminal self, we recall, delivers futural information through dreams, sudden visions, and ghostly hallucinations, all of which are mediated through unconscious processes. In *Nostradamus Factor,* Swann classifies precognition into four categories: (1) spontaneous forewarnings while asleep, (2) spontaneous forewarnings while awake, (3) communal alerts experienced by large groups of people, and (4) consciously controlled future-seeing. Much of the book is spent on exploring precognition in dreams (first category) and actively soliciting precognition through drawing mind images (fourth category).

Swann develops several novel theories of divination in identifying seventeen laws underlying future-seeing. Some of the laws assert the reality of prophecy, such as law 2: "Future events ARE foreshadowed via spontaneous NostraFac experiences."[130] Other laws revolve around the true nature of time, the chronological language of the mind, and the shortcomings of rational ideology. For example, when Swann claims that different kinds of time exist (law 3) and that all organisms experience a range of time references (law 4), he asserts that a variety of chronological modes (linear, cyclical, astrological, etc.) exist and that the human mind is capable of perceiving all of them.[131] Even if the physical body exists in one part of space-time, the mind can see into other space-time continuities as well. (We saw a strain of such thinking with Dobbs's psitron cognition in chapter 1.) When Swann writes "the independent future-seeing mind must possess its own modes of processes, its own 'language,' its own reason and logic, and its own goals or purposes," he asserts that the innate language of the mind is the *image*, which is universal, ancient, and more accurate than "rational" modes like speech.[132] This claim confirms Swann's dedication to the remote viewing ideogram as the ultimate carrier of truth. When Swann writes "preconception and future-seeing do not mix" (law 11) and "the rational-logical mind rejects the 'incredible'" (law 13), he contends that conscious ratiocination remains the biggest impediment to successful precognition.[133] Similar to how analytical overlay corrupts the purity of information from the Matrix, logical expectations and assumptions corrupt precognitive data. Only by trusting what we receive from the unconscious can precognition truly work. As we can infer, these laws affirm the same psychical terrain established in Swann's earlier novels and manuals.

However, *Nostradamus Factor* makes clear that seeing the future is a ubiquitous human trait we have an obligation to explore. Early in the text, Swann admits, "I was under the general impression that foreseeing could only be accomplished by special people who mysteriously possessed specific talents for doing so."[134] This is the medium-centric notion of the paranormal mind. However, he soon realized the only real difference between gifted and non-gifted people was the former had "developed some kind of intellect-control over it."[135] "It is the *intellectual interest* in developing the control that seems to stimulate increasingly gifted future-seeing," he contends.[136] This is a riveting claim: merely enhancing one's interest in psi enhances psi itself. Furthermore, "the *process* of our deep-level future-seeing powers remain obscured and un-located not only because they are deep-mind functions but because they are intellectually obscured by myths, misinterpretations of evidence, inappropriate intellectual assumptions, and sometimes outright denial of future-seeing phenomena."[137] The modern ideological default is skepticism of the supernatural, and it has hampered humanity's ability to future-see because it invisibly forecloses the very possibility of the marvelous. Universal skepticism leads to the universal inability to foresee. This is yet another intriguing claim wherein the power of psi worldwide is tied to levels of belief, intellectual engagement, and everyday practice. The scientification of the modern world has deeply harmed the paranormal mind; so too has the excessive rationalizing of mainstream parapsychology. Implicit in the rise of logical thinking is the decline of transcendent belief, and by extension our paranormal abilities.

Nostradamus Factor is therefore an instrument of epistemological, ideological, and phenomenological change. By disseminating the idea that "our species must possess some kind of Nostradamus-like factor," the text adds to society's store of psychical knowledge.[138] If it convinces readers of the actuality of future-seeing, then the book can spur an ideological shift on what the mind can achieve. And if those readers gain an intellectual and practical interest in psychical phenomena, then phenomenological access to future-seeing, clairvoyance, and other supposedly impossible abilities is enhanced at a species level. As such, *Nostradamus Factor* should be considered a literary technology. In the same way that remote viewing is an instrumentalization of science for military means, Swann's books are an instrumentalization of science for cultural means—tools designed to reverse psi skepticism and expand humanity's paranormal potential.

Swann's mission to achieve universal psychical advancement via paranormal pedagogy comes through most clearly in his proposal for an ESP institute, the International School for Applied Psychic Studies (ISAPS). In the early 1990s, Swann envisioned a school designed to teach the general public about telepathy, clairvoyance, and precognition. By tapping into the emerging New Age movement and lay interest in psi, Swann imagined a virtual academy of ten thousand to fifty thousand students worldwide taking classes on all varieties of psychic phenomena. In addition to teaching, ISAPS would publish psi tutorials and manifestos by Swann and other like-minded psychics. While Swann's idea never went beyond the planning stage, analyzing his unpublished prospectus is highly instructive because it showcases ISAPS as an ideological technology for enhancing the global psychic mind.

In explaining his rationale for a virtual psi academy, Swann writes, "No school of applied psychic studies has ever been founded, and in its absence Western literacy regarding psychic phenomena is not very high."[139] Widespread psychic "illiteracy" is a serious problem because it stunts humanity's full potential. Swann explains:

> Social-specific reality boundaries, however, have meaning far beyond psychic matters. Aggregate or collective social forces are very powerful. Their constructed, self-contained reality limits artificially format the manners of action, behavior, attitudes, knowledge, minds, and consciousness of those shared in the boundaries either knowingly or unknowingly. We adapt to the social experiments into which we're born, and our subsequent experiencing and formatting educational processes within those environments fully place us within the margins of their various social-specific reality limits. We either function within those boundaries in social clone-like ways, or if not, the social forces maintaining them will reject or alienate us.[140]

Several claims in this passage are worth explicating. When he argues that "collective social forces" artificially format our attitudes and behaviors, Swann speaks of ideology, and specifically the scientific ideology that has long dismissed the paranormal. Such hegemonic ways of thinking and seeing create "reality limits" on what the human race can imagine and do. For example, if society tells us telepathy is impossible, then it actually curtails one's ability to perform mind-to-mind communication. This claim reinforces a common belief in parapsychology that telepathy, clairvoyance, and precognition are strongest among true believers.[141] For Swann, the widespread skepticism of ESP produces actual roadblocks to its practice and potential. The result of antipsychical discourse is the "consciousness fragmentation of the species."[142] Swann subtly alludes that the collective mind of humanity could—and should—be united at a metaphysical level. Whether it is called the mother-sea, the Matrix, the ESP core, or the Nostradamus factor, the unconscious mind of man is linked. These links are fragile, though, and for centuries, scientific ideology has weakened them further. In contrast, strengthened belief in the paranormal would lead to new gestalts and to invigorated psi capabilities worldwide. If the general public could only break from the iron cage of scientificity and recognize the truth of the paranormal mind, then the collective psychic ability of the species would be augmented. This subtext is present in many of Swann's previous works, but his vision for ISAPS explicitly claims paranormal pedagogy must precede psychical awakening.

The solution to the problem of psi skepticism is "panoramic consciousness," or "gaining power-active participation in a larger, functional super-state spectrum of consciousness potentials."[143] In his ISAPS proposal, Swann claims all humans have superstate potentials (access to the metaphysical psychical world), which are normally impeded as a result of modern culture. If the public can bypass the dominant world views suppressing truth of mind, then more people could participate in the collective unconscious. Like his books, ISAPS serves a technological function because its purpose is cultural change. Through its proposed array of classes, lectures, and books, Swann's school becomes a vehicle for nudging humanity toward a superstate—that is, toward a "shared consciousness format, a design-proscribed format within the overall spectrum of our species indwelling consciousness-experienced potentials."[144] At stake is more than occult knowledge but access to actual superpowers like "creativity, inventiveness, ingenuity, intuition, psychic perceptions, direct observations of invisible forces, and persuasion."[145]

When placing ISAPS alongside Swann's earlier works, we can see his entire oeuvre participates in an ideological revolution designed to amplify the Matrix and the hidden powers of the mind. As such, his school and literary texts should be construed as instruments of epistemological and phenomenological transformation. This panoramic view helps us understand Project Stargate, *Star Fire,* and his other writings in a new light. Taken independently, his projects might seem like bad science, quirky entertainment, or crazed ploys for attention. Such a cynical reading misses the true scope of Swann's agenda: enhancing the shared paranormal mind of humanity. As we saw in chapter 2, viruses and bacteria may act as agents to instantiate the world mind. For Swann, his writings and theories are these agents; they are the catalysts for maximizing the global collective unconscious and pushing our awesome powers of clairvoyance, foresight, and self-discovery to ever higher stages.

Ingo Swann and the Psychic Renaissance

In his posthumously published book, *Psychic Literacy and the Coming Psychic Renaissance* (2018), Swann ties together several themes guiding his unlikely career in psychical research. "An evolutionary leap is taking place regarding psychic matters—surprisingly not because of any special breakthroughs in the familiar field of parapsychology, but because of advances in areas outside of it," he writes. "This evolutionary leap will require a new level of psychic literacy—a broader spectrum of the background information we use to make sense of psychic matters and how they interact with our daily lives."[146] These dual calls to action—to improve psi literacy and to participate in a psychic renaissance for global mind—crystallizes much of Swann's idiosyncratic work. It also provides a fitting coda to this chapter.

As we have seen, Swann played a direct role in transforming psychical research from a science to a technology in the waning decades of the twentieth century. Whereas previous researchers like Myers and Rhine once sought general principles of the paranormal mind, Swann and his military enablers desired ways to deploy it for the state. Their weaponization of the paranormal via remote viewing was a direct response to Cold War threats and the desire for enhanced governmentality, but it also captured the political priorities of the age. Through concepts like the Matrix, the signal line, and unconscious ideograms, Swann assembled different bits of psychical history (unconscious drawing, field theory, etc.) into an innovative program of militant clairvoyance. At the same time, his life's work was a democratization of the paranormal away from scientific elites who had overanalyzed it to the brink of death. He saw the potential of psi in all people, and he hoped to cultivate it, strengthen the paranormal unconscious, and spearhead a psychic renaissance. "Real foresight is always psychic in nature, and it is interesting to note that we are the inheritors of dominant social forces which skeptically denied the importance of human psychic potentials," he writes in *Psychic Literacy*. "These potentials thus have not been developed and *used,* and social outcomes have trended toward disaster as a result."[147] Through arguments like these, Swann demonstrates an unswerving belief that a cultural shift in psi awareness will transform humanity, generating new vistas of perception and unlocking powers like clairvoyance. In this sense, Swann truly was a revolutionary. He often criticized parapsychologists for being too small-minded in their research questions and experiments. There was hardly anything small about Swann's ambitions; he wanted nothing less than to change the mind, the human species, and the scientific ideologies governing us. One could easily argue that his visions were too grand because the paranormal mind never became the leading edge of military weaponry. Funding for Project Stargate dried up in 1995, with critics rebuking its operational efficacy. The paranormal mind never replaced nuclear warheads as America's greatest geopolitical weapon. But if we look at the effects of his incredible vision—in TV shows like *Stranger Things,* the ceaseless stream of Project Stargate memoirs, all the remote viewing websites peppering the internet—one could also make the case Swann succeeded in promoting the paranormal beyond his wildest dreams.

If nothing else, Swann opens our eyes to new arenas of literature we can no longer ignore. Paranormal fiction and nonfiction have been dismissed, avoided, and laughed at by scholars for decades. But psi manuals, memoirs, manifestos, novels, and intelligence reports carry within them the secret history of psychical research. Today much of that content has migrated to the internet and proliferated in unexpected ways. These unusual texts deserve our critical attention for the ways they shape scientific discourse, political history, paranormal epistemology, and cultural belief. Such an examination would reveal the breadth of parascientific networks and the processes by which marginalized science proliferates. Parascience sprawls through the interconnections of science with literature, philosophy, and myth, but it also interweaves with Cold War ideology, cartography, geodesy, and the threat of the psychic other. If we hope to understand how supernatural powers like clairvoyance evolve into surveillance instruments like remote viewing, then we must cast a wider net into paranormal science and its various literatures. Swann would undoubtedly agree, for such a renewed interest would also help unleash the coming psychic renaissance.

4 ON GHOSTS AND GHOST VISION

The Hundred Secret Senses, Comfort Woman, *and the Asian American Spirit Medium*

AROUND THE SAME TIME that it was conducting remote viewing sessions on Soviet embassies and military bases, the CIA was also monitoring the latest developments in Chinese parapsychology. In a recently declassified report, "Parapsychology in the People's Republic of China: 1978–1989," Leping Zha and Tron McConnell describe the Chinese government's preoccupation with understanding—and harnessing—psychical abilities among its citizens. At the root of this next-generation research program was the ancient concept of qi.

The term in China for paranormal powers is EFHB, or exceptional functions of the human body.[1] In the 1980s, Chinese scientists studying EFHB determined that 40 to 63 percent of children under the age of ten demonstrated some ESP capability.[2] While some researchers believed EFHB was electromagnetic in nature (psychoenergetic theory), others hypothesized it was related to qi. Translated from Chinese as "air" or "gas," qi is a form of energy in ancient Chinese philosophy and traditional medicine believed to flow through all things in the physical and metaphysical worlds. The permeation of this elusive substance throughout the universe is significant because those who hone qigong, the practice of qi cultivation, could ostensibly heal, read, and even control others' bodies. "By sending *qi* to certain parts of the body through 'meridian channels' and by practicing in certain ways, psi abilities can be attained," the CIA report asserts. "There are said to be abilities more profound than psi which can be attained by higher qigong masters who can freely control *qi* by the mind and cause it to circulate through the entire human body."[3] The CIA was particularly concerned that many high-ranking Chinese officials supported EFHB research. Dr. Qian Xuesen was an MIT- and Cal Tech–trained rocket physicist who established the military-focused 571 Institute: "psi study was their official task, and as in other top defense-related institutes, they were well-equipped and well-funded."[4] Hu Qiaomu of the Central Political Bureau is quoted as saying, "We should mobilize every unit in our society to study qigong science."[5] A third proponent was none other than Deng Xiaoping, the Communist Party leader who personally employed qigong masters to improve his personal health. When Dr. Qian makes statements like "Chinese qigong is modern science and technology," he affirms Western fears that psychic warfare is a battlefront America is doomed to lose.[6] While these national security concerns speak directly to Cold War politics, we cannot ignore the fabulous irony that an ancient vital force resides at the center of a late twentieth-century scientific research program.

The connection between literature, non-Western science, and the paranormal mind constitutes the focus of this chapter. Throughout this book we have examined the development of paranormal logic through Western sciences like quantum mechanics and systems biology. However, as EFHB and qigong suggest, the paranormal is not confined to Western culture alone. Stories and legends from nearly every culture contain references to ghosts, divine visions, and prophecies. Nor is Western culture the only one to develop a systematic interpretation of nature—a science—to comprehend the paranormal. If we hope to fully understand how paranormal concepts have evolved over the course of the twentieth and twenty-first centuries, expanding the texts, sciences, and cultural frameworks through which they have migrated is crucial. This chapter thus marks an important pivot in the book by examining the paranormal mind via non-Western epistemes and multicultural literature. Non-Western science remains a fledgling subject area in both science studies and literary history. Most STS studies approaches to literature use traditional disciplines like physics to engage with literary works; such a strict focus neglects the full range of knowledges accumulated by other cultures. In contrast, I take seriously the ways non-Western societies scientifically interpret the world and, by extension unorthodox phenomena like spectral communication.

A compelling figure in this expanded paranormal conversation is the spirit medium. In psychical research, the medium is a central figure for holding séances, calling forth ghosts, and speaking with the dead. She—for she is almost always a woman—is alternatively celebrated, feared, mocked, and revered. Along with telepathy, clairvoyance, precognition, and telekinesis, communicating with spirits has always been a distinguished occult gift—hence the preeminence of mediums across psi history and literature. She is a rare creature in this regard, a powerful female figure whose exceptional sensitivity upends the typical gender politics of androcentric science. In ethnic fiction, female mediums play crucial roles as prophets, healers, and guides; they are idiosyncratic bridge figures connecting the living with the dead. From the mystical Miranda Day in Gloria Naylor's *Mama Day* to Lourdes del Pino Puente in Cristina García's *Dreaming in Cuban*, characters who specialize in seeing and speaking with spirits abound. In Asian American fiction in particular, the female ghost seer is a recurring trope for contrasting Eastern modes of ancestral knowing with Western modernity. While such characters can invite Orientalist readings on Asia's backwardness, my goal is to resituate ghosts, ghost vision, and spirit mediums within richer frameworks of indigenous science. More specifically, I aim to elaborate on the phenomenology of spirit communication via qi and Korean shamanism—not only to destabilize Western science as the only valid framework for interpreting reality but also to make the case for utilizing multicultural science in literary analysis. One of the wider objectives of parascientific STS is to explore marginalized (sub)cultures and contexts across the history of science. The sheer dominance of Western science can be ideologically blinding, but literary critique must acknowledge that alternative modes of knowing and being exist. Recognizing these modes permits productive readings of the ghost seer as a literary archetype. Though the ethnic spirit medium is sometimes viewed as a racialized construction, analyzing her through non-Western epistemes allows for a deeper understanding of how she navigates space, time, nature, and the afterlife, as well as the circumstances through which paranormal phenomena arise. In this way, the spirit medium is reclaimed from Orientalism and emerges as a locus for reconfiguring the paranormal mind. There is a scientific logic to how Asian American spirit mediums perceive ghosts and conduct clairvoyance, and we should explore their world views on their own terms. In doing so, we can explore parascience beyond the standard arenas of Western thought.

The following pages review the politics of gender and non-Western epistemology in science before examining two of the most famous—and controversial—spirit mediums in Asian American literature. In Amy Tan's *The Hundred Secret Senses* (1995), Kwan Li is a seer who can communicate with ghosts and see into the distant past. The character has been widely criticized as an Orientalist caricature, but I reinterpret her through Chinese metaphysics and contend that her "backward" powers must be understood by her paranormal ability to see qi. More generally, her capacity to perceive the world through a non-Western scientific paradigm expands how the paranormal has been rationalized to modern audiences. I then turn to Akiko Bradley in Nora Okja Keller's *Comfort Woman* (1997), a talented ghost seer who is also Honolulu's preeminent medium for the Asian community. In examining her supernatural powers through the scientific and cultural lens of Korean shamanism,

I reconsider the novel as an epistemological text providing insights into the paranormal mind. By delivering lost histories to their Western audiences, Tan and Keller adopt the roles of mediums themselves and become bridges for Eastern ethnoscience.

This chapter serves several purposes. First and foremost, it prioritizes the role of non-Western science in the multicultural development of the paranormal mind. For too long, indigenous frameworks have been ignored in the history of science, so it is crucial to recognize how alternative views of nature have molded interpretations of psi. A second task is the reformulation of the ethnic ghost in literary criticism. Ghosts are recurring entities in Asian American, African American, and Native American literature, just to name a few fields, and they reflect long-running systems of belief involving spirits and spirituality. Too often, though, literary criticism only views ghosts as symbols for something else: collective trauma, invisibility, historical violence, and so on. I suspect metonymy has become the dominant mode of supernatural literary analysis because of the underlying assumptions in Western culture that ghosts do not exist. Ghosts are not real, so the task of the literary critic is therefore to interpret them as proxies for other, more substantial ideas. In contrast, I argue literary scholars can, and should, interpret ghosts qua ghosts. In other cultures, ghosts occupy a constitutive part of material reality, and the ability to perceive and cognize ghosts—what we might call ghost vision—represents a normal aspect of that reality. Recognizing the cultural diversity of spirit ontology bypasses the existing parochialism of ghost discourse and opens fresh spaces for literary interpretation.

My third point is related to the second insofar that a parascientific approach reclaims Asian American ghosts from the sphere of Orientalism. By looking at qi naturalism and shamanism as coherent systems of knowledge following certain rules, we can appreciate ghosts and ghost vision within a rigorous social context instead of dismissing them out of hand as primitivism. Put differently, my aim is to recognize the "eccentricities" of non-Western world views as part of global lived experience. A fourth function is to bring gender politics into the parascientific conversation. Parascience foregrounds disenfranchised ideas, movements, and groups in science. Women have been excluded from modern science for much of its history, which is why the female medium is such an outlier. As a central figure in a patriarchal field, the woman medium demonstrates how the paranormal upsets the scientific order of things in more ways than one. Finally, this chapter expands the range of paranormal literature. The literary history of the paranormal mind traverses Victorian fiction, modernist literature, science fiction, New Age speculation, and psychic manuals, among other genres. All play pivotal roles in transforming and disseminating ideations of the supernatural. The same is true for ethnic fiction. Just as Western texts have deployed modern science to rationalize precognition, Asian American works can apply Eastern frameworks to explain the unexplainable. By highlighting how authors have deployed non-Western epistemes in their paranormal poesis, we can also understand how multicultural fiction participates in parascientification.

A recurring theme throughout this book has been the import of a parascientific approach to literature, blending science, philosophy, history, and myth to gain a deeper understanding of the epistemic and esoteric flows of contemporary culture. The paranormal mind is not an antiquated idea lingering in the forgotten recesses of Anglo-American literature but a complex, recurring theme that occurs in ethnic literature as well. Its presence across different groups, chronologies, and genres suggests that the paranormal is something of cultural universal. Let us give it the attention it deserves.

Beyond Western Science

To situate the epistemes undergirding the paranormal typologies found in ethnic fiction, we must first reconsider a core tenet of Western science, namely the assumption that there is only one kind of science. The dominant narrative of science is that it constitutes a methodology and body of knowledge birthed in the seventeenth century with the founding of the Royal Society. By generating hypotheses, conducting experiments, analyzing data, and emphasizing reproducibility and objectivity, humanity has derived systematic insights into nature. According to Auguste Comte, the founder of positivistic philosophy, science rejects the metaphysical and spiritual follies of classical knowledge to deliver if not ultimate truth then an inexorable migration toward truth.[7] From the humble beginnings of natural philosophers like Sir Francis Bacon and Robert Boyle has arisen a steady accumulation of facts lifting mankind from the dark ages of ignorance, magic, and totemism into the technological mastery of the world we now enjoy today. Or so the story goes.

Much of this utopian chronicle of science is infused with a latent Eurocentrism, an "internalist" epistemology assuming there is one kind of correct knowledge in the world:[8] positivist knowledge produced by the modern West.[9] Disseminated through textbooks and invoked by scientists, politicians, and the general public, internalist epistemology is the prevailing perception of scientific knowledge in contemporary society. One problem with this narrative is that Western science is conflated with science in general; it rests on the premise that only Eurocentric methods and facts are credible—a notion discounting the entirety of information that all other societies in all other historical periods before the seventeenth century have produced as misinformed, superstitious, or, at best, prescientific. One error is misconstruing the epistemic norms of the modern era as universalist and ahistoric. More troubling is the implicit acceptance of Western Europe as the world's only fount of real knowledge, which renders all other modes of world knowing as other. It is a colonization of knowledge itself.

Thankfully, there is a countermovement among various philosophers of science and social scientists challenging such epistemological hegemony. According to anthropologist Bronislaw Malinowski, every successful civilization throughout human history has amassed scientific information: "A moment's reflection is sufficient to show that no art or craft however primitive could have been invented or maintained, no organized form of hunting, fishing, tilling, or search for food could be carried out without careful observation of natural process and a firm belief in its regularity, without the power of reasoning and without confidence in the power of reason; that is, without the rudiments of science."[10] In *We Have Never Been Modern,* Bruno Latour argues that the entire notion of modernity is based on a false distinction between the "premodern" past (the sociocultural world of myth, magic, and superstition) and the "modern" present (the postsocial world of neutral, scientific knowledge).[11] If we accept Latour's assertion that science is social, then we can also see it is an amalgam of premodern, modern, non-Western, and Western discourses.[12]

Arising from this nonmodern pivot in science studies is a key question: is more than one kind of science possible? Postcolonial theorists Sandra Harding, Walter Mignolo, and Bernd Reiter all say yes. Harding has long been one of the most prominent voices calling for the decolonization of Western science. Like Malinowski and Latour, she rejects the idea of science as a superior Western episteme stripped of religion, magic, and culture. Harding instead defines science as "any systematic attempt to produce knowledge about the natural world."[13] The decisive term is *any* because it rejects Eurocentric principles (e.g., laboratory-based empiricism) and historicity (e.g., Royal Society hagiography) as the dominant narrative of scientific process. By accepting "any and every culture's institutions and systematic empirical and theoretical practices of coming to understand how the world around us works" as science, Harding disentangles scientific reasoning from a solely Western context.[14] Crucially, she also empowers global, theological, and philosophical approaches to studying nature as part of a multiparadigmatic vision of scientific epistemology.

One of Harding's central pursuits is the reconsideration of ethnoscience, non-Western knowledge systems such as indigenous knowledge, local knowledge, traditional medicine, and oral histories—techniques that cultures worldwide have used to navigate reality for millennia.[15] Different societies inherently ask different questions about nature, and as a term, *ethnoscience* recognizes the scientific and technological achievements of non-Western cultures while also taking their frameworks seriously as complex, rational endeavors.[16] We can see a clear correspondence with Mary Midgley's many-windows perspective, discussed in chapter 2, in which the borderland knowledge of non-Western groups provides additional insights into the way humanity has historically understood the world.

Other postcolonial scholars have since advanced Harding's agenda by calling for a sea change in the ways we interpret and deploy multicultural epistemology. For instance, Walter Mignolo has called for a *pluriverse* to replace the monolithic uni-verse of knowledge controlled by Europe. The pluriverse is a multicultural, multiepistemic approach to epistemology that "consists in seeing beyond this [Western] claim to superiority, and sensing the world is plurally constituted," which ultimately produces alternative modes of seeing, knowing, and interpreting reality.[17] Bernd Reiter furthers Mignolo's call by seeking out a successor science that reflects the varieties of practices and technologies that all cultures have used to thrive in the world. His approach, *mosaic epistemology,* involves "the search for alternative and place-based epistemologies and approaches" reflecting the actual diversity of global sciences without conferring special stature to Western science alone.[18] "What we think is real certainly is not the only reality out there," he writes, "as different people access the same reality from difference places and thus either see, or experience, a different slice of the same reality, or they perceive a different reality altogether."[19] Ethnoscientific, pluriversal, and mosaic epistemologies all demand a scientific multiculturalism cognizant of how Western and non-Western societies interpret the world.

These critics provide a robust theoretical foundation from which I now wish to push into a more parascientific direction. Much of postcolonial science studies has focused on how non-Western thought has preceded and engaged with established arenas in Western science, like indigenous medicine and farming practices. Far less research has examined the ways native science engages with phenomena like telepathy; this is the sort of supernaturalism that modern (in a Latourian sense) critics have typically dismissed in non-Western cultures as folkloric belief. However, decolonial approaches allow us to reconsider the paranormal as part of a globally interconnected world;

multicultural STS expands paranormality itself when literary texts extrapolate ethnoscientific principles in increasingly heterodox directions. As we have seen time and again, the paranormal mind is many things because every scientist, writer, and theorist seems to modify it in his or her own way. What I therefore propose undertaking in this chapter is a pluriversal approach to ethnic fiction that explores how indigenous science can produce ever stranger iterations of psi.

Ethnic fiction is a rich source for investigating the entanglements between non-Western science and the supernatural. Ghosts, prophecies, clairvoyance, and communication with the ancestral dead are common themes in multicultural fiction. Whether we look to the mythic transformations in Charles Chestnut's *The Conjure Woman* or the shamanistic hauntings of Louise Erdrich's *Tracks*, American ethnic fiction is replete with tales of spiritualistic and esoteric phenomena. As such, it is a productive site for interpreting how different cultures interpret—and rationally explain—what the modern West identifies as paranormal. Most literary critique of multicultural supernaturalism tends to bypass epistemological context in favor of metaphor. One of the most popular approaches for analyzing ethnic ghosts is Kathleen Brogan's concept of cultural haunting. Brogan argues that ghosts in the Western literary tradition typically symbolize individual anxieties and fears (e.g., the ghost of Banquo haunting Macbeth), whereas ghosts in ethnic fiction represent the cultural transmission of communal traumas—the injuries, injustices, and lingering memories of an entire group.[20] For example, the titular ghost in Toni Morrison's *Beloved* is a (meta)physical manifestation of collective suffering experienced during the Middle Passage.[21] Avery Gordon similarly contends that literary ghosts are not dead people so much as "social figures" standing in for something lost, missing, or banished.[22] Conceptually, they signify the forgotten subjects of history.[23]

This predilection for metaphor is an intriguing scholarly move because it bypasses ghost ontology for symbology. Standard literary analysis of the supernatural may acknowledge the existence of ghosts in non-Western culture but immediately shunts it aside to foreground intergenerational trauma, collective silencing, and so on. To be clear, I have no issue with this approach because ghosts in ethnic fiction obviously serve important thematic functions when highlighting themes of social injustice. My problem is that too often, this is the only way literary scholars analyze ethnic supernaturalism. When scholars default to the symbolic as a matter of course, we miss the cultural specificity of ghosts. Moreover, I contend that we do ourselves a disservice by ignoring the fact that spirits play a central role in many cultural interpretations of reality. Around the world and throughout history, ghosts participate in people's lived experience. Humans, ancestors, spirits, and demons all constitute aspects of (meta)physical existence. A fuller reading of the supernatural ought to take this into account, which means understanding how ghosts fit into the greater schema of nature. When David Lloyd claims "aesthetics naturalizes representation," he describes the way Western culture can write colonized and BIPOC peoples out of existence by failing to acknowledge their ways of thinking and being.[24] If literary critique fails to recognize the ghostliness of ghosts, it negates both them and the world views they inhabit. This is precisely why R. A. Judy and Nahum Chandler value *paraliteracy*, which they define as "the inhabitation of thought and culture arising by way of its double position as simultaneously within and outside of the 'mainstream.'"[25] Paraliteracy identifies a pluriversal approach that allows us to place Western symbology and non-Western ontology in hermeneutic partnership; in doing so, we deepen our multicultural comprehension of ghostliness.

Asian American fiction helps us examine the parascientific and paraliterary contexts of ghosts. Many canonical works in Asian American literature prominently feature undead spirits, including Maxine Hong Kingston's *The Woman Warrior* and Amy Tan's *The Joy Luck Club*. Other ghostly narratives include Aimee Liu's *Face*, Larissa Lai's *Salt Fish Girl*, Therese Park's *A Gift of the Emperor*, Chang Rae Lee's *A Gesture Life*, and Viet Thanh Nguyen's *The Sympathizer*. Remarkably little criticism analyzes how these ghosts fit within their respective epistemes. By providing an ethnoscientific grounding, a parascientific approach to Asian American supernaturalism not only liberates critique from purely metaphorical readings but also pushes back against hasty charges of Orientalism. In this regard, such fictions act as a Foucauldian counterdiscourse where the scientifically marginalized can speak for themselves.[26]

Gender, Science, and the Female Medium

Before turning to the Eastern sciences of ghost vision, let us quickly review the technics and politics of Western mediumship. The spirit medium has attracted the interest of occult, gender, and STS scholars for years. As a distinguished female figure, the medium is certainly an oddity in modern scientific history. As feminist scholars have long observed, the social structure of science systematically disempowers women: "many of its applications and technologies, its modes of defining research problems and designing experiments, its ways of constructing and conferring meanings are not only sexist but also racist, classist, and culturally coercive."[27] Most obviously, modern science resisted providing education, credentials, and jobs to women for centuries.[28] When women began to enter the scientific workforce in the 1820s, underrecognition and segregated employment (low-paid positions like technicians, teachers, etc.) ensured power remained with men.[29] Even the cultural connotation of rigorous, analytical male scientists served to debilitate women as delicate, soft, and emotional in Western society.[30] This is why the female medium is so peculiar. She leveraged the so-called hypersensitivity of her biology to gain power in spiritualist and psychical research circles.

The concurrent rise of psychological research, spiritualism, and telecommunication technologies at the end of the nineteenth century transformed women's social roles. It was well established during Victorian times that women suffered from "fine nerves," leading to episodes of hysteria, fainting, and mental overstimulation.[31] Consequentially, many businessmen, occultists, and scientists believed the sympathy and empathy inherent to the fairer sex could be harnessed for electronic and even psychic communication: "Feminine nerves and temperaments were seen as crucially distinct from men's: only women generated what we phrase as sympathetic excess—an affective or spiritual quality—that could transform mediating apparatus into the carrier of intentional self-to-self communication."[32] According to Jill Galvan, it was no accident that practically all stenographers and telegraph operators in the United Kingdom and the United States were women; the scientific ideology of the time posited they were sensitive enough to "feel others," permitting them to accurately receive and relay information.[33] Nor was it an accident that the famous Fox sisters were sisters; women were better attuned to ghostly voices for the persnickety task of spiritual telegraphy.[34] As Galvan notes, mediums, telegraphists, and telephone operators alike were "tools" for gathering and distributing information.[35]

This made the female medium a vital figure for conducting psychical research. William Crookes notes that in "some parts of the human brain may lurk an organ capable of transmitting and receiving other electrical rays of wavelengths hitherto undetected by instrumental means . . . in such a way, the recognized cases of thought-transference and many instances of 'coincidence' would be explicable."[36] If the subliminal self explained most every kind of paranormal phenomenon, then finding mediums like Eusapia Palladino and Leonora Piper was mission critical. Unlike most women in Western science, female mediums were integral for the everyday practice of psi and therefore highly valued. Without them, it would be nearly impossible to plumb the inner workings of the unconscious, to prognosticate the future, or to contact spirits from the other side. At a structural level, psychical research depended on women in ways few other scientific disciplines did.

There were two psychical theories for explaining mediumistic communication: the survival hypothesis and the living agent hypothesis. Both required sensitive women. The survival hypothesis, which was championed by spiritualist-leaning researchers, claimed that "knowledge [was] acquired by the medium via communication with a discarnate or deceased individual."[37] This view assumes that some aspect of soul or consciousness exists beyond physical death as a ghost and that certain sensitives can see and speak to them. The medium was "well attuned to the subtle cues, sometimes described as vibrations, by which the spirits expressed themselves," writes Galvan.[38] Thanks to "the idiosyncrasies of female neural biology," particularly her fine nerves, she could successfully see and speak to the dead.[39] In contrast, the living agent hypothesis claimed that "knowledge [was] acquired by the medium telepathically by scouring the mind of the sitter or loved one of the disincarnate or other living people."[40] This view, supported by science-leaning researchers, contended that ghosts did not actually exist. Any extraordinary information the medium gathered was telepathically retrieved from the mind of another living person in the room. The existence of actual ghosts is therefore unlikely, and ghost vision can be explained away as telepathic image transfer. Living agent psi is related to Edmund Gurney's theory of phantasms: ghosts are not dead people existing in perpetuity but telepathic hallucinations sent from the mind of one person (agent) to another (percipient).[41] Alternatively, a haunted location might "produce in the brain that effect which, in its turn, becomes the cause of a hallucination."[42] Whether obtaining their data from the living or the dead, women were indispensable to the practice of mediumship, and by extension psychical research.

Mediumship politics and theory have long captured the interest of scientists, academics, and the public. Nevertheless, this overview remains relatively incomplete because non-Western interpretations of ghosts and mediums are not well represented in the conversation. The works of Amy Tan and Nora Okja Keller should help in this regard.

The Hundred Secret Senses and Chinese Metaphysics

One of the most important spirit mediums in Asian American fiction is also its most notorious: Kwan Li of *The Hundred Secret Senses*. Amy Tan's third novel holds a peculiar place in Asian American letters. Tan is one of the most popular—if not the single most popular—contemporary writers of Asian American literature. Her debut work, *The Joy Luck Club* (1989), was a best seller that introduced Asian American fiction to mainstream audiences and established many tropes now synonymous with the genre: immigration, the American dream, mother–daughter relationships, model minorities, tiger mothers. However, her work has also elicited critiques of Orientalism rooted in Eastern inferiority. According to Ruth Maxey, Tan often portrays China as a land of Confucian patriarchy, oppression, and folk legends—a premodern backwater drowning in superstition.[43] In contrast, America is depicted as the space of feminism, equality, opportunity, and futurity—a Manichean relationship neatly fitting into Edward Said's Orient/Occident dyad.[44] The plethora of ghost stories shared by Jing Mei Woo, Lindo Jong, and other members of the Joy Luck Club feeds directly into this criticism. If such supernaturalism was viewed as problematic in her first novel, though, they are trifling compared to the plot and themes of *Secret Senses*.

The Hundred Secret Senses is easily the most paranormal of Tan's works. The novel revolves around two character pairs who exist across two different geographies and historical periods. The main story line transpires in modern-day San Francisco and involves half-sisters Olivia Laguni and Kwan Li. Raised by a white mother after the death of her Chinese immigrant father (Jack Yee), Olivia sees herself as a normal American fully assimilated into mainstream culture. When she is seven years old, Olivia is joined by Jack's daughter from a previous marriage; twelve years older than her half-sister, Kwan Li quickly assumes the role of Olivia's primary caretaker and becomes the maternal figure in Tan's standard mother–daughter schema. More importantly for our purposes, she becomes a personal guide into the world of Chinese esotericism. As it turns out, Kwan possesses several unusual powers, the most fantastical of which is the ability to recall her previous lives on Earth. This unusual ability instantiates the novel's second story line, which takes place during the Boxer Rebellion (1899–1901), an anti-imperialist conflict that pitted the so-called Boxers (Chinese nationalists) against Christians, missionaries, and foreign spheres of influence during the Qing dynasty. In this reincarnation narrative, Kwan's previous embodiment on Earth was as Chinese peasant Nunumu (Ms. Moo for short), who is personal servant to and best friend of Nelly Banner, a white American living among Christian missionaries in the rural village of Changmian. Incredibly enough, Nelly Banner is Olivia Laguni's earlier manifestation on Earth. Over the course of the novel, Kwan helps Olivia navigate several romantic difficulties and cultural misadventures that mystically parallel their experiences as Ms. Moo and Nelly, respectively. In particular, the troubled relationship between Olivia and Simon Bishop (half white–half Asian writer) echoes the ill-fated romance between Nelly and Yiban Johnson (half white–half Asian translator), and the overall conceit of "unfinished business on Earth" powers the novel's ghostly subtext. Eventually the two story lines converge when Olivia and Kwan travel to Changmian and visit the very places where they had lived together over a century earlier. Throughout the novel, the tensions between Eastern and Western identities, Chinese antiquity and American modernity, and spiritualism and science catalyze the improbable narrative.

From the outset, Olivia and Kwan exhibit two wildly different world views when interpreting reality. For most of the novel, Olivia sees the world through the paradigm of Western science, or what we might call Occidental eyes. For her, only the rules of positivistic logic as explained by physics, biology, and chemistry can properly explain one's perception of nature. Occidental eyes is in fact how most readers raised in the modern West interpret reality. It is what Latour would call the modern world view and what Mignolo would call the parochial universe of Western knowledge. This mind-set is why Olivia casually dismisses Kwan's paranormal abilities for so long. It is best demonstrated when six-year-old Olivia first learns that Kwan can see ghosts; she promptly informs her parents, and Kwan is sent to a mental institution. At the hospital, "the doctors diagnose Kwan's Chinese ghosts as a serious mental disorder" and attempt to cure her via electroshock therapy.[45] This techno-scientific solution to a pseudoscientific problem is Occidentalism in action: expunging supernaturalism through the most advanced means possible. While she feels guilty for instigating her sibling's medical torture, Olivia also rationalizes it: "in a way, she [Kwan] brought it on herself."[46]

In contrast, Kwan views the world through "yin eyes," or what we might call Oriental eyes.[47] Rather than perceiving things through the paradigm of post-Enlightenment science, Kwan experiences a distinctively *premodern* reality teeming with spirits, ghosts, mystic energies, and mixed chronologies. As such, because she is unconstrained by the limitations of modern science, she can literally see and do things most Western readers would consider impossible. Kwan's immersion in this occult world is a major reason why she has been widely criticized as an Orientalist foil, a comic sidekick whose pidgin English and folk wisdom clearly mark her as inferior to her Americanized half-sister.[48] Worse, her paranormal powers prefigure her as a retrograde character defined by esoteric myth. "By rendering the Chinese as simultaneously animalistic and divine," writes Sheng-Mei Ma, "Tan in effect becomes an invisible creator [of self-Orientalism] whose creatures enact the Orientalist fantasies of her massive 'mainstream' following."[49] Ma's criticism is valid, especially when the story movies to Changmian village for the novel's final chapters. "This representation of China is that of a strangely *hyperexoticized* country, distant from civilization," observes Lina Unali. "We are taken to isolated, formerly unvisited places, where ravines with caves are occupied by bats, where the winds howl eerily, where one may lose the sense of gravity, where some people may get temporarily lost while others will be forever lost."[50] Tan is certainly guilty of fetishizing China's alterity, but I suggest a parascientific reading of Kwan's abilities that is both more generous and more generative. It is more generous because it does not assume Tan disdains her own cultural heritage. It is more generative because it draws from Chinese metaphysics to explain how an Asian American medium sees the world while also expanding the types of sciences literature can deploy to understand the paranormal.

The occult is not a Western phenomenon. In recent years, many scholars have explored the intersection between spiritualism, mesmerism, Theosophy, and Magick with Hinduism, Buddhism, and African thought. For instance, Franz Anton Mesmer might have asserted that an invisible, magnetic *fluidum* permeated the cosmos back in the 1770s, but his idea overlaps with the ancient Indian force of prana.[51] First appearing in the *Chandogya Upanishad* (eighth-sixth century BCE) and *Brihadaranyaka Upanishad* (700 BCE), *prana* means "breath," "vital force," or "life principle" in its Sanskrit context.[52] Dominic Zoehrer notes that early occultists immediately connected prana and the *fluidum* as two names for the same unifying force: "The attractiveness of the notion of *prana* among the mesmerists and within the emerging occultist movement was not least due to this intermediating function as a *tertium datur* between the immaterial and material realms."[53] Madam Blavatsky and Henry Steel Olcott were not the only ones claiming the *fluidum,* also known as "*pran,* od, aura, electromagnetism, or whatever else you prefer to call it," could cure disease.[54] Vivekananda, the early popularizer of modern yoga, outlined a "unified theory of alternative healing" in which prana is the elemental basis of *fluidum.*[55] For him, prana is a "vibrating energy" that manifests in forces like gravity, magnetism, nerve currents, and thought.[56] The connection between Western occultism and ancient Asian science is crucial because prana also shares several characteristics with qi.

This chapter opened with the idea of qi powering China's psi programs, and in fact the same vital energy powers Tan's eccentric novel. Chinese metaphysics, or *xing er shang xue,* refers to the "true nature of reality, its content and structure, to place human beings with the cosmos in relation to other kinds of things."[57] In this respect, it resembles the natural philosophy—that is, the early sciences—of thinkers like Aristotle and Plato. A fundamental difference between Western and Chinese metaphysics is the lack of distinction between material and immaterial reality; whereas the former generally views the mental world of abstract forms as categorically different from the actual world of concrete stuff, the latter holds these oppositions in tight formation within a greater whole.

The yin-yang symbol (☯) is representative of this pan-unified view of nature. In ancient Chinese philosophy, yin connotes darkness, femininity, and softness, whereas yang stands for light, masculinity, and hardness. Yin is everything yang is not. Each concept cannot stand alone, for it only accrues meaning as a complementary half of reality. "The concept of yinyang," explains Robin Yang, "involves viewing the world as a web of interconnected forces simultaneously and spontaneously to form the reality that we know."[58]

What makes the dynamic connection between yin and yang possible is qi. In Chinese, *qi* translates literally as "air" and figuratively as "energy flow," "life-force," or "spirit." It is similar to the eighteenth-century vitalist idea of the élan vital insofar it can be considered an ethereal essence flowing completely throughout nature.[59] However, qi is more comprehensive than the vital force because its presence in Chinese metaphysics does not distinguish the living from the nonliving. Qi is omnipresent and naturally occurring, flowing in and through all things in the universe. In this regard, it is perhaps more akin to India's prana or Europe's ether (the continuous substance natural philosophers once believed filled all the space in the universe). In Chinese metaphysics, qi constitutes both yin and yang. It can take solid form (yang) to construct material reality as well as abstract form (yin) to construct immaterial reality. As such, qi is the elemental medium through which all things manifest, as JeeLoo Liu describes:

> In Chinese *qi*-cosmology, the universe is seen as self-existing and self-sufficient, constituted by *qi* alone. Chinese *qi*-cosmology interprets the origin of the universe as the result of various transformations of *qi*'s operations. There is no supernatural entity that operates in any way on things in the natural world, and there is no divine creation or intervention. The prevailing assumption behind this worldview is that the basic element of the universe is *qi*, which is a continuous form of energy that can be manifested in both material and spiritual forms. Life and death, or existence or annihilation, are simply various states of *qi*. The world of nature and world of men, equally constructed by *qi*, are both governed by the principle of *qi*.[60]

This is a radically different paradigm of science from the Western tradition. With qi as the singular element of the universe, its various forms represent the totality of seen and unseen reality: it is Earth, sky, water, mountains, stars, animals, past, present, future, light, dark, and everything in between. Four general principles characterize the traditional scientific system that Liu calls qi-naturalism or qi-cosmology:

1. All elements of nature are configurations of qi: "The foundation of the universe is simply the transformation of *qi* in different stages, and the universe is nothing but the totality of *qi* in its various forms."[61]
2. The realm of qi is a closed system. "All existent things are produced by the integration of *qi*. . . . There is no causal influence for any entity outside the realm of *qi*."[62]
3. Qi explains all phenomena: "Everything can be reduced to *qi* and all events can be explained in terms of *qi*'s operations."[63]
4. Qi operates at macro and micro levels: "Global developments of *qi* externally and relationally influence the internal distribution of *qi* within each single entity."[64]

These principles not only illustrate the universality of qi in all aspects of existence but also highlight the systematic thinking governing this particular ethnoscience. Western science valorizes itself as the only logical system of knowledge, but these qi frameworks indicate it is a rational system as well. "Chinese *qi*-cosmology explicates the way the world functions with a conceptual scheme vastly different from the scientific exploration offered by the natural sciences," writes Liu, but difference does not mean error.[65] As Mignolo has argued, a pluriverse allows multiple knowledge paradigms to coexist. For all its differences from modern science, qi-naturalism is a science in its own right, and for centuries, the Chinese developed complex classifications, medicines, and practices around its tenets and navigated their lived realities according to its laws.

The rules of qi also produce several intriguing consequences. For example, the fact that qi connects all dimensions of existence suggests that it operates across time and space as well as life and death, which spawns many possibilities falling under the Western category of the paranormal. One striking prospect is the existence of ghosts as a normal part of nature. According to neoConfucian Zhang Zai, Chinese philosophy does not ascribe to dichotomies of matter and spirit. Because the "realm of *qi* covers the mechanistic, the organic, and the spiritual dimensions of existence," qi links life with death and the human with the spectral.[66] "Spirits and ghosts are simply different functions of *qi*," explains Zai. "When *qi* integrates into concrete things, it is called heaven's 'magnificent transformation' *(shenhua)*; when the *qi* that constitutes a living thing has disappeared or merged with the vacuous *qi*, it is called the return of the realm of ghosts *(you)*."[67] In this ethnoscientific paradigm, qi erases the Western distinction between living and dead.

Another consequence is that anyone with special awareness or consciousness of qi would have insights into its manifestations and movements beyond what is visible in the material world. In fact, within the scientific context of qi-naturalism, a spirit medium is basically someone who can see qi in all its multitudinous forms. As such, the Asian ghost seer is not a freak of nature but a savant whose superior vision reveals the true totality of qi's ubiquity. This, we can conclude, was the underlying logic behind China's EFHB research: just as psychical researchers sought women sensitive enough to receive ghostly vibrations, so did the Chinese government desire people particularly sensitive to qi. If we allow ourselves to go even further, the presence of qi as universal substance means a talented medium can see and hear ghosts. "The words ghost and spirit only signify the passage of coming and going or the process of expansion and condensation," Zai contends.[68] Because a medium can observe condensed (human) and expanded (ghostly) forms of qi, then she can communicate with spirits as well as she does with humans; she is supernatural insofar she navigates material and immaterial realms with equal aplomb.

Following the rationale of qi-naturalism to this end, it stands to reason a ghost seer can see and know things that are impossible for most anyone else, which is to say she possesses paranormal knowledge. "An individual's existence may be transient, but the stuff that makes up the individual—qi—is nonetheless indestructible," writes Liu. "Qi itself is ever-present and will never be annihilated."[69] The "never" is significant because it authorizes retrocognition as an epiphenomenon of qi-naturalism. In psychical research, retrocognition is the ability to have "knowledge of the past, extending back beyond the reach of ordinary memory."[70] The mirror image of precognition (seeing the future), retrocognition (seeing the past) is defined by the acquisition of impossible knowledge. In chapter 3, we learned of several Western theories involving a psychic reservoir of knowledge: William James's mother-sea, Ingo Swann's Matrix, and so on. In a Chinese context, qi serves as that marvelous reservoir. Qi never disappears, so a medium can ostensibly "see" traces of the past in ways most of us only see the present; after all, historical forms of qi can still reside in the present, and will do so into perpetuity. This means that distant events can remain observable eternally. The skilled practitioner of qi can recognize its prior instantiations and observe ancient events, reconstruct history, and remember what is technically impossible to recall. Retrocognition also gains import in a qi-naturalist paradigm because it maps so well onto reincarnation. The belief that living beings are born again and again on Earth is common to many Asian cultures. Qi ethnoscientifically explains such enduring existence. In a world of eternal qi, retrocognition means that a person could recall previous incarnations on Earth; past events seen in dreams are not random but focalized through the experiences of a former self. Paranormal abilities consequently emerge as logical by-products of a rational system undergirded by qi's infinitude.

This returns us to Kwan Li and *Secret Senses*. In the novel, Kwan's powers spring from a childhood encounter with death when she nearly drowned in a flash flood. After lying "dead" in a coffin for two days, she suddenly awoke, carrying back to the land of the living a special connection to the land of the dead. Henceforth Kwan possesses what she calls "yin eyes," which is a short-hand reference to yin-yang eyes in Chinese folklore.[71] Akin to having the third eye in Western occultism, yin-yang eyes means having the ability to see both yin and yang aspects of nature. Implicit in this phrasing is the idea that most people, especially those raised in the technoscientific West, only see the yang half of reality; they can see only the hard, physical world privileged in modern scientific discourse. This is what Olivia perceives with her Occidental eyes. Thanks to her near-death experience, Kwan can see the yin aspects of abstract, metaphysical reality as well. Given that qi is the fundamental element constituting yin and yang, we can infer that Kwan's powers originate from her mediumistic ability to perceive qi's totality. While the modern West is blinded to half of existence, Kwan's Oriental or yin eyes are open to the infinite permutations of reality's fundamental element.

Kwan boasts three powers, each of which showcases how ethnic literature expands scientific approaches to the paranormal: healing powers, ghost vision, and retrocognition. All three fit into the scientific framework of qi-naturalism, and focusing on them will help us reconceptualize *Secret Senses* as an epistemological novel operating beyond Western science. More precisely, it is a pluriversal novel that positions Western and non-Western sciences on an epistemological par, and in so doing, it strengthens a paraliterary approach to the paranormal.

In terms of healing, Kwan can miraculously sense maladies in people she has never met. According to Olivia, Kwan "can tell when she shakes hands with strangers whether they've ever suffered a broken bone, even if it healed many years before. She knows in an instant whether a person has arthritis, tendinitis, bursitis, sciatica—she really good with all the musculoskeletal stuff—maladies that she calls 'burning bones,' 'fever arms,' 'sour joints,' [and] 'snaky leg.'"[72] Kwan is also a psychic healer who can "release" those whom she touches from worry: "when she puts her hands on the place where you hurt, you feel a tingling sensation, or a thousand fairies dancing up and down, and then it's like warm water rolling through your veins."[73]

While Kwan's therapeutic prowess is an unimportant power Olivia only mentions to highlight Kwan's radical otherness, qi-healing is in fact an integral part of Chinese medicine. According to traditional Chinese medicine, qi circulates throughout the human body along certain pathways known as meridians, which distribute yin and yang into organs to produce good health.[74] As John Longhurst explains, "Disease occurs when there is an excess or deficiency of either *yin* or *yang*," or disruptions, blocks, or imbalances caused by improper movement of qi.[75] A traditional Chinese healer can identify yin-yang imbalances and recommend certain foods, physical training, or acupuncture for treatment. A qi master like Kwan can do more, though. As the passage above suggests, she is so in tune with qi that she can touch someone's body, gain knowledge of their meridians, move qi with her mind, and restore the proper yin-yang balance.

Kwan also has the power to see ghosts, which likewise stems from the logic of qi-naturalism. As Kwan tells a young Olivia, "I have yin eyes. I can see yin people . . . they are those who have already died."[76] When the frightened Olivia asks if there are ghosts in the room with them at that very moment, Kwan says yes. "Oh yes, many. Many, many good friends. Don't be afraid, Libby—ah, come out. They're your friends too. Oh see, now they're laughing at you for being so scared."[77] While this too is a minor scene in the novel, it provides insights into qi's specific operations. For Kwan, ghosts are not unusual entities but part and parcel of the qi-universe. Her nonchalant recognition of the ubiquity of ghosts speaks to the ontology and phenomenology implicit to qi-naturalism. As Zhang Zai points out, humans and ghosts are merely different manifestations of qi. At death, a person's qi adopts a softer or looser form invisible to the Occidental eye, but still recognizable to Kwan. The mechanics of ghost vision is thus rationalized by qi's status as an etheric substance existing beyond life and death, as well as by the medium's capacity to track its permutations. Communicating with ghosts—seeing their faces, speaking with them, laughing along with them—confirms the idea that ghosts are simply people in different qi forms. It is worth acknowledging that Kwan's ghost vision drastically departs from psychical research's phantasms. The ghosts of *Secret Senses* are not telepathic hallucinations. Nor are they symbols of collective trauma. They are ghosts according to the spirit hypothesis: disincarnate spirits of the dead. More precisely, they are manifestations of qi, just like you and me, but visible only to those with the special sensitivity (and ethnoscientific world view) to perceive them.

Kwan's most powerful qi-based paranormal talent, which drives the story's dual timelines, is her ability to peer into distant history and remember past lives. "For most of my childhood, I thought everyone remembered dreams as their past lives," explains Olivia. "Kwan did. After she came home from the psychiatric ward, she told me bedtime stories about them, yin people; a woman named Banner, a man named Cape, a one-eyed bandit girl, a half and half. She made it seem as if all these ghosts were our friends."[78] For Olivia, Kwan's dreams are just elaborate fictions. Clearly, the Western mind cannot comprehend the actuality of Kwan's retrocognition. This is why Olivia explicitly calls Kwan's visions "dreams." For most of the novel, the tales of Nelly Banner and Ms. Moo are unreal. In contrast, Kwan knows she is witnessing history. "I can't say exactly how long ago this happened," she explains. "Time is not the same between one lifetime and the next. But I think it was the during the year 1864. Whether this was the Chinese lunar year or the date according to the Western calendar, I'm not sure."[79] The exact date of these memories may elude her, but Kwan never doubts the factuality of her former existence.

Her certainty derives from what she calls "the hundred secret senses."[80] Olivia describes them as "the senses that are related to primitive instincts, what humans had before their brains developed language and the higher functions—the ability to equivocate, make excuses, and lie. Spine chills and musky scents, goose bumps, and blushing cheeks—these are the vocabulary of the secret senses."[81] Olivia's description is problematic insofar it deploys Orientalist assumptions that the "primitive" Eastern mind can access truths obscured to the "higher" functions of Western cognition. Such "power of the primitive" rhetoric is not new; for example, in 1877 Blavatsky highlighted the supernatural mental prowess of the "sages of the Orient" in works like *Isis Unveiled*.[82] In fact, Kwan is exactly the kind of Orientalist sage who awed occultists at the end of the nineteenth century with their knowledge transcending Western science.

That said, Tan's investment in qi-naturalism falls into a familiar psychical terrain where the unconscious has greater access into the reality of things than consciousness. "Secret sense not really secret," explains Kwan. "We just call secret because everybody has, only forgotten. Same kind of sense like ant feet, elephant trunk, dog nose, cat whisker, whale ear, bat wing, clam shell, snake tongue, little hair on flower. Many things, but mix up together."[83] The secret senses are instinctual, a kind of cognition more animalistic and innate than rational. In "Dreams and Occultism" (1933), Sigmund Freud writes, "One is led to the suspicion that this [telepathy] is the original, archaic method between individuals and that in the course of phylogenetic evolution it has been replaced by the better method of giving information with the help of signals which are picked up by the sense organs."[84] Here, Freud posits psi-related secret senses as our aboriginal form of communication and knowledge transfer. Kwan shares this position when she claims all humans possess such senses and that we have simply forgotten about them. Like Swann's beliefs in chapter 3, humans possess otherworldly abilities that have been lost as a result of modernity and hyperrational thinking.

It would be a mistake, though, to construe these overlaps with psychoanalysis and psychical research as Tan's subscription to Western paranormality. Indeed, her investment in qi is what makes *Secret Senses* so productive as an epistemological text. Kwan makes it clear that something binds all things together, whether living person to living person, living person to the dead, or present mind to ancient events. "Memory, seeing, hearing, feeling, all come together, then you know something true in you heart," Kwan tells Olivia. "You use you secret sense, sometimes can get message back and forth fast between two people, living, dead, doesn't matter, same sense."[85] The fluidity between all the senses, as well as the fluidity between physical and metaphysical states, is a clear indicator that Kwan is really referring to qi. Because qi is the fundamental essence of reality, it is also the medium that makes possible communication between ghosts and humans as well as information transfer between past and present. Qi is the invisible glue holding the entire narrative of *Secret Senses* together. As the novel's paranormal substance par excellence, it is the universal energy connecting past and present lifetimes, humans and spirits, the living and the dead, and material and immaterial worlds. Implicit in the qi-naturalistic world view is that epistemes beyond Western science exist, and if the rest of us could just open our eyes, we too could see the pluriverse as Kwan does.

One of the major developments in the novel occurs when Olivia finally acknowledges the facticity of Kwan's visions, which is to say she acknowledges the epistemic validity of qi-naturalism. Early in *Secret Senses,* Kwan reveals key details about her past life in 1864 China:

> That was the year I gave Ms. Banner the tea. And she gave me the music box, the one I once stole from her, then later returned. I remember the night we held that box between us with all those things inside that we didn't want to forget. It was just the two of us, alone for the moment, in the Ghost Merchant's House, where we lived with the Jesus Worshippers for six years. We were standing next to the holy bush, the same bush that grew the special leaves, the same leaves I used to make the tea.[86]

Kwan's retrocognition is both comprehensive and compelling. We soon learn Nelly Banner did in fact live with Christian missionaries (Jesus Worshippers) in a compound where a businessman (Ghost Merchant) had mysteriously died. Further, the music box at the center of the reincarnation memory actually reappears late in the novel as material proof of Kwan's paranormal sight.

At the end of the novel, Kwan and Olivia journey to China on a travel-writing assignment. On arriving in Changmian village, Olivia experiences a peculiar sense of déjà vu: "I feel like I've seen this place before," she admits—because she had indeed lived in Changmian before, as Nelly Banner.[87] Olivia and Kwan's previous lives are subsequently verified via the music box. In the Boxer Rebellion story line, several Chinese servants in the Ghost Merchant's House commit suicide by opium pills rather than face execution in the aftermath of General Cape's murder. Nelly and Ms. Moo collect their knickknacks in the music box as personal mementos: "For Dr. Too Late, it was a little bottle that once contained his opium pills. For Miss Mouse, a leather glove she always clutched when in fear. For Mrs. Amen, the buttons she popped off her blouses when she sang out loud. For Pastor Amen, a travel book. And for Lao Lu, the tin with leaves from the holy tree. [Miss Banner] placed these things in the box, along with the album where she wrote her thoughts."[88] Ms. Moo then hides the music box in a nearby cave complex for posterity. Over a century later, Kwan rediscovers the music box when she and Olivia visit the caves. "This my box I hide long time ago. Already tell you this. This box always want show you."[89] When Olivia opens the box, all the nineteenth-century keepsakes are revealed once more: "I glance at the section of the box that is now exposed. It's a knickknack drawer, a catchall for loose buttons, a frayed bobbin, an empty vial. . . . Kwan is examining a kidskin glove, its fingers permanently squeezed into a brittle bunch."[90] Pastor Amen's travel book, *A Visit to India, China, and Japan* by Bayard Taylor, appears, along with tea from the holy bush.[91] Olivia even reads Nelly's journal and finds an entry on Ms. Moo: "Further the topic of suicide, Miss Moo informed me that it is strictly forbidden among Taiping followers, unless they are sacrificing themselves in the battle for God."[92]

The greatest discovery in this scene is not the music box itself but rather what it represents, namely that qi is real and that the hundred secret senses exist. The mementos merely serve as concrete proof of Kwan's retrocognition, and they convince Olivia that her half-sister does in fact possess powers transcending Western science. "In spite of all my logic and doubt, I can't dismiss something larger I know about Kwan: that it is not in her nature to lie," admits Olivia. "What she says, she believes is true. . . . I believe her. I have to."[93] This is a remarkable shift in the novel because for thirty-odd years, Olivia has categorically disavowed Kwan's paranormal abilities because they do not make scientific sense. Now armed with incontrovertible evidence that Ms. Moo's dreams are actually true, Olivia finally acknowledges there is more than one way to view the world. As Jiena Sun argues, the discovery of the music box transforms Olivia from disbeliever into believer.[94]

Olivia's acceptance of yin eyes leads to a second transformation regarding epistemology. When she first arrives at Changmian, she announces, "And being here, I feel as if the membrane separating the two halves of my life has finally been shed."[95] One way to interpret this sentence is through the Asian American bildungsroman, in which the reconciliation of one's ancestral and modern ethnic identities is a key checkpoint en route to subject formation.[96] But an equally productive reading suggests the barriers of Western epistemological domination have finally given way to a pluriverse of knowledges. When the hegemonic "membrane" of modern scientific ideology disappears, Olivia can accept the epistemological equality of Western and non-Western world views. It also means that going forward, she can place the worlds of yin and yang on an epistemological par, Western and Eastern paradigms fully equivalent. In Chandler's terms, she gains paraliteracy. This is no small achievement for Olivia. Western science positions itself atop the epistemic hierarchy, a form of Eurocentric supremacy that has blinded Olivia to the possibilities of qi-naturalism. In dissolving the membrane and erasing her cultural bias, she acknowledges that qi takes many forms; that modern science only makes half the world visible; that a rich secondary metaphysical world exists; and that her sister has always been one of its psychic explorers. Western epistemology may have kept her blind, but no more. Her eyes and mind are now wide open. As a stand-in for the modern Western reader interpellated into scientific modernism, Olivia serves as a model whereby the knee-jerk skepticism of Eastern knowledge is replaced by pluriversal acceptance.

To be clear, Olivia's entrance into mosaic epistemology does not mean that she too can see the world of yin. She is not a spirit medium, after all. What she does gain, though, is an openness to alternative scientific frameworks desired by twenty-first-century decolonial scholars. Olivia once considered herself a modern—completely scientific and rational—but she ends the novel as a Latourian nonmodern ready to admit the universe consists of mixed physical and metaphysical content. When Olivia's partner, Simon (Yiban in his former life), gets lost in the Changmian caves, Olivia helps bring him out; during the process, she accepts that we have never been modern. "We stumble into the open, where the light is so hard I can't see. I blindly pat Simon's face, half expecting that when I can see the world again, he'll be Yiban and I'll be wearing a yellow dress stained with blood."[97] Later she thinks, "I have to believe it's not too late to tell Kwan, I was Miss Banner and you were Nunumu, and forever you'll be loyal and so will I."[98]

Benzi Zhang argues, "Though the relation of yin and yang worlds, Tan attempts to present a broader vision of the various dimensions of life."[99] This broadening occurs on several levels. For one, I disagree with scholars who interpret Kwan's yin eyes as an Orientalist feature expressly designed to disparage the East. Instead, I view yin eyes as an important literary device promoting epistemological equality by contending that there is more than one scientific way to view the world. For Kwan and Olivia, modern science and qi-naturalism are two equal paradigms for navigating the world. Kwan also helps us understand ghosts as ghosts, as entities manifesting an alternative type of qi. Her insights as a female medium align with the longer history of women in psychical research. Most importantly, Kwan becomes an avatar for expanding the paranormal mind through non-Western science. Her phenomenological perception through Chinese metaphysical frameworks highlights new rules, theories, and laws that explain paranormal phenomena in ways we have not encountered before. Her ability to see the totality of qi helps readers understand past lives, traditional medicine, spectral communication, and retrocognition from beyond Eurocentric approaches. For all the critical shade cast on her character, Kwan is an undeniably useful ambassador for diversifying the parascientific space around ghosts and ghostly visions.

Korean Shamanism and *Comfort Woman*

Another famous spirit medium in Asian American letters is Akiko Bradley of *Comfort Woman*. Since its publication in 1997, Nora Okja Keller's novel has earned copious praise for its unique portrayal of World War II sex slaves and their resulting traumas. From 1938 to 1945, the Japanese Imperial Army coerced between three hundred thousand and four hundred thousand Asian women—primarily from Korea, China, and the Philippines—to serve as "comfort women" for soldiers at military bases. Sometimes kidnapped from their homes, sometimes lured away from their villages with the promise of factory employment, comfort women faced excruciating violence at the hands of Japanese soldiers, including daily rapes, beatings, and starvation. Many committed suicide during and after World War II. Others were shamed by their own families as damaged goods, a postwar blow compounding their wartime sufferings. While the existence of military sex slaves was initially suppressed for cultural and political reasons, both testimonials and research on comfort women have emerged in the closing decades of the twentieth century and shed light on this brutal episode in history.[100]

Nora Okja Keller's *Comfort Woman* explores the trauma of comfort women through the braided narratives of Korean émigré Akiko Bradley and her daughter, Beccah. In the present story line, Beccah Bradley is a second-generation Korean American in Honolulu, Hawai'i, coping with the recent death of her mother. For most of her life, she viewed Akiko as an embarrassment, a crazy woman who wore funny clothes, performed strange rituals, sacrificed chickens, and spoke with ghosts. However, as Tina Chen observes, Beccah "misreads" Akiko because she is ignorant of Korean history and culture.[101] During the funeral preparations, however, Beccah uncovers her mother's hidden past as a military prostitute and finally recognizes Akiko's sufferings, sacrifices, and, most importantly, her inner strength—not only in rebuilding her life but by creating a better one in America for her family. In the story line set in the past, Akiko recounts her personal experience as a comfort woman, including being sold by her own sister to a Japanese military camp at the age twelve, her physical and sexual abuse at the hands of soldiers, her harrowing escape to a Christian mission, an unhappy marriage to Reverend Richard Bradley, and finally her troubled life in America. Given that she is already dead at the start of the book, Akiko essentially speaks to readers from beyond the grave; she is a ghost revealing the traumas of Korean colonial history. This is thematically resonant because Akiko, when alive, was the most celebrated spirit medium in Honolulu, and her duties included conveying stories of the dead to the living. In channeling her mother's ghostly experiences for the rest of us to hear, Beccah emerges as a spirit medium in her own way as well.

As one of Asian American literature's preeminent fictions, *Comfort Woman* has received its fair share of ghost scholarship, nearly all of which defers to metaphorical readings of ghosts as trauma symbols. The text clearly works well for this purpose. Akiko is haunted throughout the novel by several spirits, particularly the ghost of Induk, a murdered comfort woman who epitomizes the myriad sufferings of World War II Japanese sex slaves. By manifesting their hidden histories and suppressed traumas, Induk emerges as a quintessential example of cultural haunting. Jodi Kim has argued that spirits like Induk are "metaphoric manifestations of violent histories that have been occluded and are symptomatic of society's failure to reckon with the psychic and material consequences of such histories."[102] Patricia Chu similarly views Induk as an imaginary construct that Akiko uses to construct her own subjectivity: "Neither a real person nor a goddess seen by anyone in the story but Akiko. . . . Induk is the manifestation of Akiko's iron will and her search for personal integrity in the absence of social support."[103] As we can see, ghosts in *Comfort Woman* criticism—and in most Asian American criticism more broadly—are inevitably construed as symbolic for injustices, forgotten voices, and imagined states beyond the specter itself. Janna Odabas captures the metaphorical dominance of ghost scholarship in *The Ghosts Within,* contending that Asian American literary spirits have evolved through three phases: (1) ghosts as "haunted" Asian American identities (1970s–1980s); (2) ghosts as instruments of cultural difference (1990s); and (3) ghosts as self-reflexive figures linked to social history (2000s).[104] Implicit in her thesis is the idea that the ghostliness of ghosts matters less than their political referents.

One exception is the work of Bonnie Winsbro. In *Supernatural Forces,* she analyzes spectral figures in Asian American, African American, and other ethic American fictions as vehicles for personal individuation.[105] "According to the reality represented by these works," she writes, "the world is inhabited by a host of spirits and deities, all of whom possess the power to influence or control events that effect humans."[106] Importantly, she recognizes that the worlds her characters inhabit are neither Eurocentric nor modern in the Latourian sense; rather, they "represent a view of reality that has been 'disproved' by science."[107] "Although science has methodically disproved the objective reality of spirits, deities, and supernatural forces," she writes, "it has failed to destroy belief in the supernatural because that belief provides a logical answer to questions that have plagued humans since the awakening of their cognitive powers."[108] While Winbro's project ultimately asks different questions than mine, her pluriversal approach recognizes different societies experience reality in different ways based on their customs, lived experiences, and sciences.

In a similar vein, I contend that going beyond metaphor is crucial for understanding the supernaturalism of *Comfort Woman,* including the typology, behavior, and historical contexts of its ghosts. Indeed, Akiko is not just any type of medium but a culturally specific one: a traditional Korean shaman. While other critics have analyzed religiosity in the novel as a form of political resistance,[109] I propose a more epistemological approach that investigates the science of shamanism and ghostliness in indigenous Korean culture. By exploring the deeper context of shamanism—its history, role in spiritual life, and ethnoscientific principles—we can see how Keller updates the mechanics of ghost ontology and communication according to native science, and in doing so, she expands the cultural logic of the paranormal mind. This method allows us to reconceive the entire novel as a séance, a channeling of spectral visions, conversations, and revelations to widen our view of the paranormal.

Shamanism is Korea's oldest and only indigenous religion.[110] While sharing certain characteristics with Buddhism, Confucianism, Taoism, and Christianity, which all migrated into the Korean peninsula over the centuries, it maintains a distinct cosmology, structure, and organization allowing us to view it as a scientific system in its own right. Originating before the tenth century BCE, Korean shamanism views humanity, spirit, and nature as part of a fluid and eternal coexistence.[111] "Each person has been in this world since the beginning of life and remains here after death," writes Hahm Pyong-Choon.

"Man is so closely intertwined with the terrestrial forms of nature, in fact, that man without nature would be impossible and nature without man inconceivably irrelevant."[112] Both qi-naturalism and Korean shamanism share the unity of all things as a central tenet. While the former has a defined substance— qi—at its core, the latter focuses on interconnection itself. "A continuum exists in everything in nature," explains Pyong-Choon. "All the parts of man's environment are necessary for his survival, and man shares his existence with nature as an inseparable part of his being."[113] Such unity is found in the shamanistic view of time, for example. In the West, time is unidirectional and evanescent. It moves from past toward future, and once a moment has passed, it is gone forever. In contrast, "the shamanistic man views time itself as a continuum. He is reluctant to segment time into such clearly defined portions as the past, present, and future. Instead of a unidirectional linear dimension, time is understood to be a combined mixture of change and repetition. Since the rhythm of life is both changing and repeating, and since past, present, and future all contain aspects of one another, the segmentation of time into three distinct parts would be an oversimplification."[114] The Korean shaman experiences a different phenomenological sense of being because past and future blur into the present.

Such fluidity is also found in the interpretation of ghosts. Traditional Western distinctions between living and nonliving things matter little in Korean shamanism. "To the shamanistic man, death is as much a part of existence as birth," writes Pyong-Choon. "Man continues to exist not only in the lives of his kindred—especially his offspring—but also in his own body. Death signifies neither the annihilation of human life, nor the commencement of nonhuman life in another world."[115] In Korean shamanism, humans do not transition into another plane of existence after death but continue existing in the material realm in alternative forms, a difference of degree rather than kind. Like different kinds of qi in qi-naturalism, there is and always has been a constancy of being. Pyong-Choon goes on to explain that in most instances, "the spirit may wander away from the buried body, but so long as the dead body remains comfortable and has no reason to seek vengeance among the living, the spirit remains with the body; it is unobtrusive, peaceful, at ease."[116] However, "if the dead person has intense ill-will at the time of his death, or if the body has been mistreated, the wandering spirit can wreak destruction on the living."[117] There is a clear cause-and-effect logic in Korean shamanism. A so-called good death, such as being surrounded by friends and family while being at peace in one's final moments, leads to a normal afterlife in which the posthuman spirit remains harmless and invisible. In contrast, a bad death involving murder, suicide, rage, or sorrow produces an afterlife characterized by restless and troubled spirits. This is similar to the logic of Western ghostliness, where murder victims (e.g., Hamlet's father) or hit-and-run victims (e.g., a phantom hitchhiker) supposedly become the subjects of ghostly afterlives. That said, Korean shamanism has a fuller taxonomy of spirits reflecting its epistemic complexity.

Within the shamanistic hierarchy of spirits, gods *(sillyong)* are higher than ancestral spirits *(chosang)*, although both are viewed as beneficial beings whom people look to for help in the same way "a Korean child looks to a grandparent for small indulgences."[118] Ancestor spirits are generally benign, though they sometimes carry resentments *(han)* and "cry for all they have missed in life."[119] Examples here include grandparents upset they could not hold a baby before dying and men who toiled all their lives without enjoying the fruits of their labor.[120] More problematic are troubled entities like restless ancestors *(chosang malmyong)* and anonymous ghosts who died violently *(yŏngsan, chapkwi).*[121] The ghosts of the murdered, poisoned, and drowned are particularly dangerous: "because they died unsatisfied, they wander angry and frustrated, venting anguish on the living." Following cause-and-effect logic, their unhappy deaths lead to "ceaseless craving" and the subsequent haunting of households where they emerge as unwanted intruders.[122]

While ancestral ghosts tend to linger close to home, other ghosts will wander widely in search of food and company. Their geographic range is vast. Kendall writes, "The world outside the home gate is filled with ghosts. Those who travel about in the world acquire, over time, an entourage of baleful ghosts."[123] Certain spaces are also ghostly; for example, red disaster *(hongaek)* is a contagious "accretion of misfortune" that engulfs people who contract it from visiting the scene of another's misfortune.[124] If you visit a location where a murder has transpired, you might bring a vengeful *hongaek* home with you. Similarly, death humors *(sangmun)* are spirits that attack those in contact with death, such as those who enter a house where an infant's corpse is kept.[125] Such ghosts follow the logic of supernatural contagion where past misfortune begets future misfortune as a result of proximity. As we can begin to see, Korean specters fit within a causal system explaining supernatural phenomena.

The shaman plays a fundamental role within this system by acting as a bridge between spirit and human realms. According to Alan Covell, the shaman (or *mansin)* is "a specially endowed person who can communicate between the two worlds, whom the spirits have 'called' to serve for their lifetime as a messenger of inter-communication between human beings and the world of the spirits."[126] Shamans, who are almost always women, are able to do this by summoning spirits before them—even into their bodies—where they can speak to the living. "The *mansin,* as shaman, claims the power to see the gods and ancestors in visions to call them down to speak though her lips. The *mansin* makes a dynamic link between her clients and their household supernatural. She determines the appetites of gods and ancestors through the visions she receives in divination sessions."[127] The shaman thus acts as a translator by summoning gods and ancestors to her, allowing their possession of her body and voice, and issuing their directives.[128] By resolving conflicts between the two realms, the shaman can help remove bad luck or illnesses from her clients' lives. In short, the Korean ghost seer speaks for the divine and the dead, thereby acting as "the source of Korean people's spiritual energy."[129] This is a privileged position; within a patriarchal system, the female *mansin* provides services that "remain a vital component of Korean ritual life."[130] In many respects, the Korean shaman serves a similar role as the European medium. The main difference is the scientific paradigm in which she operates. For the former, ghosts are a ubiquitous part of the natural order, and therefore the shaman is not an anomaly so much as a talented nexus connecting human and posthuman spheres, a paranormal investigator of our interconnected being.

This brief background into Korean shamanism helps us interpret *Comfort Woman* in fresh ways, particularly around Akiko's relationship with Induk and what it reveals about non-Western ghosts and ghost vision. What should be obvious is how the narrative structure of Keller's novel mirrors the Occidental eyes/Oriental eyes binary found in *Secret Senses.* Whereas the Americanized daughter sees the world according to the positivistic lens of Western science, the Asian mother figure sees the world according to indigenous science. This sets up the same epistemological problematic each novel sets out to resolve. While the ghost of Induk in *Comfort Woman* is typically interpreted as a mental fabrication or a cultural haunting, she should be understood first and foremost as a *yŏngsan,* which not only explains her peculiar behaviors but also sheds light on the ethnoscientific operations of Akiko's paranormal vision.

As mentioned earlier, Induk was a sex slave killed for protesting her egregious treatment by Japanese soldiers. Her ghost follows Akiko for the rest of her life, alternatively nurturing, berating, and haunting her as she travels from Korea to America. Critics generally view Induk as an imagined fiction ("intense fantasies rather than visitations from the dead")[131] or a symbol of female suffering ("the spirits of the foremothers who endured the various inequities enforced through patriarchal discourse").[132] However, reading Induk through Korean shamanism helps us understand her as a "character" in her own right. To begin, it is important to note that Induk experiences a bad death in *Comfort Woman.* On the last day of her life, Induk refused to quietly submit to rape:

"She denounced the soldiers, yelling at them to stop their invasion of her country and her body. Even as they mounted her, she shouted: I am Korea, I am a woman, I am alive. I am seventeen, I had a family, just like you do. I am a daughter, I am a sister."[133] For her brazen resistance, Induk is brutally murdered, "skewered from her vagina to her mouth," and tied to a pole as a harsh example for the other comfort women.[134] The gruesomeness of her death matters. Akiko describes Induk's desecrated body several days after the fact: "Hair tangled through and around maggoty sockets and nostrils. Gnawed arms ripped from the body but still dangling from the hands to the skewering pole. Ribs broken and sucked clean of marrow. Flapping strips of skin stuck to sections of the backbone."[135] The defilement of her body foreshadows an angry afterlife. It should therefore be no surprise that Induk returns as a very specific type of spirit: a wrathful *yŏngsan* haunting the location of her murder. "Ghosts of the drowned, ghosts who were shot, ghosts who died of carbon monoxide poisoning, maiden ghosts, bachelor ghosts"—these are spirits whose lives ended violently, writes Kendall. "Because they died unsatisfied, they wander angry and frustrated, venting their anguish on the living."[136] In addition, Pyong-Choon notes, "if the dead person had intense ill will at the time of his death, or if the body had been mutilated, the wandering spirit can wreak destruction on the living."[137] Induk checks all the unfortunate boxes associated with bad deaths: a violent end, seething anger in her heart, bodily mutilation, and a lack of proper funeral rites. According to the shamanistic paradigm, she is predestined to live on as a vengeful, mournful wraith. Induk herself confirms this hypothesis: "No one performed the proper rites of the dead. For me. For you. Who was there to cry for us in kok, announcing our death? Or to fulfill the duties of yom: bathing and dressing our bodies. Combing our hair, trimming our nails, laying us out? Who was there to write our names, to even know our names and to remember us?"[138]

The elements surrounding Induk's life and death explain her erratic behavior. In shamanism, ghosts can follow the living back home and stay with them for extended periods. In Induk's case, she serves across two continents as Akiko's guide, midwife, and sexual partner. She first appears to Akiko soon after her escape from the military base, in the same woods where she herself had been killed. Akiko visited a haunted locale, so Induk could be considered a red disaster or death humor eager to spread misfortune. Surprisingly, Induk treats Akiko as a fellow sister-in-arms rather than a host to torment, which demonstrates more *chosang* behaviors than *yŏngsan* ones. This suggests Induk is not simply a type but a complex entity with a will and a consciousness, which is to say a former human who has moved into the next phase of life.

Induk's first act is to nurse Akiko back to health. Knowing that her fellow comfort woman will die without help, she passes along the name of a nearby shaman, a "fox spirit who haunted the cemeteries of deserted villages, sucking at the mouths of the newly dead in order to taste their otherworldly knowledge."[139] "This is Manshin Ahjima," she tells Akiko. "Old lady of ten thousand spirits. Go to her, and she will prepare you."[140] When Akiko accidentally encounters the shaman and speaks her name, Ahjima is nearly frightened to death: "E-yah! . . . The dead knows me!"[141] Akiko should have no knowledge of Ahjima's name or location. In fiction, there is pleasure when miracles happen. But in shaman cosmology, it follows that the spirit of Induk can travel about, know the local geography, and recognize the people traveling through it. By taking on the role of guardian spirit, Induk offers kindness to a young woman who is suffering as she once had.

Later, when Akiko is trapped in a loveless marriage to Reverend Bradley, Induk takes on the role of a sexual partner. Instead of enjoying sex with her colonizer husband, Akiko prefers intimacy with Induk: "She licks at my toes and fingertips, sucking at them until my blood rushes to greet her touch. I feel her fingers wind through my hair, rubbing my scalp, soothing me, while her mouth caresses my chin and neck. My body prickles."[142] When Akiko becomes pregnant with Beccah, Induk appears in the delivery room like a guardian angel: "She stood beside me, shadowed by mask, gown, and a halo of light."[143] A critical reader might question Induk's shape-shifting role as homoerotic partner and midwife. Indeed, Chu's argument that Induk is a by-product of Akiko's imagination is quite convincing. However, if we accept Akiko's interpretation of Induk's reality, then this ghost does indeed fit into a shamanistic science. In the shamanistic world view, ghosts are agents with their own desires, needs, and impulses. After all, they were once human too. Such ghosts crave company, food, and the lives they once led. More broadly, this would suggest the very human activities we witness in the novel, like providing aid, expressing outrage, and experiencing sexual pleasure, are all normal ghostly behaviors too. To this end, I contend that Induk should not be construed exclusively as a function of Akiko's mind or a metaphor for suffering but also a spectral entity acting on its own desires. Such a paraliterary reading honors Korean shamanism and allows us to accept Akiko on her own ethnoscientific terms rather than via a Westernized schema. Shamanism explains Induk's ability to predict the future; ghosts can tap into the fluidity of time. It justifies her preternatural knowledge of Manshin Ahjima. It explains her ability to follow Akiko from Korea to America, as ghosts will attach themselves to people. It resolves Induk's ambiguous actions because ghosts maintain the contradictions of human behavior in posthuman form. As such, Induk has always appeared real to Akiko because she *is* a being who exists within Akiko's ethnoscientific understanding of reality. In sum, Induk is a multidimensional figure whose existence is not limited to symbolic functions alone, and if we take the time to understand the Korean shamanistic world view, we gain insights into Akiko's phenomenological and ontological world.

Shamanism also informs our reading of Akiko. Just as Olivia believed Kwan was "insane," Beccah believes her mother is "crazy." If we accept shamanism as a scientific paradigm structuring the novel, though, Akiko is well within her rights to see ancestral spirits, gods, and hungry ghosts wandering the material world. She simply perceives and navigates reality according to a different set of rules. The task for Beccah—and for us as readers—is to meet the shaman halfway and understand those rules.

The shamanistic system underlying non-Western ghost communication in *Comfort Woman* is illustrated through two spiritual ceremonies, or *kuts*. A *kut* is a summoning ritual designed to cure an illness or misfortune by following a cause-and-effect framework.[144] Angry spirits cause sickness and misfortune, so the shaman's role in a *kut* is to call them forth via dancing and incantation, identify the source of the problem, and help resolve the issue. "A kut honors a basic structure, a progress through the house wherein gods and ancestors appear in place and in approximate sequence," writes Kendall.[145] During the ceremony, the family catches positive influences and secures them within the home while the shaman locates wandering ghosts and noxious influences and expels them. Through this stepwise process, the house is cleansed of misfortune, and all the parties involved can go forth in peace.

The first *kut* takes place in the novel when the character Aunt Reno visits Akiko's apartment and finds her in the middle of a trance. "Arms flailing, knees pumping into her chest, my mother danced without music," describes Beccah. "She danced away from me, hearing music I could not hear, dancing and dancing until her rasping breaths filled the air."[146] Beccah thinks her mother has lost her mind, but she fails to realize Akiko has entered a possession state in which the shaman has summoned a spirit into her body and *become* that person: "The invocation of spirit is a necessary condition of the kut and the shaman transforms herself into the supernatural spirit."[147] Hence, when Akiko reprimands Reno for stealing a scarf in a heavy patois, it is not Akiko speaking but rather Reno's deceased mother: "You, Baby Reno, you always wanted dis scarf. So did your sister, but I nevah wanted for you two to fight over um. 'Bury it wit me,' I told you. You make me one promise, you good-for-nuttin', and still you wen tell yoah sister I gave um to you."[148] While this scene mostly functions to certify Akiko's extraordinary abilities—Reno immediately recognizes Akiko's gift and sets her up to become Honolulu's premier psychic—it also reinforces the way ghost vision operates in Korean shamanism. Ghosts are wandering figures in this ethnoscientific system. Whether hiding in forests of Korea or wandering the streets of Honolulu, spirits are mobile figures who can declare their presence to humans or possess the bodies of shamans during a trance. In the latter case, ghosts speak with their own voice, remember with their own memories, and see with their own eyes, which is why in the scene above Akiko speaks with an accent not her own and knows things she cannot possibly know. As a bridge between the living and the dead, Akiko fulfills her shaman role by invoking a spirit (Baby Reno's mother), identifying the problem between the angry ghost and human (conflict over a scarf), and resolving the issue (being honest with sister, returning the scarf, etc.). Instead of seeing metaphors or telepathic hallucinations, shamans recognize ghosts as actual entities in the world with a full range of emotions, memories, and capabilities. Ghost vision in a shamanistic episteme signals the ability to see the full spectrum of human and posthuman life, which is to say insight into the true totality of being.

This ethnoscientific definition of ghost vision also transpires in a second, invisible *kut* taking place in the novel. I call it invisible because the ceremony is embedded in the structure of the novel itself, namely in the manner that Akiko and Beccah each take turns narrating portions of the larger story that is the mother–daughter relationship. We must recall two important facts: first, Akiko has been dead the entire novel, and second, she never revealed the details of her tragic past to Beccah. At the end of *Comfort Woman,* Beccah discovers a box left by Akiko that contains much of her family history in song. The casual reader might assume Beccah has reassembled Akiko's past and is merely ventriloquizing her mother's life story for the audience. But a closer examination reveals the sheer level of detail found in Akiko's story line resists such an explanation. It is better to assume, then, that Akiko's narrative is not a transcript read to us by Beccah but actually Akiko herself speaking to us. We know in Korean shamanism that ghosts cannot speak directly to humans; they need an intermediary, someone who serves as a connector between spiritual and human realms. When we put these various elements together, it makes thematic and structural sense that Beccah is a medium lending her body and voice to her Akiko, who now communicates directly to us from beyond the grave. This aligns with the novel's plot because Beccah is an obituary writer, a figurative speaker for the dead, who is the daughter of a medium, a literal speaker for the dead. By uncovering her mother's secret past, Beccah has also become more acquainted with her cultural heritage; for example, she gains newfound appreciation for ghosts like Induk and Saja the Death Messenger, and she performs the proper funeral rites for Akiko in the novel's final chapter. This evolution from skeptic to believer echoes Olivia's transformation from Western to non-Western thinker in *Secret Senses*, and it foreshadows Beccah's final transformation into a Korean shaman (of sorts) who invokes Akiko's spirit so she may finally speak truth to power. Beccah's symbolic and aesthetic evolution into a ghost seer is the final twist of her character, and it allows her mother to do what she never could do while living: give voice to forgotten traumas repressed by history. Other critics have ably addressed these elements of historic violence.[149] I contend that the very act of performing a *kut* is equally powerful because it reconfigures our entire understanding of the text. In broad terms, half the novel is a paranormal episode in which a spirit medium channels the spirit of her dead mother and unveils clairvoyant and retrocognitive truths. In addition, the reader is implicated in this supernatural practice. We are participants in this *kut,* and the world that unfolds through Akiko's eye is the world of the female Korean shaman, of the ghostly. As such, we are not passive observers of the non-Western paranormal but active community members privy to its production, insights, contexts, and mechanisms. We are brought into the fold to experience the epistemological and phenomenological world view of the shaman, and through this process, we gain a novel perspective of how her paranormal mind works and what she can achieve. As Chen writes, "In offering Korean shamanism as a way of reading and responding . . . Keller proposes a hermeneutics that capitalizes on what we can learn from historically and culturally global frameworks."[150] Korean shamanism is a radically different episteme than modern science, and even Western occultism; it correspondingly produces different kinds of paranormal phenomena. Furthermore, it shows how the parascientific entanglements of literature and indigenous science lead to a wide array of ghosts and ghost visions that most readers have not encountered before.

Ghosts and Epistemologies

The two spirit mediums covered in this chapter—Kwan Li and Akiko Bradley—challenge existing ways of thinking about literary ghosts as well as the typical approach to analyzing paranormal phenomena in literature. Kwan and Akiko both navigate material worlds chock-full of spirits: hungry ghosts, laughing ghosts, angry ghosts, benevolent ghosts, household ghosts, traveling ghosts—the list goes on. For them, spirits are not just symbols for forgotten traumas but entities fitting into the larger tapestry of physical and metaphysical being, figures with complex feelings, desires, and passions because not long ago, they were human too. Although it is important for literary critics to make connections to social history, it is equally important to address epistemic contexts and to recognize ghosts play a significant role in the ethnoscientific understanding of reality, whether it is qi-naturalism, Korean shamanism, or some other local knowledge system. By interpreting ghosts qua ghosts, we gain insight into the frameworks undergirding the phenomenological world views of characters who do not see the world through Western scientific paradigms. In many ways, the paranormal is normal for non-Western cultures. Ghosts in traditional Korean and Chinese culture are not aberrations of nature but part of a larger spectrum of existence including humans, ancestral spirits, reincarnated lives, vengeful demons, viral ghosts, and haunted locations that create material effects in the world.

In addition, ghost seers help us address the politics of multicultural science and gender in parascience. Parascience gives name to the epistemic boundaries occupied by concepts, beliefs, and groups historically excluded by mainstream science. Unfortunately, non-Western science falls into this area. *Secret Senses* and *Comfort Woman* are counterdiscursive texts—epistemological novels seeking to expand our understanding of mind, metaphysical realities, and ghostliness beyond a strictly Western lens. It is significant that the Americanized protagonists in each work, Olivia Laguni and Beccah Bradley, change over the course of their respective narratives to eventually accept the existence of ghosts and the veracity of non-Western modes of knowing. As proxies for the Western reader, Olivia and Beccah subtly advance a multiepistemic program asserting that several forms of knowing exist in the world, that alternative frameworks can be used to navigate reality, that modern science alone does not constitute the uni-verse of our collective knowledge. Their shared conversion into what Walter Mignolo calls the pluriverse and what Bernd Reiter calls mosaic epistemology matters because it points to a new way of interpreting the supernatural through global contexts.

It is also meaningful that the non-Western mediums in each novel are women. Women play a key role in paranormal science, and these characters are no exception. In psychical research, women leveraged the dominant scientific ideology of hypersensitivity and fine nerves into a form of epistemic power; their "biological weakness" became a strength as people like Mary Craig Sinclair emerged as leading figures for studying the paranormal. Kwan Li and Akiko Bradley perform similar functions for non-Western mediumship. Their stories are not concerned with scientific power per se, but they are nonetheless larger-than-life figures whose sensitivity to qi and ghosts provide deep insights into Chinese and Korean ethnosciences. Like European mediums, their unique gifts implicitly make them scientific pioneers. If our inquiry into parascientification shines a light on forgotten players in scientific history, we would do well in focusing on other important figures like the female medium.

Along these lines, we can even interpret Amy Tan and Nora Okja Keller as mediums as well. At a basic level, a medium links two incommensurate spheres. The spheres discussed here mostly concern human–ghost and material–immaterial binaries, but the divide between Western and non-Western science is a central motif as well. As authors of Asian American fiction, Tan and Keller bridge Asian and Eurocentric frameworks as equal epistemes worthy of investigation and extrapolation. They mediate their cultural, historical, and scientific differences for us, and in doing so, they help us perceive what we could not see before. By connecting contemporary Western thought with the specter of Eastern science, they help resolve the ongoing problem of internalist epistemology.

We can also credit Tan and Keller with expanding the breadth of paranormal literature. Scholars have usually sought the paranormal in genres like modernism and, to a lesser extent, science fiction. But retrocognition, ghost vision, and precognition exist widely in ethnic fiction as well. If we recognize supernatural phenomena as part of different scientific paradigms, then the range of paranormal fiction greatly amplifies. It also expands our understanding of what the paranormal is and what it means. When we move beyond Victorian psychology into areas like qi-naturalism and shamanism, the sciences available for interpreting the natural/supernatural boundary increase exponentially. This chapter is just a start in such an epistemic expansion, as there are works in African American, Latinx, and indigenous fiction that similarly draw from other scientific epistemes to explore the unexplainable. I see the spirit medium as a revolutionary figure because she points to a pluriversal approach to parascience largely overlooked in cultural analysis. In the epistemic networks constituting parascience, we need not limit ourselves to Western paradigms or Eurocentric works of literature. Many cultures across time and space differentially perceive the ghostly and the human. Kwan and Akiko have helped open our eyes to this reality.

5 THE MULTIVERSE AND THE MIND

The Alternative Chronologies of Atomik Aztex *and*
A Tale for the Time Being

On March 11, 2011, at 14:46 JST, a magnitude 9.0 earthquake struck forty-three miles off the eastern shore of Japan. One of the most powerful earthquakes ever measured in the modern era, it shifted the island of Honshu eight feet to the east, triggered hundred-foot tsunamis that engulfed seaside towns, and led to the partial meltdown of the Fukushima Daiichi nuclear power plant. Over fifteen thousand deaths and $200 billion in damage were recorded in the aftermath of what is now called the Great Tōhoku earthquake. In Ruth Ozeki's *A Tale for the Time Being,* one of the imaginary fallouts of this very real natural disaster is a copy of Marcel Proust's *À la recherche du temps perdu,* which is carried to an island off British Columbia where a Japanese American writer— also named Ruth—discovers it several years later among other tsunami debris. As we soon learn, the inside of the book has been co-opted by the diary of a sixteen-year-old girl named Nao Yasutani, who has inscribed within its pages her thoughts on Zen Buddhism, high school bullying, ancestral ghosts, and the fluidity of space-time. What makes this hybrid object truly remarkable, though, is not just what it says but what it does: it forges a supernatural connection between Nao and Ruth. Nao's personal understanding of Zen telepathically leaks into Ruth and provides the older woman critical insights into the true nature of time and consciousness. Ruth in turn has clairvoyant visions of Nao's past and even begins altering the material conditions of that already elapsed reality. In this regard, Proust's *In Search of Lost Time* could be construed as a parascientific object blurring Eastern, Western, mythic, scientific, and literary elements to invoke novel variations of impossible phenomena. By the end of the narrative, thanks to Nao's diary, Ruth has become a wholly paranormal being; she performs telepathy, telekinesis, and astral travel, among other feats. What makes her entrance into psi particularly fascinating is the element of time. She grows from a passive figure grounded in the strictures of Western linear time into an active agent traversing the open landscapes of Eastern nontime via backward glimpses into the past and sideways jumps into alternative realities. The novel/diary at the core of *Time Being* thus produces a paranormal mind that is both different and more powerful than psychical theory can explain.

In this chapter, I build on chapter 4 by focusing on the interconnections between the paranormal, non-Western science, and speculative ethnic fiction. In the book's opening chapters, I examined the intellectual development of the paranormal mind through the lens of Western science and literary genres like science fiction, which followed a more traditional STS approach where Western modes of knowing (e.g., physics, biology, statistics) converse with cultural texts. In contrast, chapter 4 analyzed the paranormal via the indigenous frameworks of qi-naturalism and Korean shamanism. The canonical ghost narratives of Amy Tan and Nora Okja Keller illustrated how native sciences can expand the epistemology and phenomenology of paranormal discourse. Now, the following pages will advance this method by exploring how contemporary ethnic speculation updates the paranormal, as well as how elements of fantasy, science fiction, magical realism, and indigenous cosmology inform epistemological politics.

It is worth acknowledging that I am reading the paranormal in a field that, until fairly recently, has emphasized the real. As Madhu Dubey observes, most writers of color have historically used realist literature to address the racist, colonialist, and capitalist past.[1] From eighteenth-century slave narratives to semiautobiographical encounters with racism in contemporary Asian America, realism and realpolitik have long dominated mainstream portrayals of ethnic fiction. Even the supernaturalism common to the genre is subsumed within the larger narrative that multicultural fiction is a manual for comprehending "authentic" (i.e., weird) minority cultures,[2] which reveals not only an ethnographic gaze but the dominance of realist aesthetics.[3] Thankfully, the recent explosion of Afrofuturist, Chican@futurist, techno-Orientalist, and other related genres has proven speculation an equally generative mode for analyzing issues of race and power. Alondra Nelson argues that such writings "look backwards and forwards" by asking "what was *and* what is?"[4] Cathryn Merla-Watson and B. V. Olguin contend that radical defamiliarizations of the real authorize a fresh reckoning with the "ghosts of colonialism, modernity, globalization, and neoliberal capitalism" by liberating readers from what has actually played out contra the possibilities of what might have and could have been.[5] Similarly, Ramón Saldívar identifies "speculative realism" (also known as historical fantasy) as an emergent subgenre of postmodernism that self-consciously deploys speculation to grapple with race.[6]

If speculative ethnic fiction is a burgeoning space for critiquing the political past and future, then it can also serve as an instrument for examining the ethnoscientific past and future. As addressed in chapter 4, postcolonial approaches like Walter Mignolo's pluriverse and Bernd Reiter's mosaic epistemology seek to understand local, indigenous, and non-Western epistemes as scientific systems in their own right. Tan and Keller's ghost narratives demonstrated how the paranormal is normalized when viewed through alternative scientific paradigms. Speculative ethnic fiction accelerates the pluriversal process through extrapolation: Beyond explaining how native sciences explain psi, what else could they potentially allow? What new psychical feats emerge when the paranormal mind is decoupled from Eurocentric science? What becomes possible when Aztec and Japanese sciences take imaginative flight? And how does this change the existing cultural discourse of the paranormal?

This chapter answers these questions through telepathy, or mind-to-mind communication. It begins by reviewing Western interpretations of telepathy, which early psychical researchers viewed as the foundation for unlocking the mysteries of the paranormal mind. I then segue into the political impact of speculative ethnic fiction on the paranormal. Critics like Kandice Chuh, R. A. Judy, and Ramón Saldívar have all linked aesthetics with social justice, and I riff on their ideas to further my own claims about epistemological justice. Finally, I turn to two speculative realist texts, Sesshu Foster's *Atomik Aztex* (2005) and Ruth Ozeki's *A Tale for the Time Being* (2013), to illustrate how indigenous science conjures new forms of telepathy exceeding the limits of Western consciousness. Both novels use a radical type of chronology to retheorize the paranormal. Chronology typically refers to the sequencing of events in time, but I reconsider it as a study or science of time—an exploration of temporality, its progression, its directionalities, and its ontological connections to space, mind, and being. Moving past linear, Eurocentric chronology allows us to see how various cultures throughout history have understood the nature of time. In pre-Columbian Mesoamerica, time had circular and spinning properties that *Atomik Aztex* extrapolates into multiple loops of space-time through which the protagonist, Zenzontli, mentally communicates with his multiverse doppelgängers. Such telepathy across infinite universes differs from what we might expect in Western science fiction because it draws from a completely different scientific episteme. The Zen Buddhism underpinning *Time Being* also acts as a chronology because it reformulates the mind's relationship to space and time. In contrast to the Western conception of human consciousness as an epiphenomenon of brain function, the novel treats the mind as an entity fully unified with space-time, a merging that enables passage into past, present, and future. In this way, Zen chronology becomes a tool for telepathy, clairvoyance, and even astral travel. By extrapolating these temporal ethnosciences, Foster and Ozeki drastically reimagine the paranormal. At the same time, *Atomik Aztex* and *Time Being* allude to other Western sciences—particularly physics— to explain their mind-bending circumventions of space-time. As such, they are exemplary works of parascience in amalgamating Western and non-Western sciences to naturalize the mind's fantastical capacities.

Speculative ethnic fiction plays a key role in parascience by introducing four major shifts in paranormal discourse. First, it brings additional literary genres under the umbrella category of paranormal literature. Chapter 4 showed that canonical ethnic fiction is filled with ghostly and mediumistic episodes that are not generally considered paranormal. Speculative ethnic fiction broadens this aesthetic purview and helps us recognize the ubiquity of supernatural experiences in world culture. Second, it participates in the larger decolonial project decentering Eurocentric science as the only real source of knowledge in the modern world. Mesoamerican and Zen chronologies are wildly extrapolated in *Atomik Aztex* and *Time Being*, and collectively, they suggest that there is more than one way to experience space-time. Third, it contributes to novel retheorizations of the paranormal mind. Even more so than the works discussed in chapter 4, multicultural fictions that are fantastical, magical realist, and science fictional help us cognize non-Western sciences as active frameworks; they are not dead bodies of knowledge illustrating how things used to be for premodern societies but are instead dynamic epistemes other cultures have (and still) use to explain ghosts, divine visions, and telepathy. Fourth, speculative ethnic fiction helps remove some of the paranormal's negative stigma. The paranormal is ideologically abnormal because it fails to cohere with the universalizing logic (Walter Mignolo's uni-verse) of Eurocentric science.[7] When speculative texts by authors of color extrapolate how other cultures have historically understood the fluidity of space, time, mind, and energy via ethnoscientific systems on their own terms, contemporary readers can reconsider the paranormal in wider ways that accept the marvelous as part of the regular structure of scientific life rather than anomalies to be expunged from it. A running theme in this book is that the paranormal mind is ubiquitous across multiple contexts, and speculative ethnic fiction helps attend to that ubiquity by highlighting how different peoples from different periods have conceived the transcendent possibilities of human consciousness.

In harking back to premodern, theological, and non-Western systems of thought to theorize the conscious and unconscious, speculative ethnic fiction utilizes new logics for explaining how the human mind works, perceives time, and ultimately transcends the scientific limitations imposed by the West. Like the traditional science fictions and ghost stories preceding it, these texts act as parascientific agents producing new kinds of paranormality for uptake into mass culture. Our pivot toward ethnic fiction also returns us, ironically enough, to the beginnings of modern occultism. The mysticism at the root of the Occult Revival traces back to Orientalist notions of Asia. Madam Blavatsky claimed all her esoteric knowledge ultimately derived from the Mahatmas, the keepers of "the Secret Doctrine of the East" who had assembled the ancient Egyptian, Chaldean, Gnostic, and Buddhist sciences lost to the modern world.[8] At first glance, ethnic literature engaging with native sciences might seem to draw from the same cultural biases that inspired fin de siècle occultists. However, I contend that such fictions do not purposely portray the East as a mystical other by which to reconstruct the West; rather, it decolonizes indigenous science as a valuable approach for reinterpreting the mind. Their aim is not to estrange the premodern but to parse its logic and push its limits, to reconceive the boundaries of the scientifically possible. Speculative ethnic fiction is a site of epistemological production because it theorizes and normalizes the paranormal within mass culture. By building from twentieth-century occult modernism, science fiction, and canonical ethnic fiction, twenty-first-century multicultural literature becomes a vibrant space where the paranormal is continually rekindled and reborn.

Telepathy and Early Psychical Research

After its founding in 1882, the SPR studied a wide array of supernatural phenomena, including mesmerism, haunted houses, ghosts, mediumship, and divining rods. Psychical researchers quickly latched onto telepathy as the most important of these—the one paranormal ability that might explain all the others. In fact, they made it the subject of their first major study, *Phantasms of the Living* (1886). Defining telepathy as "the phenomenon of transmission of thought or sensation without the agency of the recognized origins of sense," Frederic Myers contends that years of field research led to the inexorable conclusion that mind-to-mind communication stood at the very center of the mind's paranormal processes: "direct action of mind upon mind has at least a generality which makes it possible that, like the laws of atomic combination in chemistry, it may be a generalization which, though grasped at first in a very simplified and imperfect fashion, may prove to have been the essential pre-requisite of future progress."[9] Telepathy was potentially the scientific bedrock on which all other paranormal phenomena was based. If the Victorian psychical researcher could understand its principles and parameters, he could in turn elucidate every other type of supernatural power.

Phantasms of the Living claimed that the preponderance of paranormal activity in nineteenth-century Britain proved the existence of telepathy.[10] In the follow-up text *Human Personality and Its Survival of Bodily Death* (1901), Myers elaborated on this thesis by establishing the general principles of the subliminal self. As covered in the introduction, the subliminal self was a pre-Freudian theory of the unconscious viewing the mind as a "porous" entity capable of sending and receiving thoughts, feelings, urges, sounds, and images to other minds, both living and dead. It was this marvelous ability to instantaneously transfer information across vast distances that authorized other occult powers. For example, in chapter 4 we learned how the subliminal self's ability to receive far-flung telepathic data from someone in distress could explain why people saw (or thought they saw) ghosts. "Telepathy," Myers argues, "looks like a law prevailing in the spiritual as well as in the material world."[11] The word choice "law" captures the import of telepathy in psychical discourse—its status as first among equals—because in its biological pathways and psychological functioning lie the skeleton key for unlocking the paranormal mind more broadly. Myers did not delve into such particulars in *Human Personality*, though; that would be the monumental task for psychical research going forward.

If telepathy is the primary law of the paranormal, one implicit corollary is that mind-to-mind communication takes place under a discrete unit of space-time. In hundreds of episodes collected in *Phantasms of the Living, Human Personality*, and the *Proceedings of the Society for Psychical Research,* thought transference occurs in a period ranging from instantaneous to several hours. For instance, a woman might feel a sudden "cloud of anxiety" at 7 PM one evening, which is the precise moment that her sister receives a telegram announcing the illness of her brother-in-law.[12] Or a woman has a strange feeling of dread about her childhood home in Scotland and learns her sister died in Edinburgh four hours later.[13] In these anecdotes, one mind exchanges information with another mind during a distinct time frame within the same plane of reality (our material world). In psychical research, mental communication does not transpire between alternate dimensions, or between multiverse versions of yourself, or by moving backward through years. For men of science like Gurney and Myers, that would have been *too* weird. I suspect this was because psychical research was predicated on Western sciences favoring linear, straightforward approaches. Aberrations such as those listed above would strain the credulity of a field that intensely desired scientific respectability.

As we shall soon see, a paranormal mind deriving from native sciences gains flexibility in terms of space and time. When liberated from Western scientificity, telepathy follows different rules and assumptions, and it consequently manifests in different ways. Psychical research may have developed the formal conceit of telepathy, but authors of color have begun to hybridize it with non-Western modes of knowledge, and in doing so have irrevocably changed it.

Epistemology, Power, and the Ethnic Imagination

Over the past decade or so, several scholars have argued that ethnic literary representation matters more than ever in the twenty-first century. Their rationale can help us understand the function of ethnic speculation in both paranormal discourse and parascience. Kandice Chuh, for instance, makes a compelling case for the necessity of ethnic fiction in *The Difference Aesthetics Makes*. Criticizing Western human liberalism for reproducing social inequity, she calls for an "illiberal humanities" based on minoritized discourses. Such an aesthetics is not representation for representation's sake or some other weak multiculturalism, but rather literature rooted in Foucauldian power/knowledge. The illiberal humanities, she writes, "acts as a counterhegemonic point of entry into illuminating the relationship of knowledge practices to structure and relationships of power."[14] For Chuh, ethnic fiction reflects and reinforces ways of being and knowing Western culture has too often attempted to extinguish. Non-Western science is a prime example of such smothering: how much does the general public know about Sumerian mathematics, Chinese astronomy, or indigenous agriculture except that Europe took the "best" parts, radically improved it, and thankfully discarded the rest?

R. A. Judy advances both Chuh's critique of the West and the political need for more ethnic speculation. He argues that Martin Heidegger, so central to Western humanistic thought, limited Dasein to *white* Being in his writings. Ethnic minorities, as epitomized by the Negro, fall into the subaltern category of primitive Dasein, in which notions like past and memory are vague abstractions limited to the "latest hunt."[15] For Heidegger and the Western tradition, "the primitive has not yet achieved an ontological understanding of Being, has not yet properly objectified things in the world as distinguishable from the world."[16] This is a grave assertion with even graver consequences, implying that non-Western, nonwhite peoples cannot signify, cannot distinguish myth from reality, and cannot conduct science, let alone fathom what a science is. It debases the non-Western subject as "nontheoretical, prephenomenological, and concomitantly, preontological"—so far below the West as to render their world views incomprehensible.[17] This is why Judy claims that the ethnic imagination matters so deeply. He redefines the term *poesis* as "the species-activity of actualizing in discrete material forms any given conceptualization of being-in-the-world, in accordance with a specifiable set of practices-of-living."[18] His definition is useful because it recasts non-Western aesthetics (perceptions, reflections, accounts of Being, etc.) in stark political terms. The Black novel does not simply share the nature of Black experience but pushes against a Western tradition that has historically viewed such experiences as unfathomably subaltern. By the same token, a qi-cosmological novel (discussed in chapter 4) does not merely teach us about qi but validates the sciences, world views, and values of an othered culture. Such works, Judy notes, "articulate appositionally, opening up infinities of other ways of being human."[19]

Ramón Saldívar elaborates on the idea of infinities when discussing speculative realism. Connecting the various writings of Salvador Plascencia, Charles Yu, Larissa Lai, and others, he defines speculative realism as a nascent postmodern form that hybridizes genre elements (fantasy, science fiction, Gothic, etc.) with classic ones (modernist, romantic, realist, etc.) to focus on racial politics.[20] Why this kind of literature at this historical moment? "At the beginning of the twenty-first century," Saldívar writes, "changing relationships between race and social justice, race and identity, and race and history now require American writers of color to invent a new imaginary for thinking about the nature of a just society and the role of race in its construction."[21] New imaginaries are needed because Western literary realism is insufficient to capture the multiverse of experiences located in minority experience: the rich cosmologies, knowledges, beliefs, values, gods, traditions, and everyday practices of the so-called primitive Dasein. Accomplishing this requires speculation—radical speculation, in fact. And while Saldívar views social justice as the goal of speculative realism, I see epistemological justice as an equally worthy telos. With epistemological justice, I return to the decolonial STS approach from chapter 4 in which we gain access to a range of scientific and philosophical traditions—none ipso facto subordinate to Western science, but contextualized to times and places to offer an enriched, global view of nature. It suggests a movement away from the Eurocentric universe of scientific knowledge toward multiepistemic sciences, as well as the cultural impetus to resist prioritizing Western knowledge above all others.

All these various aspects of speculative ethnic fiction—illiberal resistance, minority poesis, epistemological justice—matter in parascience because scientific knowledge is power. Parascience names a site of unorthodox knowledge circulation, but it also speaks to the nature of power relations in the epistemological economy. Speculative works by authors of color can shift those relations. This is a positive and necessary development. We should be able to see reality as other cultures have seen it, to understand its hidden mechanics as they understood it, to extrapolate its possibilities as others may have extrapolated it. Such a literary mode, such a methodology, not only enrichens our understanding of nature but also reveals new dimensions of the supernatural.

Atomik Aztex: Mythology, Chronology, Telepathy

Sesshu Foster's *Atomik Aztex* is an excellent entry point into the ethnoscientific transformation of telepathy because of its complete immersion in Mesoamerican culture, religion, and technology, as well as its totalizing poesis. On its surface, this speculative realist novel is a revisionist tale that reimagines contemporary life if the Aztec civilization had never perished. Like Ernest Hogan's *High Aztech,* it is an alternative history exploring what the world might be like if the European conquest of the Americas had failed and Aztec society reigned in its stead. *Atomik Aztex* goes further than Hogan's cyberpunk take on pre-Columbian life as a postmodern pastiche that is "part fantasy; part hallucinatory Global south realism; part muckraking novel; part historical novel; part chronicle in the tradition of *crónicas de Indias* . . . part Los Angeles cartoon noir, and wholly science fiction alternative and counterfactual history."[22] The narrative unfolds through the perspectives of two main characters. The first is Zenzontli, the Keeper of the House of Darkness, a high-ranking Aztec military officer in a reality where the Aztecs not only thwarted the Spanish conquistadors in the sixteenth century but went on to defeat the West and its "bullshit ideology" in the modern era.[23] In Foster's words, "An Aztec warrior speaks from the other side of life about some other side of History."[24] But Zenzontli suffers from splitting headaches and horrible visions of a second world—which is our world—where a second narrator named Zenzon is an undocumented immigrant from Mexico working for minimum wage at a California slaughterhouse.[25] The narrative winds back and forth between these protagonists as world-historical battles mirror the battles of daily minority life. The ironic parallels between the Zenzons has become rich fodder for literary critics, who have argued that labor exploitation is a mode of human sacrifice, corporate capitalism is a systemized form of bloodletting, bigotry is a cultural universal, and colonization and genocide are natural by-products of unchecked power.[26] The novel undoubtedly has much to say about the destructive ideologies of religion, free markets, and racial hierarchies.

Nonetheless, it is the telepathic exchange between the two Zenzons that I find truly fascinating. While a casual reader may accept their mental correspondence as the price of admission into such a zany text, it is actually the central crux of the novel, given its investment in Aztec science and culture. As we will soon see, Aztec space-time does not operate like Western space-time, and therefore neither perceptions of reality nor narrative structures in the text correspond with Western knowledge paradigms. In the same way that Akiko Bradley and Kwan Li (chapter 4) gain otherworldly powers by occupying non-Western scientific epistemes, so too does Zenzontli acquire uncommon capabilities courtesy of his Aztec background. More specifically, the Aztec calendar constitutes a scientific framework allowing him to experience multiple space-times, or what we might call parallel realities. In true science fiction fashion, Foster extrapolates this chronology, or the science of time, to introduce telepathic communication across these realities. The novelty of such paranormality should not be overlooked; instead of transmissions from one mind to another within the same block of space-time, as understood in Western telepathy, *Atomik Aztex* showcases communication between different versions of the same person across multiple historical realities. These psychic missives break with early psychical models, but they rationally build on the guiding principles of Aztek science. Just as Alan Moore extends the psitron's power in *Watchmen* beyond Adrian Dobbs's original version (chapter 1), and just as Greg Bear augments Gaian consciousness in *Vitals* beyond Lovelock's model (chapter 2), Foster propels *Aztex* chronology in his novel to conjure bold visions of paranormal cognition. To understand how Aztec telepathy works and why it matters in a parascientific sense, we must familiarize ourselves with the Mesoamerican calendrical sciences.

As mentioned earlier, I use the term *chronology* here to reference to an actual science of time: a systematic and comprehensive study of temporality, including but not limited to its principles, movements, technological applications, and social functions. In an ethnoscientific context, chronology can be viewed as a sweeping body of knowledge as well as an instrument of temporal control. For the Aztecs and other native tribes of central and southern Mexico like the Olmecs, Toltecs, Zapotecs, and Maya, the ancient Mesoamerican calendar was the most important social and scientific object in their respective civilizations. As historian Vincent Malmstrom has pointed out, much of Aztec intellectual life—astronomy, mathematics, writing, architecture, urban planning, and so on—was bound up in the calendar and its related scientific discourses.[27] According to Elizabeth Boone, the Aztec calendar "stood for an entire body of indigenous knowledge, one that embraces both science and philosophy."[28] The Aztec world interwove physical and metaphysical existences, and the calendar was a tool for bridging these natural, supernatural, and religious realms. "By explaining the temporal and supernatural forces that shaped and governed their world as it was known to them, they [the Books of Fate explaining the calendar's nuances] articulated its universal laws," explains Boone. "Their goal was to express the unrepresentable, to provide through structured figuration an understanding of invisible forces and principles. These divinatory and religious books were thus equivalent to our books of philosophy, theoretical physics, astronomy, and astrology."[29] The comparison between modern science and the Aztec calendar is apt because the latter is a unifying structure—a paradigm—authoring the society's principles of cosmology, being, space, and time. In short, it is a fully fledged system based on rules, empirical observations, rational inferences, and ideological values constituting a body of knowledge from which the emergence of additional technologies and powers—including paranormal ones—are completely rational.

Aztec chronology constitutes multiple interlocking vigesimal (base 20) calendars. The secular *xiuhpohualli* calendar consists of eighteen months that are twenty days long, with five unlucky days added to produce the familiar 365-day solar calendar (see Figure 7).[30] More significant to the cultural life of Aztecs was *tonalpohualli*, the sacred calendar made of thirteen months of twenty days,[31] a 260-day calendar believed to correspond to the length of a woman's pregnancy, and thus symbolic of rebirth.[32] The circular shape of these calendars highlights their nonlinear progression. Aztec time is cyclical, so as one cycle ends, it immediately begins once more, signaling that what has previously transpired shall occur again and again. The cycle of the seasons (fall → winter → spring → summer → fall, and so on) provides a familiar analogy for Western audiences. A key difference here is that instead of the seasons cycling along a linear progression of months into years, Aztec time loops back to the beginning.

Every fifty-two years, though, the secular and sacred calendars synchronize to their original starting points to form the Aztec "calendar round," which

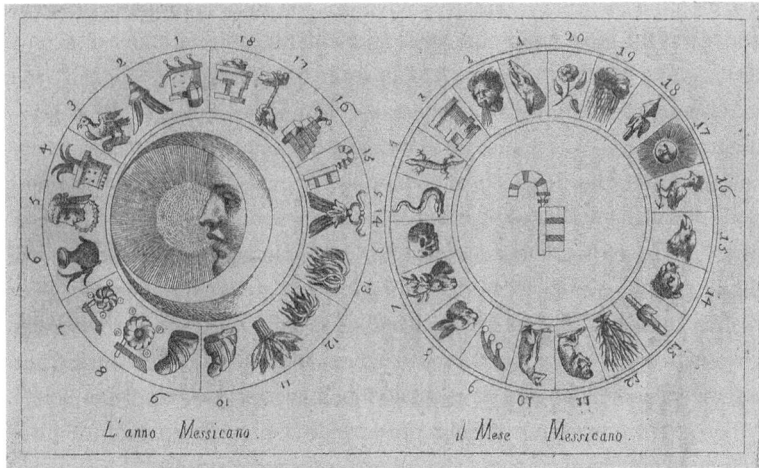

Figure 7. The Aztec year and the Aztec month. Francesco Saverio Clavigero, "L'anno Messicano and il Mese Messicano," from *Storia antica del Messico, 1780–1781,* Rare F129. C61, Dumbarton Oaks Research Library and Collection, Trustees for Harvard University, Washington, D.C., https://www.doaks.org/resources/rare-books/storia-antica-del-messico/ lanno-messicano-il-mese-messicano.

signals the end of one age and the birth of another. Unlike Western time, which extends into the future in an undifferentiated and endless manner, *the calendar round* serves a very human purpose: it provides an ontological break at which point all things begin anew. It is a dangerous time, though: "At this juncture it was thought that the cosmic forces might cease and the sun, lost in the underworld, might not summon up the vitality to rise again, thereby preventing the inception of another set of fifty-two years," writes historian Burr Brundage. "Universal death would follow."[33] In Aztec mythology, the god Nanahuatl (or Nanahuatzia) sacrificed himself in a bonfire so that the sun would continue shining on Earth, thus commencing the Fifth Age (or Fifth Sun). With *the calendar round* representing the end of an epoch, there was no guarantee the sun would return or reality would continue. According to Aztec chronology, a sacrifice was therefore needed to maintain existence itself.[34] David Carrasco explains that "sacrifice was a way of life for the Aztecs, enmeshed in their temple and marketplace practices, part of their ideology of the redistribution of riches and their beliefs about how the cosmos was ordered, and an instrument of social integration."[35] Put differently, ritualistic human sacrifice served an ontological purpose: it was a *technology* helping Aztecs assert a measure of control over time and existence. According to Brundage, the New Fire Ceremony was an elaborate ritual held at the end of each calendar round to ensure the successful birth of a new era. On that night, a procession of priests would gather at a sacred hill in Tenochtitlán to perform their existential duties:

> At the appropriate time a distinguished captive of war—a ruler or at least a great captain—was sacrificed, his heart being ripped out and offered to the god of fire. In the cavity in the dead body a special fire priest then placed the sacred fire board and fire stick and, whirling the latter between his palms, elicited the first sparks of new fire. No words were spoken until the spark had been nourished into a flame. This fire was then applied to a great pile of fagots on the temple terrace which grew into a magnificent bonfire, signaling its joyful message across the night. Into this sacred fire was thrown the body of the sacrificial victim in commemoration of the birth of the sun out of the ashes of the god Nanahuatl, who had thrown himself into the fire and perished that the sun might come into being. For many miles around in the blackness Aztecs of all walks of life had been waiting. On mountain peaks and rooftops, in some cases far off across the lake, they hailed the knowledge brought to them by the sight of the first fire of the new calendar round, the knowledge that they would be granted yet another fifty-two-year reprieve and that life would continue.[36]

As the passage indicates, human sacrifice is an ontological instrument perpetuating reality itself. Such a powerful technology is not magical so much as a logical extension of their scientific world view. As a surrogate for Nanahuatl, the distinguished captive reenacts the god's benevolent sacrifice. By submitting his life-force to the flame, he powers the birth of a new sun. The bonfire provides empirical evidence of the ritual's success, and the next morning's sunrise offers further substantiation. As everyone could plainly see, the danger had passed; time and space were renewed, and life would continue once more unabated. As technologies go, human sacrifice is incredibly powerful because it allowed the Aztecs to maintain control over reality itself.

I have elaborated on the major theories and practices of the Mesoamerican calendar to substantiate Aztec chronology as a science in Sandra Harding's expanded sense of the word.[37] While it does not follow the rules of Western science, theirs was a truly elaborate system of hypotheses and observational knowledge that explained the operations of the natural world. Moreover, it was an all-encompassing system structuring every aspect of society, from the orders of priests who were its chief scientists to the interpretation of dreams to the planting of crops according to calendar directives. Just as modern science organizes so much of daily life today, so did Mesoamerican chronology for the Aztecs. As it turns out, it also structures the logic for new kinds of telepathy in *Atomik Aztex*.

Navigating Foster's novel is a discombobulating affair. It begins with a brash introduction—"I am Zenzontli, Keeper of the House of Darkness of the Aztex, and I am getting fucked in the head and I think I like it"—which acquaints us with the high-ranking military officer in the alternative Aztec-dominated reality.[38] A few pages later, though, Zenzontli is praising his Toltec ancestors for their technological prowess when his consciousness abruptly shifts into the Zenzon on the killing floor of an abattoir: "Heads pile up on a wet concrete floor. Bloody meat slaps the wet concrete. The blade spins, zinging through the air till its tiny teeth bite into the pallid skin, the circular blade sinking out of sight underneath descending to the bone."[39] This character is a longtime employee of the real-life Farmer John slaughterhouse in Los Angeles, where his main task involves killing pigs and cutting off their limbs for further processing down the assembly line. For the next five pages or so, Zenzon provides depressing details about his job, his life, his wayward children, and other travails of being a poor, undocumented immigrant in America. Without warning, the narrative shifts back to Zenzontli, who reveals, "I was walking along the kanal outside Xochimilko reeking of human excrement and tropical vegetation" when he falls into the water because he had yet another daydream of a world where Europe colonized the planet and his people were reduced to cogs in a capitalist machine. The rest of the novel zigs and zags between these and other Zenzons: a soldier being sacrificed in Tenochtitlán, a migrant crossing the Mexico–U.S. border, an Aztec general fighting in Russia, and so on. The polyphonic effect of multiple voices constructing the narrative resembles modernist experimentation, like in *As I Lay Dying*. While William Faulkner's text amalgamates a panoply of characters' thoughts, *Atomik Aztex* is distinctive because the various consciousnesses all belong to the same character, albeit variations of himself occupying alternate space-time realities.

The invasion of multiple minds into the novel's stream of consciousness is not just a clever narratological trick for comparing Western and Mesoamerican systems of power, religion, and economics but an unorthodox rendering of telepathy as well. Telepathy, we recall, occurs when the mind of one person gains direct access to the thoughts, feelings, images, fears, and knowledge of another person in a way unexplainable to modern science. This is exactly what we see in *Atomik Aztex*. In fact, the entire novel is best construed as a polyphonic, telepathic exchange between multiple Zenzons. What makes these mental transactions so innovative is their passage across parallel realities, beyond the single block of space-time assumed in psychical research. SPR researchers observed telepathic exchanges across rooms, cities, and even countries. They theorized that telepathy occurred between the minds of the living as well as the souls of the deceased. But in all these scenarios, telepathy transpires across a shared plane of being: the world you, me, and the dead all occupy. *Atomik Aztex* extrapolates a new mode of mind-to-mind communication in which someone gains direct access to the consciousness of their other selves across infinite planes of reality.

Foster initially offers a Western scientific framework to explain the novel's multitelepathic structure. "Perhaps you are familiar with some worlds," he writes, "stupider realities amongst alternative universes offered by the ever-expanding omniverse, in which the Aztex civilization was 'destroyed.'"[40] The "ever-expanding omniverse" is an allusion to Hugh Everett's many-worlds interpretation of quantum mechanics. In 1957, Everett published "'Relative State' Formulation of Quantum Mechanics" in *Reviews of Modern Physics* and unleashed one of the most widely debated ideas in twentieth-century science. Challenging the standard Copenhagen interpretation that quantum states collapse to a particular time and space when observed, the many-worlds interpretation posits all potential quantum states branch into equally valid parallel realities. This fascinating theory claims existence is fundamentally fluid and multiple. If scaled up macroscopically, it suggests "the world we see around us is just one of countlessly many such worlds."[41] In other parallel realities, then, there are different versions of history populated by different versions of ourselves, each one as real as the world in which we currently exist. According to Everett's original theory, there are no crossovers between different realities. Contact points across the omniverse are now a common trope in science fiction, however, with Spock meeting himself in *Star Trek* (2009), superheroes fighting alternative versions of themselves in the Marvel Cinematic Universe, and Michelle Yeoh's Evelyn Wang accessing her other lives in *Everything Everywhere All at Once* (2022). In this regard, the narrative movement of *Atomik Aztex* is partially explained by this peculiar yet theoretically feasible schema.

While Everett's multiverse provides a workable analogy to Mesoamerican space-time, a more nuanced understanding of the narrative structure can be achieved by recognizing the ethnoscientific framework underlying Zenzontli's experience and by using a pluriversal approach that provides a stronger account of the novel's phenomenological design. Zenzontli alludes to this native science when he says, "Luckily we Aztex believe in circular concepts of time, cyklikal konceptions of the universe where reality infinitely kurves back upon itself endlessly so all that has existed does exist and will always exist and so forth into infinity."[42] Historian Kay Read argues that Aztec time combines both linear and cyclical elements. For example, Aztec time has passed through five sequential ages in addition to the unending cycles of *xiuhpohualli* and *tonalpohualli*.

She contends that the combination of progressive and circular temporalities results in "spinning" time, which can be pictured as a "spinning rope growing larger and larger as new threads are continually added on top of old ones."[43] This rope analogy is useful because it captures the directional and ontological multiplicities of chronology in *Atomik Aztex*. Cycles of time overlap with each other like fibers spinning together, continuously creating larger threads; these threads can double back on themselves, spiraling back to produce new piles of overlapping fabric. "Moments arranged themselves like the rope itself, in which various fibers of differing lengths are spun together, now some overlapping and others not, now others overlapping that had not meshed before," Read explains. "Time spirals and spins around and around, like the rope turning back or doubling over itself, its fibers now meshing, now not. No section of rope was ever exactly like a previous sector."[44]

The fractured and zigzagging narrative of *Atomik Aztex* snaps into coherence when reconceived through Read's spinning time, an indigenous chronology where multiple minds endlessly and telepathically intertwine, come together, fall apart, and shuttle back and forth. This is well illustrated when Zenzon meets Nita, a labor organizer, at a café. While the two discuss unionizing all the slaughterhouse employees, Zenzon sees "some guy running down the middle of the street, smoke coming out of his hair and clothes; he was on fire."[45] We eventually learn this man (a multiverse doppelgänger of Zenzon's co-worker, Weasel) participated in a diner holdup that went haywire. We also learn the robbery was foiled by the murderous Skidrow Slasher, a multiverse doppelgänger of Zenzon's other workplace pal, Zahuani.[46] And of course, Zenzon is not just a minimum-wage worker. He is also the Keeper of the House of Darkness, who may or may not have illegally crossed the Mexican border to work in the Farmer John factory. In this story, multiple characters' realities loop back and forth into intertwined piles. Different multiverse versions of Zenzontli occupy these realities, each intersecting with other multiverse characters traversing their own narrative threads. Such crossings are not quirks of the narrative but logical by-products of Aztec chronology. Just as *xiuhpohualli* and *tonalpohualli* are two aspects of time that periodically fuse at the calendar round, so too do the various selves of Zenzon, Weasel, and Zahuani traveling their own cycles of time converge with one other. If Aztec space-time is a fabric woven and warped via cyclical interconnections, then *Atomik Aztex* fastidiously represents that fabric through ethnoscientific phenomenology. Judy uses the term *para-semiosis* to describe "a dynamic constitution of the world as a fluidity of multiple enactments of referentiality whereby being human is enunciate in flow."[47] Foster's text is that rare novel capturing such fluid multiplicities—not just of non-Western humanism but also of indigenous, chronological being.

Not every Aztec character in the novel is sensitive enough to be a telepath. Some, like Zenzontli, have full psychic access to their infinite lives. Others do not and consider him mentally unstable. Still, there are others in between who possess vague, uncanny knowledge of their multiverse counterparts. For example, Zahuani unconsciously knows intimate details of his other self: "I think [the failed diner robbers] shot the waitress by accident. I think she was a friend of theirs. I think she was the true love of the shorter of the men. I think he was from Las Vegas & she was from Bakersfield. I think he lost his soulmate right then & there on the linoleum floor."[48] Throughout the novel, Foster reinforces Aztec chronology as a radical set of possibilities transcending the singular teleology of Western time. One question that naturally arises is what mechanism allows telepathic knowledge to move across the consciousnesses of these multiverse characters. What broader principles explain how Zenzon connects to other Zenzons in the ever-expanding omniverse? To comprehend this, let us delve into the secret weapon Foster develops as the ultimate achievement of Aztec science.

Ritualistic sacrifice served as an ontological technology for Aztecs to produce reality itself. However, Zenzontli slyly alludes in his opening monologue that the "Aztec Sciences of the Human Heart" have progressed rapidly since the Toltecs first unlocked the power of sacrifice.[49] In fact, instead of human sacrifice merely producing *one* reality, as seen in the New Fire Ceremony, Aztec science in the novel can now produce *multiple* realities. Just as advancing Western science allows us to squeeze more energy from a gallon of gasoline or more lamplight from a single watt of electricity, Aztec technology has improved to the point that it can wring more reality from the human heart. One of the key extrapolations in *Atomik Aztex* is that the act of cutting out human hearts *produces* the alternative realities in Zenzon's polyphonic narrative. Equally important for our purposes is that Aztec heart science maintains mental connections across those realities, which is to say it ensures telepathy across the multiverse. While Zenzontli never fully explains these psi mechanics, the subtext is that all minds in all Aztec space-times preserve a networked connection in which the consciousness from any one node can tap the consciousness of another.

This is why Zenzontli can boast, "No matter what horrible fate awaits us at the hands of some momentarily viktorious enemy, we have altered the space-time continuum of the universe through our Aztek sciences and technologies so that we shall emerge victorious on one level or another sooner or later, something which causes our enemies no end of anguish and horror once they realize their own hearts are the key, the crucial item in the orchestration of our mastery of the time-space continuum."[50] This passage underscores the swift transition from Aztec chronology to technology to telepathy to psychopolitical conquest. Whereas the Toltecs originally conjured one reality from human hearts, Aztecs now produce many. The alteration he mentions is the ability for talented telepaths like Zenzontli to simultaneously traverse the multiverse and achieve victory in one space-time or another. Domination in any one reality allows for domination elsewhere because all the Zenzon minds are interconnected.

The technological advancement of *Atomik Aztex* is twofold: ontological and phenomenological. In terms of the former, Aztec scientist-priests have mastered being itself. Read contends that Mesoamerican calendars provided Aztecs with a form of ontological control: "If things are alive, they move; if they move, they are time; if they are time, they can be calculated; and if they can be calculated, with some luck, they can even be ordered and controlled."[51] The Aztec Sciences of the Human Heart represent a new apex of humanistic dominion: the manufacture and manipulation of space-time. In terms of the latter, the Aztecs also possess a new mental technology: a telepathic connection throughout the omniverse in which every mind maintains communication with itself across every splitting branch of reality, sharing not only experiences but also "viktorious" knowledge guaranteeing success in other worlds.

Further, Foster's extrapolation of telepathy fuses and advances the Aztec concepts of *ixtli* and *teyloia*. According to Aztec culture, *ixtli* is the locus of perception situated in the head representing "complete consciousness" found in communication with the outside world; such a concept is similar to cognition in the Western sense.[52] *Teyloia* is the "force that exists beyond death," which is situated in the heart. Aztecs rulers often passed on their *teyloia* to guide future political leaders.[53] This idea is akin to Myer's subliminal self insofar it is an immortal essence persisting beyond the passing of the material body. It would seem, then, that the Aztec sciences have raised the power of *ixtli* across all potential realities such that one's mind is fully aware of itself across multiple unfoldings, loopings, and cycles. At the same time, it is enjoined with the vitality of *teyloia*, so consciousness always survives beyond death. The end result is a telepathic mind transcending any single body or reality, timeless and quasi-omniscient thanks to the interconnection between an infinite number of conversing selves. The novel provides two illuminating examples of this omnipotent paranormal mind.

The first occurs when Zenzontli prepares his Jaguar troops for a military incursion against the Nazis in an alternative battle of Stalingrad within an alternative World War II. The context for this battle is unusual because Zenzontli is a thoroughly twenty-first-century figure with knowledge of voodoo ekonomix, the 1980s war on drugs, and the music of Juan Lennon (the Mesoamerican John Lennon). Yet his military assignment is circa 1942 and involves leading his forces against Nazi Germany on the banks of the Volga River. While the unorthodox chronotope of this situation can be explained by the multitemporal spinning of the Aztec calendar,[54] he also demonstrates telepathic knowledge from another reality.

When Zenzontli stands before his troops at the outset of the Stalingrad siege, he attempts to rouse their spirits with a speech. What follows is a telepathic communiqué from beyond his known universe:

> How do we take out the Nazis? How do we win? How do we prevail in our attempt to try hard? Listen hard and listen well. . . . *SMOKING CRACK, ROCK OR FREE-BASING? Do you want to stop? Acupuncture & yoga relation techniques. Call now! 1-800-810-5551. SEXY YOUNG GIRLS EXPLORE THEIR SEXUALITY IN THEIR OWN HOME VIDEOS. Only $19.95 + $3.95 S&H. Media Vision Films 18375 Ventura Blvd., #173 Tarzana CA 91356 (818) 420-9843. CASH FOR YOUR CAR WHILE YOU DRIVE IT! Lease not a loan (888) 678-6866.* LARGER BREASTS! 100% Natural Safe & Affordable Alternative to Surgery! Fast & Guaranteed Results! Toll Free 1-877-6-BREAST. PET BEHAVIOR PROBLEMS? Solved! Call for psychic counseling.[55]

This motivational speech goes on for another full page, but we can immediately tell it does not derive from Zenzontli's Aztec-dominated reality but rather the western capitalist reality of Zenzon the immigrant. A Marxist critique of this passage might focus on capitalism's obsession with sex and beauty (all for the low, low cost of $19.95!), but my concern here is paranormal phenomenology. Zenzontli's oration resembles a stream-of-consciousness reading pulled from the back pages of a smut magazine or advertisements on a local cable station at two o'clock in the morning. In either case, this message is not of Zenzontli's world but rather noise from Zenzon's capitalist consciousness, whose own experiences slowly suffuse the Keeper of the House of Darkness.

This scene showcases the experience of Aztec hematic technology. People like Zenzontli are aware that other realities exist and can access information (like pornographic ads) from their alter egos' realities. Telepathy is not an anomalous phenomenon in this world but a rational extension of the scientific system. One way of viewing Aztec space-time is like a musical round with ever-changing harmonies: "At one time the sopranos and tenors might be singing their particular lines; at another, the tenors, altos, and bassos, and yet another, sopranos, contraltos, and tenors; and so on . . . all these ditties merge and diverge in the most complex and confusing ways, each according to the same steady, never failing metronomic beat of the days."[56] *Atomik Aztex* conjures this multivocality as a form of telepathic phenomenology, endlessly conjoining new Zenzons into the narrative structure such that we, as readers, are left unsure which Zenzon is the real Zenzon—and this is the point. All the Zenzons are equally real, occupying real worlds and relentlessly communicating between themselves thanks to advances in the chronological sciences.

A second example of Aztec telepathy highlights how memory and cognition transcend death in the multiverse. Late in *Atomik Aztex,* the perspective shifts to a Zenzontli who is battling atop the great pyramid of Tenochtitlán as part of an elaborate ritualistic sacrifice, a la the New Fire Ceremony. This Zenzontli is also a military officer in an Aztec-dominated reality, though I read him as a different figure from the other military officer who perished in Stalingrad. After killing several warriors, he is finally defeated and brought to the Tizok sacrifice stone: "Then they carried me to the chakmool, laid me out on it, opened me up and cut out my heart."[57] In another text, we might expect the narrative to end with the speaker's death. Incredibly, Zenzontli keeps talking: "*Except that never happened.* Let me make that clear, I would never let that happen. That might have happened on some alternative reality when I wuzn't looking, some fucking Other World when they didn't let me get my two cents in. But it didn't happen this time. *Cuz I didn't let it happen.*"[58] Zenzon speaks truth to power, because the narrative does in fact go on. The next chapter proceeds without missing a beat, and we learn how Zenzontli is apprenticed to a gangster, Xalatoktli, and eventually crosses into the United States.

What this unique passage suggests is that Zenzon possesses a telepathic awareness of his other selves—and his other existential possibilities—across the omniverse. He recognizes there exist other realities, other calendars of time, beyond the one in which he just died. He also possesses a mental connection to those realities so that he can cognitively jump into more promising worlds, where he will no doubt emerge victorious. This is why his ritualistic murder "didn't happen"—because the power of his *ixtli* exceeds death. His consciousness expands to all cycles of the multiverse; defeat in one cycle is only a temporary setback because his mind, and life-force, will always continue elsewhere. If we follow Read's interpretation of Aztec chronology, Zenzontli can see the spinning lines of space-time, fully aware of the present—as well as other fail-safe universes, should his current thread run out. This is why he can confidently make the following declaration: "We Aztek warriors can face our own deaths with komplete kool, knowing full well that on some place of the spirit we go on living forever and ever, knowing that whatever torture, torment and travail the present finite moment brings us we exist simultaneously in all the happiest moments of our lives and these go on shining forever like the stars."[59]

If we follow Michel Foucault's claim that knowledge is power, then Aztec telepathic knowledge is the ultimate form of power. *Atomik Aztex* imagines a hyperbolic advancement of hematic technologies of control. To know there is always another reality to experience, another space-time to apply your wisdom, another chance to get it right, another world in which victory is assured—this is a form of telepathic omniscience and immortality. At the novel's end, we finally understand that what we have read over the past two hundred pages is just a small slice of the paranormal mind at play. "Somebody played Russian Roulette with reality," Zenzon tells us. "Now it's time to start again."[60]

The paranormal mind that Foster develops in this novel is significant for several reasons. It posits a new type of telepathy because the minds involved belong to different realms of the multiverse, different *tonalpohualli* and *xiuhpohualli* calendars floating in space and time. The visions Zenzontli sees of another world in which the Aztec empire collapsed, in which he is a minimum-wage employee, in which he dies on a sacrificial stone, in which he is an unstoppable soldier, all suggest that he possesses something akin to panoptic vision over space-time. It also gives his consciousness a measure of eternity. We watch Zenzontli die over and over, yet his mind keeps trucking forward, adapting new personas, experiencing new lives, and ultimately achieving victory. The Aztec paranormal mind persists in perpetuity because it moves among and between the *ixtlis* of the multiverse. Our definition of mind expands exponentially under this system. It is no longer a defined object moving linearly through space and time but instead a parasemiotic entity traversing the various branches of the ever-expanding omniverse, linked with other minds navigating other realities. Consequently, Zenzontli's is the most expansive mind we have thus encountered; it is immortal, omnidirectional, unconquerable, and telepathic across all realities that currently and will ever exist.

Saldívar and others have written extensively about the novel's social justice aims, but I have paid special attention to its epistemological agenda because Foster so wonderfully inhabits its ethnoscientific world. Aztec science is not a dead episteme in this story world but a dynamic body of knowledge that has been cultivated by its priests and extrapolated by its greatest warriors; it is as compelling to contemporary BIPOC authors as quantum physics has been for modernist and science fiction writers in the West. I can think of no higher form of epistemological justice than twenty-first-century writers taking the Aztec Sciences of the Human Heart seriously and reveling in its endless possibilities.

Toward the end of the novel, Zenzontli relates the epic journey that made him the Keeper of the House of Darkness, but it is also the tale that divulges his illegal crossing into the United States. In the merging of these realities, he ends up paddling a boat, knowing that his mind and spirit would go on, that it was indomitable: "I was—at best—starting at the beginning of a long, difficult, trouble filled journey from which I might never return in this world, and what enabled me to paddle with feeling, to bail with immediate passion, paddle & bail & move on up the coast—slowly, inexorably the boat moved north, swell after swell, wave on wave—was the happiness that solitude amounted to in me. They'd tried to trap me & could not; they'd tried to break me & could not; they'd tried to kill me & I was alive."[61] This passage speaks to Zenzontli's personal sojourn, but more importantly, it is a microcosm of indomitable Aztec paranormal consciousness. Cycles of time, of life, of history, of past and future keep spinning and expanding, and with it, the myriad minds of Zenzontli keep marching on. Such a mind intimates the deathless subliminal self that Frederic Myers first envisioned in the pages of *Human Personality*. But it also speaks to the power of Aztec science, which transcends the ideological limitations of the West and the aesthetic urgency of minority poesis.

A Tale for the Time Being and the Science of Zen

One of the most important scholars of Mesoamerican calendrical systems was Yale anthropologist Floyd Lounsbury. Originally trained as an expert of Iroquois languages, Lounsbury eventually mastered Mayan linguistics and helped develop the methodology to translate ancient Mayan hieroglyphics during the 1950s.[62] From the 1960s to the 1990s, he broadened his research into Mayan calendars, number systems, and astronomy. This long-running fascination in indigenous chronologies may very well have played a role in the interests of his daughter, novelist Ruth Ozeki. We began this chapter with her 2013 novel *A Tale for the Time Being,* a curious blend of speculative fiction, metafiction, and transpacific fiction that has received attention for its ecological, feminist, capitalist, and globalist critique.[63] My concern is epistemological—specifically, how Zen Buddhism operates as an alternative chronology for reformulating the bounds of the supernatural. Similar to *Atomik Aztex,* the novel stretches consciousness beyond what is conceivable in the Western sciences, and it consequently reconfigures broader cultural interpretations of the paranormal.

The two interweaving story lines at the center of *Time Being* hinge on the Proust novel/journal that washes up on the British Columbia coast. The first involves a Japanese American teenager, Nao Yasutani, whose family has returned to Tokyo after her father lost his computer programming job in Silicon Valley. Nao's diary, placed within a copy of Proust's *In Search of Lost Time,* initially seems like a *shishosetsu* confessional of a teenager coping with bullying, but the plot veers into the metaphysical as Nao encounters ancestral spirits and even turns into a ghost herself.[64] The second narrative involves Japanese American writer Ruth, a metafictional stand-in for Ozeki, who discovers Nao's diary on the beach and is literally pulled into her story. Over the course of the novel, Ruth translates the journal entries, begins seeing events from Nao's life, and eventually gains the ability to alter her past.

Much of the novel emphasizes the interrelationship between two knowledge systems structuring the narrative, Zen Buddhism and quantum mechanics. The former is mostly the province of Nao's story line and the latter the province of Ruth's, which seemingly reinscribes the traditional East–West dichotomy of spirituality and science: the backward Orient understands reality through premodern practices, whereas the futuristic Occident interprets reality through technoscientific rationale. Ozeki assiduously highlights their communal features, though. Zen, one of the ancient religions of Japan, contains metaphysical teachings about the unity of space and time that approximates (and some might say anticipates) the quirks of nature modern physicists have uncovered at the smallest scales (subatomic) and highest velocities (light speed). For this reason, Ozeki conspicuously foregrounds Western science throughout the book. For example, Ruth has a cat named Schrödinger, who is lost during a lightning storm and assumed dead for the duration of the novel, only to turn up very much alive. This is obviously a reference to Schrödinger's cat, the paradoxical thought experiment on the Copenhagen interpretation in which subatomic particles simultaneously exist in two mutually exclusive states (i.e., the cat is both dead and alive).[65] Ozeki also offers a brief primer on modern physics through the appendixes at the end of the novel, where she explains quantum mechanics (appendix B), Schrödinger's cat (appendix E), and the theories of Hugh Everett (appendix F), the author of the multiverse. For the Western audience of *Time Being,* quantum physics is a (somewhat) familiar scientific system that helps explain the idiosyncrasies of the novel's plot and narrative structure.

However, to accept Western science as the central logic of Ozeki's novel is to misunderstand the novel's epistemological project. Quantum physics may be better known to the text's American readers, but Zen is the true key for comprehending *Time Being* and the paranormal mind it envisions. Indeed, if we adopt mosaic epistemology and treat Zen as an alternative science paralleling modern physics, we gain a richer appreciation of space-time and the impossible phenomena transpiring in the novel. As a unified and systematic body of knowledge that recognizes deep connections between space, time, mind, and physical being, Zen is the holistic scientific system undergirding *Time Being*. One task, then, is to understand how it produces new forms of consciousness.

Dogen was the founder of the Soto school of Zen whose writings on "being-time" (or "time-being") provide the intellectual backdrop for Ozeki's novel. Born in 1200 in Kyoto, Japan, Dogen chose the monastic life early on. After spending his early years studying at Buddhist temples in Mount Hiei and Kenninji, he traveled to China where he met the great Zen teacher Ju Ching. It was from Master Ju Ching that Dogen recognized the true power of zazen, "a traditional form of Buddhist meditation which emphasizes upright lotus posture, steady breathing, and mental freedom from attachment, desires, concepts, and judgment."[66] In 1227, Dogen returned to Japan, where he spent the rest of his life dedicated to expanding Zen by establishing new temples and writing influential texts like *Shobogenzo*.

Zen is often misunderstood in the West as an abstract philosophical doctrine or a spiritual exercise designed to achieve therapeutic calm (e.g., a practice for achieving a "Zen" state of mind).[67] Hee-Jin Kim argues that Zen's mystical aura is overblown when Dogen's system is in fact deeply analytical.[68] We should therefore interpret Zen as its practitioners do: as a comprehensive schema explaining the natural world. At its core, Zen is a systematic means of "universal liberation" that involves training the mind to recognize its true place in nature's schema.[69] It must be first understood in contradistinction from the positivistic tendencies of objectification and atomization. Like the systems thinking we saw in chapter 2, it rejects positivism's separation between parts and wholes, individuals and nature, mind and body, and so on. Describing its hybridized approach to existence, Taigen Dan Leighton writes, "The world is a site of radical, mutual interconnection of the subjective and objective, in which each event is the product of the interdependent co-arising of all things."[70] The basis of this interconnection is the human mind. Unlike Descartes, who separated mind from matter, Dogen saw the individual mind as completely—and radically—continuous with the material world in all its dimensions. In Zen, "all phenomena originate in the mind, and when the mind is fully known, all phenomena are fully known."[71] This is not a Japanese variation of solipsism; Zen does not claim the world exists within the individual. It actually makes the opposite claim. For Dogen, mind is universal; it is connected, vibrant, and coterminous with reality itself. "Mind comes into and out of being with the psychophysical activities of the mind and the creative activities of the physical universe," writes Kim.[72] "Mind is at once knowledge and reality, at once the knowing subject and the knowing object, yet transcends them both at the same time . . . ontology, epistemology, and soteriology are inseparably united."[73] Reality is thus not a solipsistic product of the mind but *enjoined* with mind in all aspects. This is a fundamental distinction; consciousness and reality, phenomena and noumena, all are in fact one. Zen dissipates all binaries into unity, and through mind, all things become accessible. When Dogen claims that "all existence is Buddha-nature," he dismisses the anthropocentric ordering of things for a universal ontology where you, me, space, and time exist in intimate connection with everything else.[74] Buddha-nature is all beings in the world and being itself, and our realization of this truth is a step toward enlightenment.

Zen philosophy is an enormous departure from classic Western phenomenology and ontology. In opposition to the "naturalistic heresy" of the West, Zen rejects a reductionist view of mind as nothing but a result of materialistic, organismic, and phenomenalistic conditions.[75] It also rejects the premise that we are all independent entities buffeted by outside forces over space and time. In Zen, we are all manifestations of an essential thusness (Buddha-nature) of which our own minds are intrinsic to its becoming. The idea of a vastly interconnected mind shares qualities with qi-cosmology and shamanism just discussed in chapter 4. It also aligns with the subliminal self, which conjoins with others' minds beyond the confines of a material body. In this respect, the paranormal mind comes full circle from the Eastern-inflected occultism of the fin de siècle back to Eastern philosophy itself.

The ontological unity of Buddha-nature has enormous implications for the fluid chronology of *Time Being* as well as for the minds traveling through it. Because all aspects of reality—including time—are ultimately connected to each other, it also means the present is intrinsically connected to past and future in ways conflicting with classical views of temporality. "Western time is seen as linear, a fleeting, current moment linked in an endless series to past moments which have gone away and future moments which are yet to come," writes critic Steven Heine.[76] Only the present is real in Western chronology, whereas the past and future are phenomenologically and ontologically unreal. Classical views of time disconnect humanity from temporality because we are subjects of a perpetually vanishing time. "Thus static, linear conceptions of time, so deeply ingrained as to be considered self-evident and beyond question or doubt, is manifested in man's preoccupation with the chronology and calculation of time in terms of hourly and seasonal sequences," writes Heine. "It also haunts supposedly more sophisticated phenomenological claims of permanence or an essential time conceived of as an abstract changeless realm outside of the limitations of contingent time."[77]

For Dogen, linear time is an illusion disconnecting us from *uji no dōri*, the truth of being time. Being-time, or *uji*, can be understood as an alternative chronology that approaches time as a "fluid, flexible, and integrated process inseparable from human activity."[78] In Zen, what is often called "nirvana" or "enlightenment" is the genuine experience of temporal reality—of being-time—existing beyond Western linear time. In *uji*, "all beings and all times are originally, perpetually, and without exception linked together each and every moment."[79] Zen "deepens" time by asserting that being-time encompasses all temporal tenses—"an ultimately equalizable and simultaneous realization."[80] According to Dogen, Zen articulates ten aspects of time: within the past, present, and future exist another subdivision of past, present, and future (the past of the past, the present of the past, the future of the past, the past of the present, etc.); these nine aspects are contained within the simultaneity of the present.[81] As this systematic breakdown suggests, Zen proffers a theory of temporal knowledge that explains an alternative experience, passage, and nature of time. Like the Mesoamerican calendar in *Atomik Aztex* and qi in *The Hundred Secret Senses*, it provides a conceptual system explaining how local cultures define nature.

Interestingly, Zen time can move in all directions—forward, backward, even sideways.[82] Dogen famously writes, "In *being time*, there is the quality of passage. That is, it passes from today to tomorrow, it passes from today to yesterday, it passes from yesterday to today, it passes from today to today, it passes from tomorrow to tomorrow."[83] Dogen rebuffs the unidirectional flow of Western time from past toward future with an Eastern chronology that is multidirectional. And because past and future are already localized within the present, they can be accessed if the Zen practitioner is prepared to perceive it. According to Leighton, what distinguishes a bodhisattva from a normal man is the ability to "enter into and inhabit time, in all its temporal aspects."[84] Through proper training and zazen, certain individuals can maintain agency, and eventually exert control, over time.

This control also holds true for physical space and matter because time and space are inseparable. In "Being-Time," Dogen writes, "The mountains are time, the oceans are time too. If they were not time, the mountains and oceans could not be."[85] By claiming that spatial entities, both biological and nonbiological, are fundamental manifestations of time, Dogen links space and time as related components of being. "Space and time are so inseparably interpenetrated, in Dogen's view," explains Leighton, "that to see one without seeing the other destroys the fundamental understanding of his thought."[86]

This returns us to mind, which enfolds the ethnoscientific principles of space, time, and being in ways paranormal to modern science but fully rational within *Time Being*. In Dogen's system, mind is "not only the totality of psycho-physical realities but also 'something' more . . . such as thusness *(tathatā; shinnyo)*, Dharma-nature *(dharmatā, hosshō)*, Buddha-nature (*buddhatā* or *buddhatva; busshō*), absolute emptiness, and so forth."[87] In Zen, mind *is* mountains, oceans, humans, time, space, eternity, and being because it is intrinsic to the simultaneity and unity of *uji*. And because the Zen mind is coterminous with all other things and temporalities, it empowers consciousness to perform the incredible feats that we witness in Ozeki's novel.

In *Time Being*, several paranormal abilities are reconceived as natural extensions of heightened Zen consciousness: ghost communication, telepathy, telekinesis, and astral travel. My intent here is not merely to find supernatural instances in Eastern philosophy but to illustrate how ethnic fiction normalizes paranormality via non-Western science. In doing so, indigenous science joins other Western sciences as a major hub within parascience.

Nao is a troubled figure for much of Ozeki's novel. Her father has become a *hikikomori*—a social recluse—who seldom leaves the house and attempts suicide twice over the course of the story.[88] This destructive home environment leaves Nao suicidal herself. At school, she is the victim of severe psychological and physical bullying, including a sexual assault that is broadcast on a *hentai* website.[89] With her home and school life spiraling out of control, Nao is sent to visit her great-grandmother, Jiko, a Zen nun living in an ancient temple at Sendai. Jiko quickly inculcates Nao into the practice of zazen meditation. By methodically breathing in and out and cleansing her mind of all extraneous thought, Nao eventually learns to liberate her mind: "It doesn't seem like such a great thing, but Jiko is sure that if you do it every day, your mind would wake up, and you will develop your SUPAPAWA!"[90] What is this superpower? Jiko reveals that to perform zazen is "to enter time completely."[91] From Dogen, we know meditation will reveal time as it truly is: no longer a linear sequence of events but a unity tied in all directions to space, being, and mind. When a Zen practitioner recognizes this higher truth, as Nao does, she is no longer restricted to the limitations of Western chronology or positivistic science. The modus operandi of *Time Being* is to literalize the extraordinary phenomenological abilities Dogen associates with zazen. It is Buddhist parasemiosis cranked up to eleven.

Nao's entry into Zen grants her the ability to communicate with the dead. Late one night at Jiko's temple, Nao encounters the ghost of her great uncle, Haruki #1, who perished during World War II. "He looked like a younger version of my dad [Nao's father is Haruki #2], only a couple of years older than me, but he sounded different, and the clothes were all wrong too. And that's when I figured it out too: if this ghost who had answered to my father's name wasn't my father, then he must be my father's uncle, the suicide bomber, Yasutani Haruki #1."[92] They converse for a while, during which time Haruki #1 admits, "I wasn't aware that I had a nephew, never mind a great niece. How quickly time flies."[93] This is the first of several Nao–Haruki interactions, and it signals the manner in which Zen has expanded Nao's consciousness. In Western occultism, mediums can communicate with the dead because they are uniquely sensitive. According to Dogen, though, our long-deceased ancestors exist in perpetuity because the past is coterminous with present and future. This is not supernatural but rather the natural order of things in being-time. As such, ghosts are always walking among us; the minds of most people are simply not prepared to see them. Just as her immersion in qi cosmology allows Kwan to see spirits in *The Hundred Secret Senses,* so does Zen practice opens Nao's eyes to this hidden truth.

Jiko later acknowledges that Nao's spectral visitations are not tricks of the mind but actual encounters. "I'm glad you met him while you were here," Jiko responds when Nao reveals her late-night discussions with Haruki #1.[94] Jiko's lack of surprise is telling for a few reasons. First, it suggests that Nao's ghost vision is genuine: spirits really do coexist with humans, and we really can communicate with them. Second, it confirms that Zen practice produces the paranormal mind. Nao had always heard about ghosts (red-faced *tengus,* hideous *nuppeppos,* etc.), but it was not until she began practicing zazen, and experiencing being-time, that she could see them.[95] To be clear, I do not mean to suggest all students of Zen can or will see ghosts. As in *Atomik Aztex,* some people are predisposed to the paranormal. Zen functions in *Time Being* as a mode of epistemological possibility beyond Western logic, and its appearance as a phenomenological superpower is an emphatic reclamation of that possibility. The existence of Haruki #1 also reinscribes the ghostliness of ghosts, as discussed at length in chapter 4. Literary criticism prefers interpreting ethnic ghosts as figures of historical trauma. Haruki #1 fits this template because he is a victim of Japanese militarist ideology: he is a World War II kamikaze pilot whose personal antiwar beliefs, and life, are sacrificed in the name of the nation's honor. But Jiko and Nao subvert Western literary symbology by substantiating Haruki #1's material existence. He is a ghost, and in the Zen world view, ghosts exist.

Zen also leads to telepathic "learning": Ruth enters being-time by reading the novel/journal. As she translates Nao's journal from Japanese into English over the course of the novel, Ruth is not only drawn into her narrative but *into her mind*. By this, I mean Ruth gains access to the knowledge, skills, and historical events that define the younger woman's life. Eventually the two become "entangled" in the Western physicalist sense, which is when "two particles can coordinate their properties across time and space and behave like a single system."[96] In physics, quantum entanglement describes an unusual phenomenon in which two or more particles maintain instantaneous communication despite separation in space and time. A classic example is when two photons with a total spin of zero are generated and separated. When one photon is measured to have clockwise spin, the other will have anticlockwise spin, which means the photons are somehow communicating at faster-than-light speeds while separated by increasing geographic space.[97] According to Albert Einstein, entanglement represents "spooky action at a distance."[98] In the novel, Ruth and Nao's transpacific correspondence, in which the former somehow knows the mind of the latter via the journal, is initially presented to the reader as a macroscopic form of quantum entanglement.

Zen confers an alternative way of interpreting their spooky telepathic connection. As Nao explains at the start of the novel, "A time being is someone who lives in time, and that means you, and me, and everyone one of us who is, or was, or ever will be."[99] *Uji* connects all things, including Ruth and Nao, and the novel affirms this link by providing the former with the Zen knowledge of the latter. Ruth, we must remember, never undergoes Zen training. All she does, really, is read Nao's journal entries *about* Zen training. It is Nao who spent months learning zazen meditation from Jiko. Yet over the course of the narrative, Nao's awareness of space-time somehow leaks into Ruth's brain such that the Canadian writer can bend the space-time continuum in ways that would awe Dogen himself. Leaky awareness is a core characteristic of the subliminal self, but here it seeps across vast stretches of space-time.

Soon after finding the journal on the beach, Ruth experiences a lucid dream in which she approaches a building in the Japanese countryside late at night. "The only illumination inside the temple came from a single room adjoining the garden, where the old nun knelt on the floor in front of a low table, leaning in toward a glowing computer screen, which seems to float in the darkness, casting its silver square of light onto the ancient planes of her face."[100] When we finally meet Jiko later in the novel, we learn her temple in Sendai really is falling apart, that her room overlooks the garden, and that she has been emailing Nao on a computer in the evenings.[101] Ruth has experienced a retrocognitive vision. This dream is not supernatural within the text's Zen logic because the past is not an inaccessible feature of temporality but something always already present. If we follow Dogen's teachings about the ten aspects of time, then Ruth has experienced the past of the present—the reality of the material past still existing in being-time.

Telepathic learning and clairvoyant sight are only minor aspects of the novel's paranormal purview. Ruth's immersion into *uji* becomes even more impressive when she begins traveling through time and changing the past. The climax of the novel occurs when a depressed Nao, knowing that her father is about to commit suicide, begins to fade out of existence at a Tokyo train station ("Nobody sees me. Maybe I'm invisible. I guess this is it. This is what now feels like").[102] Recognizing this existential crisis—despite the fact that she is reading about it several years after the fact—Ruth decides she cannot allow her entangled counterpart to die from despair. Consequently, she descends into another lucid dream with the expressed purpose of saving Nao. Incredibly, her gambit works. In the uncanny dream sequence that follows, Ruth locates Nao's great-grandmother: "Jiko holds out her glasses and Ruth takes them and put them on because she knows that she must. The murky lens smear the world, as fragments of the old nun's past flood through her: spectral images, smells and sounds, the gasp of a woman hanged for treason as the noose snaps her neck; the cry of a young girl in mourning; the taste of a son's blood and broken teeth; the stench of a city drowned in flames; a mushroom cloud; a parade of puppets in the rain."[103] This is a pivotal sequence in the novel because we witness Ruth's absorption into being-time. It is a cessation from Western linear time, a withdrawal from Western science altogether—and a complete submersion into Zen chronology. The taking up of Jiko's glasses symbolizes the exchange of a Western perspective for an Eastern one. Like Olivia Laguni, who finally accepts yin eyes as equivalent to Western science in *The Hundred Secret Senses,* and like Beccah Bradley, who finally accepts shamanism and Western rationalism as coequal world views in *Comfort Woman* (chapter 4), Ruth enacts a paradigm shift. The glasses are a powerful metaphor because they literally change the way she perceives the world. With that new paradigm and pluriversal mind-set arise formidable powers for navigating being-time. By gaining access to Jiko's vision, memories, history, pain, and consciousness, Ruth also harnesses the "supapawa" implicit in Dogen's teachings. What we witness here is the synthesis of individual selfhood into the ultimate unity of space and time. As Ruth describes it, "There is no up. No down. No in. No out. No forward or backward. Just this cold, crushing wave, this unnameable, continuous merging and dissolving."[104] In Zen, mind is coterminous with reality, and Ruth experiences that dissolution here. Moreover, the directionlessness of space-time is implied by the absence of up and down, of forward and backward progression; space and time become entirely fluid in the enlightened mindscape of Zen thusness. Through the atomization of individual being into *uji*, Ruth gains the extraordinary ability to migrate from today into yesterday.

As the dream continues, Ruth finds herself in the past and across an ocean, in a Tokyo park where she convinces Haruki #2 not to kill himself. I use the term *dream* loosely because it is not oneiric at all but rather a material event in the material world. When Ruth sits beside Nao's father at the park, he straightens and bows his head in a tentative greeting. "Are you the one I'm waiting for?" he asks.[105] Ruth knows in her timeline that Nao has died (or rather, faded away) because her journal was filled with blank pages after making her final declaration: "this is what now feels like."[106] But Ruth convinces Haruki #2 to live for his daughter's sake. This timely intervention sets off a chain reaction in which he immediately leaves the park, finds Nao at the train station, and prevents her suicide/disappearance. "This is what Nao feels like" never happens—which is to say that time is rewritten. Death is averted. History changes. Interestingly, Nao's survival produces supernatural effects in Ruth's timeline. After Ruth's climactic travel into yesteryear, the blank pages of the journal magically repopulate with Nao's words, her life.

In the Western occult tradition, astral travel is the idea that some aspect of the consciousness or soul can travel to distant parts of the universe. It was based on the same logic of Aleister Crowley's famous precept: "believe thyself to be in a place, and thou art there."[107] Such a marvelous capability, usually restricted to the esoteric elite, involved producing "a material replica of an embodied self, which left the temporal body of the occultist and journeyed at length in astral realms."[108] While Ruth's travel across geographic space certainly fits a criterion of astral travel, her hypertemporal ability to enter the distant past, change it, and consequently alter the future transcends the abilities of even the most powerful Magick practitioner. Zen provides a different rationale for this uncommon skill. Instead of conjuring a second consciousness and moving it to other physical realms, Zen views all spaces and times as accessible through the mind. Once a Zen disciple recognizes the fundamental unity of *uji*—the merging and dissolving of all things—the mind can then simply travel to other spaces and timelines. In the same way that you and I can walk into another room in everyday life, a Zen master can conceivably walk through time and across the world. The wearing of Jiko's glasses—the adoption of Zen as dominant paradigm—makes this impossibility possible. As in *Atomik Aztex*, non-Western chronology delivers a different and more powerful version of paranormality than previously seen in Western scientific and narrative discourse.

Zen consciousness also explains the strangest paradox in the novel's dynamic space-time: the reappearance of Haruki #1's wartime letters. During World War II, Haruki #1 kept a stash of personal correspondence detailing his abuse at the hands of the Japanese military. His letters were lost when Haruki intentionally crashed his plane in the ocean. However, during her immersion into *uji*, Ruth brings the letters from her present timeline, which had washed ashore with the diary, and brings them backward through space-time into Jiko's temple, knowing Nao and her father would eventually discover them. This act brings resolution to the long-standing mystery of Haruki #1's death and peace of mind to the Yasutani family. Telekinesis—the movement of objects with the mind—is one of psychical research's most debated powers, but in *Time Being*, it is just another component of being-time.[109] If mind is connected to other spaces and temporalities, then mentally moving an object—like a set of letters—from 2013 British Columbia to Sendai circa 2010 is not unlike physically moving them from the mailbox to the kitchen counter. The limitations of Western cause and effect and space-time motion becomes inconsequential from the standpoint of Zen consciousness.

This returns us, at last, to Nao's diary. Ruth, unlike Jiko or Nao, is not a practicing Buddhist. Nevertheless, the diary provides Ruth with a temporary form of enlightenment and space-time passage. In chapter 1, I described the psitron as a parascientific object because it blurs scientific, literary, and mythic elements to rationalize the paranormal for new audiences. In this regard, Nao's journal is a parascientific object as well; it is an amalgamation of Western and Eastern discourses that highlights the esoteric powers of the human mind. *In Search of Lost Time* is a meditation on phenomenology, particularly in the ways that memory is a form of time travel. Through memory, we can sojourn from the present into the past and onward to the future and back again. In his own way, Proust suggests there is *something* beyond linear time, and this something is eminently accessible through consciousness. Nao's own words replace Proust's in the novel/diary, but the central message remains the same. Drawing from Zen chronology, Nao highlights that beyond the illusion of linear time is *uji*. And that knowledge, that alternative rationalization, is what ultimately produces Ruth's paranormal powers. The journal is parascientific because it combines several unorthodox discourses to legitimize the logic of paranormal consciousness for its reader. A parascientific object is a bridge between the normal and the paranormal, the scientific and the superscientific, the possible and the impossible. Nao's diary entangles Nao's and Ruth's minds, and in the process yokes Eastern and Western interpretations of time, space, and cognition.

If we allow ourselves to take yet another step backward, we might even say that *A Tale for the Time Being* is a second-order parascientific object acting on us, the contemporary Western reader. Ozeki's novel is itself a hybrid object combining Nao's Eastern metaphysics with Ruth's quantum physics. Both epistemes acknowledge nature as inherently strange, and the former in particular authorizes telepathy, clairvoyance, astral travel, and telekinesis as hidden properties of the mind. The novel suggests that if a twenty-first-century audience can adopt a pluriversal approach and interpret the world from beyond classical science, then paranormal communication across the multiverse is theoretically possible.

Speculative Ethnic Fiction as the New Literary Phenomenology

When we analyze Aztec and Zen theological systems as chronological sciences, paranormal phenomena arise as their logical by-products. Just as Western science fiction acts as a laboratory for theoretically developing occult sciences, so too does speculative ethnic fiction become an instrument for developing the metaphysical claims of indigenous sciences in increasingly supernatural dimensions. My intention has not been to impugn these epistemes as pseudoscientific. The issue with Western intellectual history is that the paranormal has been pseudoscientific for so long that anything associated with it runs the risk of defamation. It would be safer to keep the religious and scientific realms separate, as academic criticism tends to do. Such bifurcation is disingenuous, though. As Bruno Latour and others have pointed out, the social, the religious, the political, and the scientific all intersect, and the paranormal mind has historically emerged from that admixture, in both Western and Eastern contexts. Throughout this book, many groups have argued human consciousness is capable of so much more than what we experience on a day-to-day basis: telepathy, clairvoyance, precognition, time travel, multiverse telekinesis. These ideas are continually renewed in parascience, and we would be blind to its development if we limit ourselves to Western modes of thought. Recognizing our multicultural and multiscientific histories is critical for understanding the intellectual progression of the paranormal mind in the twenty-first century.

The turn to ethnoscience over this chapter and the last has been my attempt to decolonize the paranormal mind from the Western scientific tradition. Mesoamerican and Zen chronologies exist apart from the dominant scientific epistemes of today, but they can play a role in explaining how different cultures have historically navigated the world and interpreted the human mind. Writers of color are increasingly turning to native sciences, knowledge systems, and practices to capture that totality of human experience. In doing so, they liberate the consciousness from the universality of Western science and illustrate how non-Western peoples have theorized mental functioning and capabilities. Mignolo claims the Western uni-verse closes off vital ways of thinking and doing. The pluriverse, in contrast, renounces the idea of world as a unified totality; it is a "decolonized political vision of a world in which many worlds would coexist."[110] *Atomik Aztex* and *Time Being* are paragons of this many-worlds approach. Not only are they acts of minority poesis (Judy) and social justice narratives (Saldívar), but they are agents of epistemological justice as well. They remind us that Eurocentric science, hermeneutics, and political economy are not the only ways of perceiving the world. Other cultures have a vision of reality, along with marvelous extrapolations of that reality, and we would do well to acknowledge that epistemic mosaic.

While the pluriverse is a useful approach for decolonizing the structures of Western epistemology, it is also valuable for reconsidering the paranormal mind in *Atomik Aztex* and *Time Being*. Both novels reconstruct the mechanics of telepathy through ethnoscience. For Zenzon, Mesoamerican chronology and the Aztek Sciences of the Human Heart become instruments for constructing multiple realities while maintaining psychic connections between them. For Ruth, Zen Buddhism authorizes a mental connection with all other things— teenage girls in Japan, years of zazen meditation, non-Western cognition— which produces telepathic learning and astral travel among other capabilities. In each text, indigenous frameworks of space, time, and mind unlock powers inaccessible to the average citizen of the West. Both works are deeply pluriversal. From the outset, Zenzon acknowledges that Aztec chronology echoes Hugh Everett's infinite worlds theory. Ruth juxtaposes quantum physics and Zen as coequal thought systems. Both novels decline Western universality in favor of a Western and non-Western mosaic. And there is a clear symbolism at play when recognition of pluriversal modes of knowledge bestows extraordinary powers to travel the pluriverse itself. When we open our eyes to the full breadth of multicultural sciences, we are, like Ruth, fully enlightened.

This same message rings true for the paranormal more broadly. Literary and historical criticism has often categorized telepathic, prophetic, and clairvoyant discourse as dead ends in Western science, unworthy of commentary. What is lost in this Eurocentric disregard is how non-Western cultures have understood the paranormal within their own scientific systems. Whether we speak of shamanism, qi cosmology, Mesoamerican calendars, or Zen, different cultures have developed their own sophisticated views of consciousness and its relationship to nature. As we have seen, several epistemes have explained how the so-called supernatural emerges from nature's core principles. Authors like Ozeki and Foster extrapolate ethnoscience in imaginative ways to not only retheorize the paranormal mind for contemporary audiences but also reframe the abnormal and normal. This is an important shift worth stressing. Other cultures have undoubtedly experienced divine visions, telepathic dreams, and spectral communication. Many of them have incorporated such phenomena into their scientific understanding of reality. Unlike the modern West, not every culture has marginalized the paranormal as an aberration to expunge; in contrast, they treat it as something to celebrate, to ponder, to scientifically analyze; they use it to gain agency, to connect with the past, to understand their future. The paranormal, it seems, can be a good thing. Fictions like *Atomik Aztex* and *Time Being* help us recognize that. For too long the paranormal has been cast out because it did not fit into existing knowledge paradigms. But if we expand our paradigms, either through speculative ethnic fictions or pluriversal approaches to science, we can estrange the paranormal mind from the conservative aspiration to fit it into a technoscientific box. In doing so, we can perhaps finally recognize it as an ever vital and vibrant cultural object.

6 THE STRUCTURE OF
PARASCIENTIFIC REVOLUTIONS

FEW ARTICLES IN THE HISTORY of science literature have caused as much trouble as Daryl Bem's 2011 "Feeling the Future: Experimental Evidence for Retroactive Influence on Cognition and Affect." Then again, few articles have ever attempted to stage the holy grail of parapsychological research: a fully reproducible ESP study that could be assessed and substantiated by any other laboratory in the world. Throughout the 2000s, Bem, a professor of social psychology at Cornell University, had become increasingly certain that psi phenomena did in fact exist. As a veteran academic, he also knew the old adage was true: extraordinary claims require extraordinary evidence. What he consequently revealed in the prestigious *Journal of Personality and Social Psychology* was nothing less than the largest and most rigorous set of psychical experiments ever run in the modern era: nine separate experiments conducted over a ten-year period involving over a thousand participants. The experiments were testing various aspects of premonition and precognition. For example, experiment 1 (Precognitive Detection of Erotic Stimuli) asked study participants to sit at a computer and guess which one of two sets of digital curtains on a monitor hid a pornographic image.[1] Only after the participants made their choice would the computer randomly assign the location of the image. Incredibly, 53 percent of participants correctly "foresaw" the location—a statistically significant result. Experiment 8 (Retroactive Facilitation of Recall I) analyzed the ability for memory to "work both ways" by testing whether rehearsing a set of words made them easier to recall—even if the rehearsal took place *after* the recall test was given.[2] Bem's research did not rely on obscure techniques or advanced mathematics; he only used established methods, basic statistics, and large sample sizes. Eight of the nine tests demonstrated conclusive evidence for ESP. At the end of the article, he even offered to share his data on

"Macintosh and Windows-based computers to encourage and facilitate replication of the experiments reported here."[3]

Bem's publication generated an immediate uproar. The problem was, it was "both methodologically sound and logically insane."[4] In a front-page *New York Times* article, psychology professor Ray Hyman proclaimed, "It's craziness, pure craziness. I can't believe a major journal is allowing this work in. I think it's just an embarrassment for the entire field."[5] For the Parapsychological Association, to whom Bem had presented early drafts of his research, the publication was a smashing success that vindicated their mission. For decades, they had sought the kind of smoking gun research that could provide definitive proof of the paranormal mind's existence. With a huge dataset and replicated results, Bem's experiment certainly looked like a pistol with smoke wafting from the muzzle. But for mainstream science, which had been winning the war against parapsychology for so long, Bem's research was not a gun so much as a grenade: it threatened to blow apart the very processes used to produced knowledge, facts, and truth.

As we reach the conclusion of this study, I bring up Bem's research because two aspects of this curious episode are worth highlighting for what they reveal about the nature of the paranormal mind and parascience more broadly. The first is that "Feeling the Future" ultimately did not convince establishment scientists that ESP is a real phenomenon. In fact, it triggered a backlash delegitimizing the commonplace techniques he used. In the years since its publication, Bem's study has emerged as a flash point for methodological problems prevalent in the social sciences, such as cherry-picking data or continually adding research subjects to a study until the desired effect is found. Statistician Andrew Gelman has identified Bem's ESP research as emblematic of the problem known as "researcher degrees of freedom"—the ability of scientists to look for *any* desired pattern in the cacophonous noise of data.[6] The suggestion is that if Bem found evidence of ESP in his data, then a researcher could practically prove anything. The methodological hand-wringing his work provoked would eventually snowball into the replication crisis spreading across the social, medical, and biological sciences. "If one had to choose a single moment that set off the 'replication crisis' in psychology—an event that nudged the discipline into its present and anarchic state where even textbook findings have been cast in doubt—this might be it," observes writer Daniel Engber.[7] It is telling, I think, that Engber's article in *Slate* is provocatively entitled "Daryl Bem Proved ESP Is Real: Which Means Science Is Broken." The implication is that psi is obviously impossible, so only a series of gargantuan and systemic oversights in the scientific process could have produced Bem's fantastical results. In his history of parapsychology, Brian Inglis argues that psi represented such a threat to establishment science that no amount of evidence could ever successfully prove its merits. As psychical researcher Whately Carington once lamented to J. B. Rhine, their critics would not be silenced "until and unless you have systematically closed and riveted and clinched the door of leakage until not the minutest crack remains."[8] Even then, "it was to become apparent that there were no watertight doors in parapsychological research."[9] The aftermath of Bem's research has proven Inglis correct: there is no study perfect enough and no amount of evidence massive enough to prove the existence of the paranormal mind to contemporary scientists. Sociologists H. M. Collins and T. J. Pinch identified several tactics that modern science historically deploys to impede psychical research, including refusing wholesale to believe evidence, citing experimental errors, and associating it with unscientific belief, and we observe several of them play out here.[10] We even see a brand-new one: renouncing the field's basic methodologies! Psychical research has been and will continue to remain outside the dominant paradigms of accepted science.

The second fascinating aspect is that even after all this time, we still see the same patterns of expulsion and recurrence. Despite its banishment from mainstream systems of thought, the paranormal mind continues to hector contemporary science and culture. Psychical phenomena have been attacked time and again across the long twentieth century, yet it always remains on the edges of the collective conscious. Like Rupert Sheldrake's 2013 TED talk with which I started this book, Bem's research is yet another incident reminding us pseudoscientific thinking continually shadows the safe spaces of scientific consensus. When I think of the paranormal, I am inevitably reminded of Peter Quint in *The Turn of the Screw.* In Henry James's psychological masterpiece, the young governess is haunted by the spectral presence of the deceased former valet at Bly. In his most chilling appearance, Quint appears on the other side of a window from the vast gray outside. *Appears* is not the right word, though, because it seems he has been there all along. Quint only stares at the governess for a few seconds, "but it was as if I had been looking at him for years and had known him always."[11] So it goes for science as it does for the poor governess. The ghost of the irredeemable past is always already leering at us from the outside, his evanescent visitations sufficient to shake the walls of the house of rationality. The persistence of the paranormal mind, whether morphic resonance, qi ghosts, or memory in two directions, suggests we still have much to learn about the circulations of anomalous science. It is not enough to say that psychical phenomena are relics of the misinformed past because the historical record shows us that the paranormal mind today is far removed from the subliminal self of yesteryear. Its durability and dynamism after more than a hundred years outside orthodox science speaks to both a structure and a process of knowledge production beyond scientific paradigms. Exploring that epistemological realm through the nexus of the paranormal mind has been the guiding principle of this study.

I adopted the term *parascience* early on when writing this book because I needed a way to discuss not-quite-scientific epistemology in neutral terms. *Pseudoscience* is such a loaded word that it inevitably betrays any attempt at reasonable debate, when reasonable debate is precisely what we need in a post-truth period such as ours. We cannot deny that unusual scientific ideas have existed and still exist all around us; nor can we evade our willful ignorance of them. Both Michel Serres and Paul Feyerabend have argued we need a theory of knowledge that can accommodate pseudoscientific, antiscientific, and almost scientific concepts that permeate the boundaries of science. Whether called anarchic epistemology, occulture, almost science, or parascience makes little difference in the end. What matters is that we start talking about it honestly and critically, and that we recognize there exists outside the borders of orthodox science an epistemic flood of inexplicable dreams, mystic encounters, precognitive visions, and the wild theories underlying them.

To be clear, this book is not a celebration of conspiracy theories, alternative facts, post-truth propaganda, or their ilk. As we know all too well, this kind of information has devastating material effects. From tobacco industry cancer "research" to refuting public health data during the Covid-19 pandemic, heterodoxy has been willfully deployed by politicians and organizations alike to society's detriment. These are serious epistemic problems that require serious pushback. But challenging oddball ideas requires discussing them; it demands contextualizing and historicizing by tracing origins, proponents, and patterns. While I would argue psi belongs to a different category of outsider science than, say, anti-vax campaigns, my larger point holds: such issues warrant meaningful, sustained critique. The problem is that psi is not covered in the academy enough, as if evading it will simply make it go away. For a cultural object as multifaceted and protean as the paranormal mind, understanding its longitudinal permutations informs us about the nature of anomalous epistemology. Rather than pretend pseudoscience does not exist, the time has come to study its circulation, evolution, and dissemination. Moreover, we must continue the task of estranging the paranormal, of recognizing the ubiquity of supernatural phenomena in global contexts and exploring its myriad logics.

In the introduction, I described parascience as an epistemic third space suffusing science where unproven concepts engage with Western and non-Western literature, philosophy, myth, theology, and art to hybridize into new knowledge forms. Parascience speaks to the arenas and processes involved in the circulation of pseudoscience, and via neologisms like *parascientification* and *parascientific objects,* I articulated the ways bad science is reinvented time and again. The psitron illustrated the multiepistemic transformation of precognition from a contested ability in psychical research to a superluminal thought particle passing through midcentury quantum mechanics and science fiction. An assemblage of biological, New Age, and apocalyptic writings demonstrated how telekinesis was reinterpreted as part of Gaia mind. Our investigation of Project Stargate showed how psychical research, cartography, and Cold War politics were recombined to reformulate clairvoyance into remote viewing, the ultimate geopolitical technology. By turning toward non-Western sciences in ethnic fiction, we saw how other cultures cognize ghosts, mediumship, and telepathy. A major goal throughout this project has been to showcase how the paranormal mind is not a fossil of the premodern so much as a plastic concept transmuted through an ever-expanding array of models, theories, ideologies, philosophies, literatures, and political forces.

The paranormal mind helps to reframe several pressing conversations within literary and science studies. Most obviously, it offers a broader view of occultism in twentieth-century aesthetics. The history and theory of the occult have generated some of the most exciting scholarship in modernist studies, and this book extends this cultural fascination with esoterica into the early twenty-first century. In doing so, it pushes David Lodge's argument in *Consciousness and the Novel* in a parascientific direction; not only is the novel the richest and most comprehensive record of human consciousness we have, but it is also a testament to the interdisciplinary, multicultural, and multigeneric attractions to the psychologically improbable.[12] The paranormal mind is not a blip in modern thought but an inexorable and pan-historical other. I have emphasized speculative fiction and ethnic fiction as two genres that interweave traditional and untraditional, Western and non-Western, and mythic and technoscientific sets of logic to continually renew the paranormal, but they are only the tip of the iceberg. Parapsychological journals, New Age treatises, superhero comics, psychology textbooks, essays on Chinese medicine, CIA reports, how-to psi guides, ghost stories, pop science best sellers, remote viewing protocols—where do we *not* find the paranormal in twentieth- and twenty-first-century culture? The enormity of what counts as paranormal literature boggles the mind. Rather than confine the occult to a specific time, place, or genre, the paranormal mind helps us see its multicultural sprawl, and our multiepistemic points for accessing it.

The ubiquity of the paranormal destabilizes strict binaries like natural-supernatural. This is one of the major political interventions of parascience—centralizing the marginal. By definition, parascience focalizes on that which exists beside, beyond, and next to what we have typically prioritized in science. But the history of science is also the history of power. Most people forget this truth, swept up in grand narratives of unbiased inquiry, neutral researchers, and following the data wherever it might lead. Science is not so simple, and there are aspects like funding, racism, sexism, prestige, and rival agendas to contend with. There are winners and losers in science, as psychical research knows all too well. Parascience refocuses our attention to the scientifically excluded, forgotten, oppressed, and maligned in all its forms. As this book has shown, I hope, not everything deemed pseudoscientific ought to be discarded; on the contrary, such texts often tell important stories.

Parascientific Revolutions makes the case that pseudoscientific and quasi-scientific topics deserve a place within contemporary science studies. STS has traditionally prioritized the creation and distribution of facts—the actants, cultural forces, organizations, and political contingencies that play a role in producing modernity's most powerful episteme. This has been and continues to be invaluable work. However, to ignore scientific outliers and their aesthetic legacies is to turn a blind eye to the workings of culture and the circulation of knowledge. In developing an architectonics of anomalous science, I have emphasized literature and science as catalysts for paranormal poiesis. The latter may come as a surprise to the positivists in my audience. It should not. From spiritual telegraph to SETI, science and technology have always reinforced magical thinking. This relationship is deeply entwined, even structural. Science produced the theories allowing precognition to become a mental particle and allowing posthuman consciousness to become a system property. Science can be both opponent and enabler of weird thinking.

Parascience also flips the script of women in science. Modern science has historically excluded women from receiving education, joining its professional ranks, and sharing its top accolades. As Sandra Harding notes, women in science are too often laboratory technicians rather than laboratory heads.[13] Parascientific inquiry means paying attention to fields like psychical research and shamanism where women play important roles. The occult sciences were highly dependent on sensitive women who could speak to the dead and scry the future, and figures like Annie Besant rose to the highest levels of their fields thanks to their "biological" dispositions. Likewise, characters like Kwan Li, Akiko Bradley, and Ruth dominate their narratives as primary investigators of the paranormal mind. Parascience brings us the stories of those systematically marginalized by modern power, and we are better off for hearing their voices. In addition, this study advances the Latourian demythologizing of science by including non-Western sciences and narratives within a larger constellation of twenty-first-century STS. As chapters 4 and 5 have just shown, the paranormal is not a Western phenomenon; cultures around the world have wrestled with ghosts, divine visions, and other marvelous events since time immemorial. Indeed, the West is distinctive in ostracizing the paranormal for the sin of existing outside the bounds of orthodox science. What our exploration of non-Western science and ethnic literature shows is how other cultures throughout history have explored the paranormal and integrated it into their knowledge paradigms. One result of this parascientific inquiry is naturalizing the supernatural, so abilities like telepathy are neither maligned nor ignored; instead, they become avenues for probing consciousness, memory, time, space, energy, and nature. The paranormal need not foreclose scientific thinking; it expands it. A second result is participating in the decolonization of STS. Through Bernd Reiter's mosaic epistemology and Walter Mignolo's pluriverse, we heard arguments for the expansion of knowledge beyond Western modes of thinking and being. Paranormal ethnic fiction puts the pluriverse into action by creating story worlds operating along non-Western epistemes, illustrating how we might cognize the physical and metaphysical worlds according to these alternative frameworks. Parascientific inquiry participates in epistemological justice; it decenters Western science and acknowledges the numerous ways people have interpreted ancestral spirits, retrocognition, and the afterlife for centuries. To understand science in its totality, we must take into consideration all the other discourses entangled with it.

I see parascientific analysis as an organic extension of post-Kuhnian science studies, but it does interrogate one element of the paradigm model—namely, its unwillingness to make allowances for unorthodox science. For Kuhn, life outside the paradigm is a Hobbesian affair: solitary, poor, nasty, brutish, and short. However, this project has highlighted the intellectual inertia of impossible science, the mechanisms of resisting paradigmatic closure. In some ways, we might be better served by turning back to Kuhn's intellectual predecessor, Ludwik Fleck, who identifies "thought styles" or "conceptions" as the scientific ideologies dictating the kind of knowledge a scientific collective can produce: "When a conception permeates a thought collective strongly enough, so that it penetrates as far as everyday life and idiom and has become a viewpoint in the literal sense of the word, any contradiction appears unthinkable and unimaginable."[14] Unlike the paradigm, though, Fleck contends that more than one thought style can exist at any given time. More importantly, they can persist long after they fall from favor. "Every thought style leaves remnants," usually in small and isolated communities, notes Fleck. "This explains the existence even today of astrologers and magicians: eccentrics who associate with the uneducated of the lower social classes or become charlatans because they do not share the community mood."[15] More than Kuhn, Fleck recognizes the tenacity of pseudoscience. However, even Fleck underestimates the power of the paranormal. Like most scientists, he assumes it is a remnant of the past, doomed to linger like a desiccated fly in the cobwebs of intellectual history. Our parascientific analysis of the long twentieth century reveals the opposite: science itself conjures the methods and theories for perpetuating the pseudoscientific. Together with literature, it from a matrix from which the paranormal mind is born again and again.

The parascientific exploration of the paranormal in this study is not complete in any way. Indeed, my hope is that it serves as departure point for future inquiry. There are many more disciplines and fields of thought left to uncover. For example, the golden age of science fiction (1938–46) is teeming with psi-fi that has only been briefly alluded to here. An in-depth analysis of this period would reveal much about the interconnections between fiction and border science. While many texts of the era use the paranormal to address questions of power, surveillance, or otherness, some may also rationalize how superhuman consciousness operates according to new technoscientific frameworks, and it is these narratives that can widen our parascientific purview. The overlap of government science and psi also deserves more attention. Wladimir Velminski's *Homo Sovieticus* provides an excellent account of state-sponsored telepathy ascending to the highest levels of Russian government during the Cold War. Understanding how the United States, China, and other countries applied ESP would provide key insights into governmentality and other forms of political power. This book has also introduced multicultural fiction to paranormal discourse, but there is still much to explore. While my own analyses have largely focused on Asian American supernaturalism and epistemes, the paranormal is found across African, African American, Latinx, Asian, and indigenous writings. From Pauline Hopkins's *Of One Blood* to Gloria Naylor's *Mama Day* to Tade Thompson's Wormwood trilogy, tales of telepathy and ghost vision fill the shelves of ethnic fiction. Many of them contextualize these powers within non-Western modes of science, and exploring how the paranormal operates in various global contexts would build on the pluriversal approach begun in these pages.

The biggest parascientific realm to address, of course, is the proliferation of paranormal thinking on the internet. It is perhaps unsurprising that with the rise of internet culture in the twenty-first century, the dominant medium of paranormal discourse has moved from print to online. Ghost-hunting websites, paranormal discussion communities, remote viewing sites, telepathy training modules, and parapsychological news have all proliferated in the information free-for-all that is the modern internet. The sheer volume of paranormal content available online today is stunning, and its easy access means more people than ever can participate in esoteric analysis. Whereas psychical research once attempted to delimit its disciplinary borders via trained personnel and laboratory experiments, the migration online signals a return to mass culture, civilian scientists, and hobbyists. It will be fascinating to see what media scholars and cultural critics make of this brave new world. How will digital natives push the original conceits of Frederic Myers, Edmund Gurney, and J. B. Rhine? How might they further recontextualize psi via African medicine or Eastern metaphysics? In full parascientific spirit, my hope is that these novel hybrids variations are met with fervent curiosity rather than disdain.

The remarkable variegation of paranormal minds we have, and will, encounter in fiction, movies, scientific literature, and the internet illustrates the entrenched nature of the supernatural in contemporary life. My research suggests as long as we have science, art, philosophy, or myth, we will have something approximating the paranormal. I am sure that even after reading this book there are those among us who wish to banish pseudoscience completely (although I contend that this is an impossible task). Instead, I would urge us to take a step back and attempt to understand parascientification as part of our larger epistemological economy. In the vast cycles of fact and fiction and information and misinformation flowing through daily life, we still have much to learn. When it comes to parascience, we are closer to a beginning than we are to an end.

ACKNOWLEDGMENTS

Like the paranormal mind itself, this book has a strange and circuitous history. Who could have predicted that a random article about a random book inscription found in a random archive would lead me to the Society for Psychical Research? Or that my love of science would take me down a rabbit hole of telepathic transmissions, psychic spies, and encounters with the spectral dead? It has been a fantastic journey, and so many have contributed to it.

Penn State's Department of English provided a wonderful environment for drafting the first version of this book. First and foremost, I am indebted to Susan Squier for her mentorship and brilliant feedback at every stage of this project. Her enthusiasm, unrelenting support, and personal writing motto— "Onwards!"—still inspire me today. Mark Morrisson taught me it was OK to geek out on weird science. Jonathan Eburne is a trailblazer of the arcane who showed me how to intellectualize outrageous ideas. Tina Chen provided so much good advice over the years that this book would not exist without her. Special thanks to Hester Blum, Claire Bourne, Robert Caserio, Chris Castiglia, Claire Colebrook, Richard Doyle, Sean Goudie, Kit Hume, Rosemary Jolly, Janet Lyon, Carla Mulford, Ben Schreier, and Garrett Sullivan. To my fellow members of Lit Lab—Sara DiCaglio, Bethany Doane, Michelle Huang, and Krista Quesenberry—thank you for the vital early feedback and for being such amazing people.

My colleagues at Wake Forest have been wonderful interlocutors, and our many conversations added clarity and depth to this work. Dean Franco provided encouragement and terrific comments on every chapter. Alex Brewer, Andrea Gómez Cervantes, Carla Hernández Garavito, Allison Kim, Luca Provenzano, and Siddarth Srikanth helped to sharpen this project in its closing stages. Special thanks to Gail Adams, Lucy Alford, Rian Bowie, Erin Branch,

Chris Brown, Reggie Brown, Amy Catanzano, Amy Clark, Meredith Farmer, Ebony Flowers, Matt Garite, Jennifer Greiman, Susan Harlan, Omaar Hena, Sarah Hogan, Jeff Holdridge, Melissa Jenkins, Scott Klein, Danielle Koupf, Zak Lancaster, Kevin MacDonnell, Judith Madera, Aimee Mepham, Barry Maine, Kaitlin Moore, Herman Rapaport, Jessica Richard, Joanna Ruocco, Gale Sigal, Jeffrey Solomon, Olga Valbuena, Corey Walker, Anna Willis, and Eric Wilson.

I am grateful to the University of Minnesota Press for bringing this book into the world. Thank you to the Proximities series editors, David Cecchetto and Arielle Saiber, for championing this project. I am indebted to Doug Armato for his enthusiasm and dedication as he shepherded it through the publication process. Special thanks to Zenyse Miller, Michael Stoffel, and all the copyediting, marketing, and design folks working behind the scenes. I am particularly indebted to my book's reviewers, Alicia Puglionesi and Peter Schwenger, for lending their considerable expertise.

My research was made possible by numerous institutions and awards. At Penn State, my work has been supported by the Center for Humanities and Information, the Institute for the Arts and Humanities, and the Department of English. At Wake Forest, this book was sponsored by the Humanities Institute with support made possible by a major grant from the National Endowment for the Humanities, the Office of the Provost, the Z. Smith Reynolds Foundation, the Sterge Faculty Fellowship, the Undergraduate College of Arts and Sciences, and the Department of English. I would like to thank the Ingo Swann Research Fellowship at the University of West Georgia for allowing me to explore Swann's papers, particularly Blynne Olivieri Parker, Brian Lord, and the Ingram Library staff. Special thanks to *Science Fiction Studies* and the R. D. Mullen Postdoctoral Research Fellowship for supporting me while I completed additional research at the University of West Georgia and Georgia Tech's science fiction collection. I am indebted to Elly Flippen, the Ingo Swann Estate, and the special collections staffs at Dumbarton Oaks and the University of West Georgia for their permission to reprint archival materials.

Several portions of this book reached early audiences. The background section on the Occult Revival and psychical research in the introduction first appeared in "The Man in the Macintosh and the Science of the Occult" from *James Joyce Quarterly* 55, no. 3–4 (2018): 115–37. A shorter version of chapter 5 was published as "Postquantum: *A Tale for the Time Being, Atomik Aztex,* and Hacking Modern Space-time" in *MELUS* 45, no. 1 (2020): 1–26. Many thanks to the journal editors and staff for publishing these essays. I am grateful to the conference organizers and my copanelists at the Society for Literature, Arts, and Science; the European Society for History of Science; and the Big Ten

English Scholars Lecture, where I presented different parts of this project over the years. Special thanks to Christina Skolnik and Samuel Stoeltje, my fellow paranormal enthusiasts at several of these venues.

This book was written in the company and conversation of good friends. Hoi Ning Ngai has always been there to lend an ear. Van Jensen's ghostly encounters inspired me to chase the paranormal to the strangest of places. Paul Kim's passion for speculative comics, cinema, and culture lurk behind these pages.

Finally, this book would not be possible without my amazing family. Some of my earliest memories involve Quilly Lee regaling me with stories of spirits, demons, and past lives. Paul Lee supported me every step of the way. Stephen Lee introduced me to ghost-hunting technologies (talcum powder on the windowsill!) when we were little kids. To the entire Poulsen clan, thank you for your unflagging encouragement. Fiona and Quinn, you bring me so much joy every day. And to Melissa Poulsen, thank you for your love, your laughter, and the light you bring into my life.

NOTES

Introduction

1. TEDx Whitechapel, "Theme: Visions for Transition," TED.com, January 12, 2013, https://www.ted.com/tedx/events/5612.
2. Rupert Sheldrake, "The Science Delusion—Banned TED Talk," YouTube video, November 29, 2014, https://www.youtube.com/watch?v=hO4p3xeTtUA.
3. Sheldrake.
4. The Skeptical movement is a broad social movement that attempts to investigate fringe topics and determine if they are supported by empirical research and sound methodologies as determined by science.
5. "Graham Hancock and Rupert Sheldrake, a Fresh Take," TED Blog, March 18, 2013, https://blog.ted.com/graham-hancock-and-rupert-sheldrake-a-fresh-take/.
6. Sigmund Freud, "The Uncanny," in *The Standard Edition of the Complete Psychological Works of Sigmund Freud,* vol. 17 (Hogarth Press & Institute of Psychoanalysis, 1953), 241.
7. Jeffrey Kripal, *Authors of the Impossible: The Paranormal and the Sacred* (University of Chicago Press, 2010), 23.
8. Pew Research Center, "Many Americans Mix Multiple Faiths," December 9, 2009, http://www.pewforum.org/2009/12/09/many-americans-mix-multiple-faiths/#ghosts-fortunetellers-and-communicating-with-the-dead.
9. Wladimir Velminski, *Homo Sovieticus: Brain Waves, Mind Control, and Telepathic Destiny* (MIT Press, 2017), 71–88.
10. The internet has clearly emerged as the deepest and most important reservoir of paranormal discourse in the twenty-first century, with hundreds, if not thousands, of websites and active online communities dedicated to subjects like psychic experiences, astral travel, clairvoyant methodologies, and ghostly encounters. However, given my priority here on paranormal developments across the long twentieth century, the internet and internet culture are not focal points in this book.

11. See Roger Luckhurst, *The Invention of Telepathy: 1870–1901* (Oxford University Press, 2002); Leigh Wilson, *Modernism and Magic: Experiments with Spiritualism, Theosophy, and the Occult* (Edinburgh University Press, 2013); Helen Sword, *Ghostwriting Modernism* (Cornell University Press, 2002); Mark Micale, *The Mind of Modernism* (Stanford University Press, 2004); Lawrence Sutin, *Do What Thou Wilt: A Life of Aleister Crowley* (St. Martin's Griffin, 2000); Martin Booth, *A Magick Life: A Biography of Aleister Crowley* (Hodder & Stoughton, 2000).

12. The chronological start and end dates of literary modernism are constantly moving targets. Here I use the traditional 1890s to 1940s time frame associated with Anglo-American modernism because most literary studies on the occult and psychical research are similarly situated within the scientific, literary, and cultural frameworks of early twentieth-century Britain and America.

13. Roger Luckhurst, "Pseudoscience," in *The Routledge Companion to Science Fiction*, ed. Mark Bould, Andrew M. Butler, Adam Roberts, and Sherryl Vint (Routledge, 2009), 404.

14. I want to highlight from the outset that I use *pseudoscience* in specific ways throughout this study. For me, the term designates nonnormative, outsider, border, and failed scientific ideas, but without the negative connotations typically attributed to it. My aim is not to denigrate ideas like telepathy but to acknowledge its marginality within Western scientific discourse. Pseudoscience is also different from parascience, the epistemic zone in which it circulates.

15. Egil Asprem, *The Problem of Disenchantment: Scientific Naturalism and Esoteric Discourse, 1900–1939* (State University of New York Press, 2014), 3–4.

16. Seymour Mauskopf, "Marginal Science," in *Companion to the History of Modern Science*, ed. R. C. Olby, G. N. Cantor, J. R. R. Christie, and M. J. S. Hodge (Routledge, 1990), 872–73.

17. Mauskopf, 875.

18. Jonathan Eburne, *Outsider Theory: Intellectual Histories of Unorthodox Ideas* (University of Minnesota Press, 2018), 10.

19. Auguste Comte, *The Positive Philosophy of August Comte* (Batoche Books, 1896), 28.

20. Andreas Sommer, "Psychical Research in the History and Philosophy of Science. An Introduction and Review," *Studies in History and Philosophy of Biological and Biomedical Sciences* 48 (2014): 44.

21. Alex Owen, *The Place of Enchantment: British Occultism and the Culture of the Modern* (University of Chicago Press, 2004), 148; Micale, *Mind*, 2; Mark Morrisson, *Modern Alchemy: Occultism and the Emergence of Atomic Theory* (Oxford University Press, 2007), 106.

22. Owen, *Place of Enchantment*, 4.

23. Madam H. P. Blavatsky, *The Secret Doctrine: The Synthesis of Science, Religion and Philosophy* (Theosophical Publishing House, 1888), 42, 45.

24. Jason Josephson Storm, *The Myth of Disenchantment: Magic, Modernity, and the Birth of the Human Sciences* (University of Chicago Press, 2017), 3.

25. Owen, *Place of Enchantment*, 19.

26. Owen, 13.

27. Alan Gauld, *The Founders of Psychical Research* (Shocken Books, 1968), 2.

28. Janet Oppenheim, *The Other World: Spiritualism and Psychical Research in England, 1850–1914* (Cambridge University Press, 1988), 136; Owen, *Place of Enchantment*, 33.

29. Owen, *Place of Enchantment*, 170.

30. Seymour Mauskopf and Michael McVaugh, *The Elusive Science: Origins of Experimental Psychical Research* (Johns Hopkins University Press, 1980), 6.

31. Owen, *Place of Enchantment*, 171.

32. Owen, 170.

33. Micale, *Mind*, 2.

34. Edmund Gurney, Frank Podmore, and Frederic W. H. Myers, *Phantasms of the Living*, 2 vols. (Scholars Facsimiles and Reprints, 1886), 1:457.

35. Frederic W. H. Myers, *Human Personality and Its Survival of Bodily Death*, 2 vols. (Longman, Green, 1903), 1:14.

36. Myers, *Human Personality*, 2:4.

37. Gauld, *Founders*, 296.

38. Micale, *Mind*, 9.

39. David Seed, "'Psychical Cases': Transformations of the Supernatural in Virginia Woolf and May Sinclair," in *Gothic Modernisms*, ed. Andrew Smith and Jeff Wallace (Palgrave Macmillan, 2001), 50.

40. For more on James Joyce's interest in occultism and psychical research, see Sword, *Ghostwriting*, chap. 3; Charles Ko, "Subliminal Consciousness," *Review of English Studies* 59, no. 242 (2007): 740–65; Erik Schneider, "'Welcomers': James Joyce and Frederic W. H. Myers," *Journal of Modern Literature* 38, no. 2 (2015): 59–70; Derek Lee, "The Man in the Macintosh and the Science of the Occult," *James Joyce Quarterly* 55, no. 3–4 (2018): 347–69.

41. Wilson, *Modernism and Magic*, 1.

42. Luckhurst, *Invention of Telepathy*, 2.

43. In 1923, the conflict between ASPR spiritualists and scientists led to a split in the organization. When spiritualist Frederick Edwards took control of the ASPR presidency and loosened the group's scientific requirements, Walter Prince and his scientific allies left and formed the Boston SPR in 1925. The two groups remained separate until merging in 1941. For more, see Arthur Berger, "Walter Franklin Pierce," in *Lives and Letters in American Parapsychology: A Biographical History, 1850–1987* (McFarland, 1988), 90–93.

44. Brian Inglis, *Science and Parascience: A History of the Paranormal* (Hodder & Stoughton, 1984), 215.

45. Inglis, 341.

46. Inglis, 314.

47. Alicia Puglionesi, *Common Phantoms: An American History of Psychic Science* (Stanford University Press, 2020), 4.

48. Two short-term academic fellowships with private funding preceded Duke University's Parapsychology Laboratory. The Stanford Fellowship in Psychical Research at Stanford University (1912–17) was supported by businessman and spiritualist Thomas Stanford. Psychologist John Edgar Coover performed ten thousand trials with one hundred subjects on telepathic card guessing and found no difference between the target (self-proclaimed psychics) and control groups. The Richard Hodgson Memorial Fund at Harvard University (1916–26) funded experiments by Leonard Thompson Troland (double-blind lamp clairvoyance), Gardner Murphy (telepathic image transfer), and George Hoben Estabrooks (telepathic card guessing and the "fatigue effect"). For more, see Asprem, *Problem of Disenchantment*, 355–66.

49. J. B. Rhine, *Extra-sensory Perception* (Bruce Humphries, 1934), 18.

50. Rhine, xxix.

51. Rhine, 50.

52. Rhine, 176.

53. Mauskopf and McVaugh, *Elusive Science,* 298.

54. Luckhurst, "Pseudoscience," 410.

55. Luckhurst, 410.

56. The term *psi-fi* has been previously used to describe fiction engaged with paranormal culture, psychology, and the unconscious. In a 2011 blog post, Jeffrey Kripal calls attention to "the incredibly messy, 'loopy' ways in which popular culture informs paranormal events, which in turn inform popular culture." In 2017, Bill Adams defines psi-fi as "psychological fiction that explores human consciousness in a technological context." I agree that having a fixed category for literature engaging with mind and paranormality is highly generative, and one of the explicit tasks of this book is expanding what counts as psi-fi. Jeffrey Kripal, "Psi-fi: Popular Culture and the Paranormal," Boing Boing (blog), January 26, 2011, https://boingboing .net/2011/01/26/psifi.html; Bill Adams, "Introducing Psi-fi: A New Genre of Fiction," Psi-fi (blog), March 9, 2017, https://www.psi-fi.net/intro-psi-fi/.

57. Mauskopf and McVaugh, *Elusive Science,* 302.

58. Mauskopf and McVaugh, 259.

59. Mauskopf and McVaugh, 253.

60. C. E. M. Hansel, *ESP: A Scientific Evaluation* (Scribner, 1966), xvi.

61. Mauskopf and McVaugh, *Elusive Science,* 274.

62. Mauskopf and McVaugh, 248, 253, 302.

63. Mauskopf and McVaugh, 304.

64. Inglis, *Science and Parascience,* 243.

65. Berger, *Lives and Letters,* 300.

66. Sommer, "Psychical Research," 43.

67. Gardner Murphy, *Challenge of Psychical Research: A Primer of Parapsychology* (Harper & Brothers, 1961), 285.

68. Asprem, *Problem of Disenchantment*, 12.

69. Myers, *Human Personality*, 1:1.

70. Inglis, *Science and Parascience*, 339.

71. Inglis, 319.

72. H. M. Collins and T. J. Pinch, "The Construction of the Paranormal: Nothing Unscientific Is Happening," *Sociological Review* 27, no 1, suppl. (1979): 237–70.

73. Thomas Kuhn defines the paradigm as "a body of rules competent to constitute a given normal research tradition" as well as "the fundamental unit for the student of scientific development." Implicit in these definitions is the paradigm's intrinsic ability to shape a scientist's entire world view. For example, the geocentric model of the universe in astronomy assumes the centrality of God and man in nature, whereas the heliocentric model does not. Thomas Kuhn, *The Structure of Scientific Revolutions* (University of Chicago Press, 2012), 11, 44.

74. Kuhn, 37.

75. Steve Fuller argues that Kuhn's *Structure of Scientific Revolutions* offers a fundamentally conservative view of scientific progress reflecting American Cold War anxieties. Ian Hacking notes that Thomas Kuhn's training in physics (theory driven) led to a paradigm model ill-suited for biology (technique driven), which is the new queen of the sciences. Steve Fuller, introduction to *Thomas Kuhn: A Philosophical History for Our Times* (University of Chicago Press, 2000); Ian Hacking, "A Q&A with Ian Hacking on Thomas Kuhn's Legacy as 'The Paradigm Shift' Turns 50," interview conducted by Gary Stix, *Scientific American,* May 27, 2012, https://www.scientificamerican.com/article/kuhn/.

76. Kuhn, *Structure*, 17.

77. Mauskopf, "Marginal Science," 870.

78. Bruno Latour, *We Have Never Been Modern* (Harvard University Press, 1993), 10.

79. Randall Styers, "Bad Habits, or How Superstition Disappeared in the Modern World," in *Magic in the Modern World: Strategies of Repression and Legitimization,* ed. Edward Bever and Randall Styers (Penn State University Press, 2017), 24.

80. Christopher Partridge, *Reenchantment of the West: Alterative Spiritualities, Sacralization, Popular Culture, and Occulture* (T&T Clark International, 2004), 4.

81. Partridge, 4.

82. Massimo Pigliucci, *Nonsense on Stilts: How to Tell Science from Bunk* (University of Chicago Press, 2010), 25.

83. Pigliucci, 24.

84. Pigliucci, 24.

85. Pigliucci, 31.

86. Pigliucci, 229.

87. Eburne, *Outsider Theory*, xvi.

88. Nahum Chandler, "The Problem of the Centuries: A Contemporary Elaboration of 'The Present Outlook for the Dark Races of Mankind,' circa the 27th of December, 1899, or At the Turn of the Twentieth Century," in *"Beyond This Narrow Now": Or, Delimitations, of W. E. B. Du Bois* (Duke University Press, 2022), 253.

89. Pigliucci, *Nonsense*, 1.

90. R. A. Judy, *Sentient Flesh: Thinking in Disorder, Poiesis in Black* (Duke University Press, 2020), 20.

91. Thomas Gieryn, *Cultural Boundaries of Science: Credibility on the Line* (University of Chicago Press, 1999), 6.

92. *Counterdiscourse* refers to a discourse positioned against dominant powers, be they political regimes, colonialism, or Western science. It is "a space in which the formerly voiceless might begin to articulate their desires—to counter the domination of prevailing authoritative discourse." Mario Moussa and Ron Scapp, "The Practical Theorizing of Michel Foucault: Politics and Counter-discourse," *Cultural Critique*, no. 33 (1996): 88.

93. Inglis, *Science and Parascience*, 12.

94. Jeremy Stolow, "Salvation by Electricity," in *Religion: Beyond a Concept* (Fordham University Press, 2008), 677.

95. Susan Squier, *Babies in Bottles* (Rutgers University Press, 1994), 11; Susan Squier, *Liminal Lives: Imagining the Human at the Frontiers of Biomedicine* (Duke University Press, 2004), 3.

96. Sherryl Vint, *Bodies of Tomorrow: Technology, Subjectivity, Science Fiction* (University of Toronto Press, 2007), 20.

97. Gary Westfahl, *The Mechanics of Wonder: The Creation of the Idea of Science Fiction* (Liverpool University Press, 1998), 194.

98. Michel Serres, *Hermes: Literature, Science, Philosophy* (Johns Hopkins University Press, 1982), 83.

99. Paul Feyerabend, *Against Method* (Verso, 2010), 48.

100. Jeffrey Kripal, *Mutants and Mystics* (University of Chicago Press, 2011), 1.

101. For a good overview on the relationship between literature and the various psychological sciences, see Leon Surette, *The Birth of Modernism: Ezra Pound, T. S. Eliot, W. B. Yeats, and the Occult* (McGill-Queen's University Press, 1993); Micale, *Mind*; Paul Peppis, *The Sciences of Modernism: Ethnography, Sexology, and Psychology* (Cambridge University Press, 2014); Maud Ellman, *The Nets of Modernism: Henry James, Virginia Woolf, James Joyce, and Sigmund Freud* (Cambridge University Press, 2010); George Johnson, *Dynamic Psychology in Modernist British Fiction* (Palgrave Macmillan, 2006); Judith Ryan, *The Vanishing Subject: Early Psychology and Literary Modernism* (University of Chicago Press, 1991).

102. See George Mills Harper, *Yeats's Golden Dawn* (Macmillan, 1974); George Mills Harper, *W. B. Yeats and W. T. Horton: The Record of an Occult Friendship* (Humanities Press, 1980); George Mills Harper, *The Making of Yeats's "A Vision": A Study of the Automatic Script* (Southern Illinois University Press, 1987); Surette, *Birth of*

Modernism; Sword, *Ghostwriting;* Wilson, *Modernism and Magic;* John Bramble, *Modernism and the Occult* (Palgrave Macmillan, 2015).

103. David Treuer, *Native American Fiction: A User's Manual* (Graywolf Press, 2006), 165; Madhu Dubey, "The Future of Race in Afro-futurist Fiction," in *The Black Imagination: Science Fiction, Futurism, and the Speculative,* ed. Sandra Jackson and Julie E. Moody-Freeman (Peter Lang, 2011), 16.

104. Kathleen Brogan, *Cultural Haunting: Ghosts and Ethnicity in Recent American Literature* (University Press of Virginia, 1998), 4.

105. *Ethnoscience* (also called *native science, indigenous science, traditional knowledge,* and *non-Western science*) refers to the knowledge and practices of science as construed by non-Western and nonmodern societies. Like the term *ethnic fiction,* it is a problematic label because it differentiates multicultural representations of science as other to Western science. It also clusters a wide variety of global cultural systems as a single block of alternative knowledge. Nevertheless, I find it and its synonyms useful terms for specifying non-Western scientific discourse, and part of this project's aim is to recuperate its meaning. Sandra Harding has already begun this process by identifying Western science as just one, albeit powerful, type of ethnoscience. Sandra Harding, *Is Science Multicultural?* (Indiana University Press, 1998), 18–19.

106. Sandra Harding, *The Science Question in Feminism* (Cornell University Press, 1986), 21, 63.

107. Puglionesi, *Common Phantoms,* 10.

108. Puglionesi, 11.

109. Harding, *Is Science Multicultural?,* 18–19.

110. Walter D. Mignolo, "On Pluriversality and Multipolarity," foreword to *Constructing the Pluriverse: The Geopolitics of Knowledge,* ed. Bernd Reiter (Duke University Press, 2018), x.

111. Karl Popper defines *falsifiability* as the ability to test whether a scientific hypothesis is true or false. A strong scientific research program creates theories that can be tested and (one hopes) confirmed. For example, Einsteinian physics made several predictions about matter and light speed that could be tested (and eventually confirmed), so it proved to be scientific. In contrast, Freudian psychoanalysis did not offer testable hypotheses that could be proven or disproven, so the field is unscientific. See Karl Popper, *Conjectures and Refutations: The Growth of Scientific Knowledge* (London, 1962), chap. 1.

112. Karl Popper, *Logic of Scientific Discovery* (Routledge, 2002), 4.

113. Bruno Latour, *Science in Action: How to Follow Scientists and Engineers through Society* (Open University Press, 1987), 41.

114. The knowledge deficit model of science communication rests on two explicit premises: first, scientific understanding is essential for good decision-making on scientific matters; and second, if people have scientific literacy, then their attitude to science will be positive. Implicit in these assumptions are three things:

(1) science has epistemic priority over other forms of knowledge; (2) the duty of scientists is to teach the public; and (3) the proper flow of knowledge is from science to society. See W. J. Grant, "The Knowledge Deficit Model and Science Communication," in *Oxford Research Encyclopedia of Communication* (Oxford University Press, November 22, 2023), https://oxfordre.com/communication/dis play/10.1093/acrefore/9780190228613.001.0001/acrefore-9780190228613-e-1396.

115. Bernd Reiter, introduction to *Constructing the Pluriverse*, 1.

116. Michel Serres, *The Troubadour of Knowledge* (University of Michigan Press, 1991), 37.

117. Michel Serres, *Conversations on Science, Culture, and Time* (University of Michigan Press, 1995), 133.

118. Serres, 148.

1. Toward a (Meta)physics of Precognition

1. Alan Moore and Dave Gibbons, *Watchmen* (DC Comics, 1986), 390.

2. Luckhurst, "Pseudoscience," 405.

3. Myers, *Human Personality*, 2:979.

4. Myers, 2:979.

5. Myers, 2:978.

6. Myers, 2:984.

7. Myers, 2:985.

8. Myers, 2:986.

9. Murphy, *Challenge*, 154.

10. C. D. Broad, "The Notion of Precognition," in *Science and ESP*, ed. J. R. Smythies (Routledge, 1967), 191.

11. Melissa Littlefield, "Matter for Thought: The Psychon in Neurology, Psychology and American Culture, 1927–1943," in *Neurology and Modernity: A Cultural History of Nervous Systems*, ed. Laura Salisbury and Andrew Shail (Palgrave Macmillan, 2010), 267.

12. William James, "A Plea for Psychology as a Natural Science," *Philosophical Review* 1 (1892): 153.

13. Littlefield, "Matter for Thought," 268.

14. *Logical atomism* is a philosophical belief popularized in the early twentieth century by Bertrand Russell in *The Philosophy of Logical Atomism* (Monist, 1919). It holds that the world consists of logical facts (or atoms) that cannot be broken down any further; moreover, only by first understanding these ultimate facts can the larger holistic system become fully known.

15. Sigmund Freud, "Project for a Scientific Psychology," in *The Standard Edition of the Complete Works of Sigmund Freud*, vol. 1 (Hogarth Press & Institute of Psychoanalysis, 1996), 295.

16. In 1887, Santiago Ramín y Cajal used Golgi stain to visualize individual neurons in the brains of birds. His discovery was seminal in proving the neuron doctrine.

17. Littlefield, "Matter for Thought," 272.

18. Warren S. McCulloch, *Embodiments of Mind*, rev ed. (MIT Press, 1988), 37.

19. William Marston, *Emotions of Normal People* (Harcourt, Brace, 1928), 52.

20. Stanley Weinbaum, "The Ideal," in *"A Martian Odyssey" and Other Science Fiction Tales: The Collected Short Stories of Stanley Weinbaum* (Hyperion, 1974), 225.

21. Littlefield, "Matter for Thought," 280.

22. Whately Carington, *Thought Transference* (Creative Age Press, 1946), 164.

23. Carington, 200.

24. Carington, 214.

25. Leonid Leonidovich Vasiliev, *Experiments in Mental Suggestion* (Hampton Roads, 1962), 5.

26. John Eccles, *The Neurophysiological Basis of Mind* (Oxford University Press, 1953), 265.

27. Eccles, 87.

28. John Eccles identifies the particles of "mental will" as psychons in his 1994 text *How the Self Controls Its Brain*. In *Basis of Mind,* he simply refers to them as subatomic particles of mentation (87). Regardless of their label, both follow the atomistic logic of fundamental thought particles.

29. Eccles, *Basis of Mind,* 277.

30. Eccles, 280.

31. Eccles, 279.

32. Arthur Eddington had hypothesized that the mind could potentially control the behavior of matter within the limits defined by the Heisenberg uncertainty principle, but he ultimately rejects this idea since the mass of the neuron is far larger than what is permissible in subatomic physics. Eccles builds on this concept by claiming that dendrons are small enough to operate at the subatomic level. See Eccles, *Basis of Mind,* 278; and Arthur Eddington, *New Pathways in Science* (Cambridge University Press, 1935).

33. Eccles, *Basis of Mind,* 284.

34. Exotic matter is matter that does not conform to the behavior of normal baryons, like protons and neutrons. Examples of exotic matter include dark matter, particles of negative mass or imaginary mass, Bose-Einstein condensates, and quark-gluon plasma.

35. Allen Thiher, *Fiction Refracts Science: Modernist Writers from Proust to Borges* (University of Missouri Press, 2005), 22.

36. For a strong introduction to the effect of quantum mechanics on literature, see Thiher, *Fiction Refracts Science;* and Susan Strehle, *Fiction in the Quantum Universe* (University of North Carolina Press, 1992). For its impact on art, see Gavin Parkinson, *Surrealism, Art, and Modern Science: Relativity, Quantum Mechanics, Epistemology* (Yale University Press, 2008); and Thomas Vargish and Delo Mook,

Inside Modernism: Relativity Theory, Cubism, Narrative (Yale University Press, 1999). For its effects on philosophy and culture, see Peter Lewis, *Quantum Ontology* (Oxford University Press, 2016); Robert Crease and Alfred Goldhaber, *The Quantum Moment* (Norton, 2014); and Stephen Kern, *The Culture of Space and Time, 1880–1918* (Harvard University Press, 2003).

37. Owen, *Place of Enchantment,* 19.
38. Arthur Eddington, *The Nature of the Physical World.* (Cambridge University Press, 1946), xi.
39. The neutrino had been theorized by Wolfgang Pauli in 1930 to explain the conservation of energy, momentum, and spin in the phenomenon known as beta decay.
40. Arthur Koestler, *Roots of Coincidence* (Random House, 1972), 63.
41. Koestler, 62.
42. The search for superluminary communication between entangled photon pairs emerged as a major research program in post-1970s physics and would eventually give rise to quantum computing. For more information, see David Kaiser, *How the Hippies Saved Physics* (Norton, 2011), chap. 9.
43. Kaiser, *How the Hippies Saved Physics,* 2.
44. Koestler, *Roots of Coincidence,* 78.
45. Koestler, 79.
46. The standard model of particle physics is a theory describing three of the four known fundamental forces: electromagnetism, weak force, and strong force.
47. Koestler, *Roots of Coincidence,* 140.
48. Adrian Dobbs, "Time and Extrasensory Perception," *Proceedings of the Society for Psychical Research* 54, no. 197 (1965): 342.
49. Adrian Dobbs, "The Feasibility of a Physical Theory of ESP," in *Science and ESP* (Routledge, 1967), 247.
50. Dobbs, "Time and Extrasensory Perception," 305.
51. Dobbs, 255.
52. Dobbs, 342.
53. Dobbs, 343.
54. The Soal-Shackleton experiments refer to a famous set of laboratory experiments run by Samuel Soal and conducted from 1941 to 1943 in which Basil Shackleton demonstrated strong statistical evidence of precognitive ability. See Murphy, *Challenge,* chap. 5. *Spontaneous cases* is SPR's term for cases in which percipients "suddenly" see ghosts or experience visions.
55. Dobbs, "Feasibility," 250.
56. Dobbs, "Time and Extrasensory Perception," 330.
57. Dobbs, "Feasibility," 250.
58. Dobbs, "Time and Extrasensory Perception," 340.
59. Dobbs, "Feasibility," 252.
60. Dobbs, 248.
61. Mauskopf and McVaugh, *Elusive Science,* 6.

62. Arthur Hastings, preface to *Experiments in Mental Suggestion* (Hampton Roads, 1962), xxviii.

63. Inglis, *Science and Parascience*, 127.

64. Lawrence Sutin, *Divine Invasions: A Life of Philip K. Dick* (Harmony Books, 1989), 210.

65. Philip K. Dick, *The Exegesis of Philip K. Dick* (Houghton Mifflin Harcourt, 2011), 271.

66. Erika Dyck, *Psychedelic Psychiatry: LSD from Clinic to Campus* (Johns Hopkins University Press, 2008), 69.

67. Dick, *Exegesis*, 7.

68. Dick, 7.

69. Sutin, *Divine Invasions*, 213.

70. Sutin, 213.

71. Sutin, 225.

72. For more on the supposed causes of 2-3-74, see Sutin, *Divine Invasions*, 221; Kyle Arnold, *The Divine Madness of Philip K. Dick* (Oxford University Press, 2016), 4; and Gabriel McKee, *Pink Beams of Light from the God in the Gutter: The Science Fictional Religion of Philip K. Dick* (University Press of America, 2004), 22.

73. Dick, *Exegesis*, 8.

74. Arthur Koestler, "Order from Disorder," *Harper's*, 1974, 60.

75. Philip K. Dick, *VALIS (VALIS and Later Novels)* (Library of America, 1981), 177.

76. Laurence Rickels, *I Think I Am: Philip K. Dick.* (University of Minnesota Press, 2010), 22.

77. See Umberto Rossi, "The Shunts in the Tale: The Narrative Architecture of Philip K. Dick's *VALIS*," *Science Fiction Studies* 39, no. 2 (2012): 243–61; John Garvey, "A Real Gnostic Gospel: The Fiction of Philip K. Dick," *Commonweal Magazine*, April 30, 2007, 13–16; Lorenzo DiTommaso, "Gnosticism and Dualism in the Early Fiction of Philip K. Dick," *Science Fiction Studies* 28, no. 1 (2001): 49–65; Darko Suvin, "Goodbye and Hello: Differentiating within the Later P. K. Dick," *Extrapolation* 43, no. 4 (2002): 368–97; Roger Stilling, "Mystical Healing: Reading Philip K. Dick's *VALIS* and *The Divine Invasion* as Metapsychoanalytic Novels," *South Atlantic Review* 65, no. 2 (1991): 91–106; and Christopher Palmer, "Postmodernism and the Birth of the Author in Philip K. Dick's *VALIS*," *Science Fiction Studies* 18, no. 3 (1991): 330–42.

78. Dick, *VALIS*, 189.

79. Dick, 186.

80. Dick, 320.

81. Dick, 187.

82. Dick, *Exegesis*, 11.

83. Jill Galvan, *The Sympathetic Medium: Feminine Channeling, the Occult, and Communication Technologies, 1859–1919* (Cornell University Press, 2010), 30.

84. Dick, *Exegesis*, 11–12.

85. Dick, *VALIS,* 189.

86. Blavatsky, *Secret Doctrine,* 6.

87. Dick, *VALIS,* 340.

88. Dick, 343.

89. Dick, 343.

90. Dick, 178.

91. Dick, 389.

92. *Watchmen* scholarship grows every year. For a good introduction to this subfield, see Mark White, ed., *"Watchmen" and Philosophy: A Rorschach Test* (Wiley, 2009); Sara Van Ness, *"Watchmen" as Literature: A Critical Study of the Graphic Novel* (McFarland, 2010); and Andrew Hoberek, *Considering "Watchmen": Poetics, Property, Politics* (Rutgers University Press, 2014).

93. Alan Moore, "Magic Is Afoot: A Conversation with Alan Moore about the Arts and the Occult," interview conducted by Jay Babcock, *Arthur,* no. 4 (May 2003), https://arthurmag.com/2007/05/10/1815/.

94. Moore and Gibbons, *Watchmen,* 140.

95. See David Barnes, "Time in the Gutter: Temporal Structures in *Watchmen,*" *Krono-Scope* 9, no. 1–2 (2009): 51; and Adnan Mahmutovic, "Chronotope in Moore and Gibbon's *Watchmen,*" *Studies in the Novel* 50, no. 2 (2018): 271.

96. Christopher Drohan, "A Timely Encounter: Dr. Manhattan and Henri Bergson," in White, *"Watchmen" and Philosophy,* 115.

97. Moore and Gibbons, *Watchmen,* 297.

98. Moore and Gibbons, 389.

99. Moore and Gibbons, 126.

100. Barnes, "Time in the Gutter," 54.

101. See the following essays in White, *"Watchmen" and Philosophy:* Christopher Robichaud, "The Superman Exists, and He's American: Morality in the Face of Absolute Power," 5–18; Arthur Ward, "Free Will and Foreknowledge: Does Jon Really Know What Laurie Will Do Next, and Can She Do Otherwise?," 125–36; and Andrew Terjsen, "I'm Just a Puppet Who Can See the Strings: Dr. Manhattan as a Stoic Sage," 137–53.

102. Sutin, *Divine Invasions,* 1.

103. Hoberek, *Considering "Watchmen,"* 3.

104. Sultan Tarlaci, "A New Electronic Journal and a New Word. Neuroquantology: Two Sides of the Same Coin," *NeuroQuantology* 1 (2003): 1.

105. Tarlaci, 1.

106. See Gao Shan, "A Possible Quantum Basis of Panpsychism," *NeuroQuantology* 1 (2003): 4–9; and Donald Watson and Bernard Williams, "Eccles' Model of the Self Controlling Its Brain: The Irrelevance of Dualist-Interactionism," *NeuroQuantology* 1 (2003): 119–28.

107. Syamala Hari, "Eccles's Psychons Could Be Zero-Energy Tachyons," *NeuroQuantology* 6, no. 2 (2008): 156.

108. Hari, 153.
109. Hari, 159.
110. Hari, 154.
111. Manuel DeLanda, *A Thousand Years of Nonlinear History* (Zone Books, 1997), 15.

2. Gaia, the Paranormal Planet

1. Many writers typically refer to Gaia as a female entity (*she, her,* etc.), in the mythological tradition of Earth's being the first mother goddess. In this chapter, I occasionally adopt similar terminology to match the ongoing Gaian discourse, although I recognize the problems inherent to such gendered rhetoric. Biological discourse has historically coded the cellular and natural environment as female and passive compared to the active, male agents that make things happen. New Age writings regularly assume women possess intuition, sensitivity, and maternal powers granting them unique access to Earth. Gender unfortunately infuses both scientific and cultural life, and I acknowledge this history even as I adopt its narrative tropes. For more on gendered language in biology, see Evelyn Fox Keller, *Refiguring Life* (Columbia University Press, 1995), chap. 1.

2. Pepper Lewis, *Gaia Speaks: Sacred Earth Wisdom* (Light Technology, 2005), xxi.

3. James Lovelock, *Gaia: A New Look of Life on Earth* (Oxford University Press, 1979), x.

4. James Lovelock, *Homage to Gaia: The Life of an Independent Scientist* (Oxford University Press, 2000), 241.

5. Daisyworld is a computer simulation of a hypothetical world that James Lovelock and Andrew Watson developed in 1983 to illustrate the viability of the Gaia hypothesis. The model tracks two daisy populations (black and white) and shows how their competition for solar energy creates feedback mechanisms that result in optimal environmental conditions for future daisy growth.

6. See Isabelle Stengers, *In Catastrophic Times: Resisting the Coming Barbarism* (Open University Press, 2015); Donna Haraway, *Staying with the Trouble: Making Kin in the Chthulucene* (Duke University Press, 2016); Bruno Latour, *Facing Gaia: Eight Lectures on the New Climatic Regime* (Polity Press, 2017); and Bruce Clarke, *Gaian Systems: Lynn Margulis, Neocybernetics, and the End of the Anthropocene* (University of Minnesota Press, 2020).

7. James Lovelock, "The Quest for Gaia," *New Scientist,* February 6, 1975, 304.

8. Michael Ruse, *The Gaia Hypothesis: Science on a Pagan Planet* (University of Chicago Press, 2013), 44.

9. Ruse, 44.

10. Plato, *Timaeus* (Harvard University Press, 1999), 57.

11. Ruse, *Gaia Hypothesis,* 50.

12. Mary Midgley, *Gaia: The Next Big Idea* (Demos, 2001), 20.

13. Midgley, 19.

14. Midgley, 19.

15. Midgley, 18.

16. Midgley, 19.

17. Latour, *We Have Never Been Modern*, 47.

18. Manfred Drack, Wilfried Apfalter, and David Pouvreau, "On the Making of a System Theory of Life: Paul A. Weiss and Ludwig von Bertalanffy's Conceptual Connection," *Quarterly Review of Biology* 82, no. 4 (2007): 1.

19. The law of exponential growth shows that the growth rate of an entity increases or decreases at a known exponential rate defined by the formula $xt = xo(1 + r)^2$, in which xt is the value of variable x at time t; xo is the value of x at time 0; and r is the growth rate. Other examples of general system laws include Boulding's first law, Malthusian (exponential) law, and Volterra's theory of mathematic biology. For more on general system principles, see Ludwig von Bertalanffy, *General System Theory* (George Braziller, 1968), 47–48.

20. Bertalanffy, *General System Theory*, 18.

21. Bertalanffy, 45.

22. Drack, Apfalter, and Pouvreau, "On the Making," 9.

23. Bertalanffy, *General System Theory*, 35.

24. Bertalanffy, 45.

25. Bertalanffy, 14.

26. Bertalanffy, 45.

27. Drack, Apfalter, and Pouvreau, "On the Making," 14.

28. According to critic Bruce Clarke, first-order systems theory is largely grounded in the cybernetics of Warren McCulloch, Norbert Wiener, John von Neumann, and their colleagues studying mechanical and computational systems. In contrast, second-order systems theory foregrounds the work of Gregory Bateson, Heinz von Foerster, Niklas Luhmann, and others who explored the broader areas of autopoiesis, "order-from-noise" principles, and system–environment interactions. Bruce Clarke, *Posthuman Metamorphosis* (Fordham University Press, 2008), 5–6.

29. Ilya Prigogine and Isabelle Stengers, *Order out of Chaos: Man's New Dialogue with Nature* (Bantam Books, 1984), xv.

30. Prigogine and Stengers, 14.

31. Humberto Maturana and Francisco Varela, *Autopoiesis and Cognition: The Realization of the Living* (Reidel, 1973), 79.

32. Ray Kurzweil, *The Singularity Is Near: When Humans Transcend Biology* (Viking, 2005), 21.

33. Lovelock, *Gaia*, 2.

34. Lovelock, *Homage*, 241.

35. Lovelock, 241.

36. Lovelock, *Gaia*, 10.

37. Dorion Sagan, "Life on a Margulisian Planet," in *Earth, Life, and System: Evolution and Ecology on a Gaian Planet* (Fordham University Press, 2015), 19.

38. In 1929, Walter Cannon defined his new theory of *homeostasis* as the process of adjustments in an internal environment (like a living body) that allowed for the self-maintenance of physiological variables within that environment.
39. Lovelock, *Gaia*, 45.
40. Lovelock, 44.
41. Lovelock, 75.
42. Lovelock, 81.
43. Lovelock, 9.
44. Lynn Margulis, "On the Origin of Mitosing Cells," *Journal of Theoretical Biology* 14, no. 3 (1967): 226.
45. Andreas Schimper, Konstantin Mereschkowski, and Ivan Wallin all posited that eukaryotic life stemmed from prokaryotic symbioses. For more, see Jan Sapp, "Too Fantastic for Polite Society: A Brief History of Symbiosis Theory," in *Lynn Margulis: The Life and Legacy of a Scientific Rebel,* ed. Dorion Sagan (Chelsea Green, 2012), 54–55.
46. Lichen is a composite organism made up of cyanobacteria and fungi. The fungi benefit from the carbohydrates produced by the photosynthetic cyanobacteria, and the cyanobacteria benefit from the environment conditions and nutrients gathered by the fungi.
47. Lynn Margulis and Dorion Sagan, *What Is Life?* (Simon & Schuster, 1995), 26.
48. Margulis and Sagan, 26.
49. Lynn Margulis, *Symbiotic Planet: A New Look at Evolution* (Basic Books, 1988), 115.
50. Margulis and Sagan, *What Is Life?*, 26.
51. The literature describing the scientific collaboration of James Lovelock and Lynn Margulis is extensive, so I will not rehash it here. Several books I have already mentioned—*Gaia: A New Look at Life on Earth, Homage to Gaia,* and *Symbiotic Planet*—discuss their work over the years. For their early thoughts on reconceptualizing life, see James Strick, "Exobiology at NASA: Incubator for the Gaia and Serial Endosymbiosis Theories," in *Earth, Life, and System: Evolution and Ecology on a Gaian Planet,* ed. Bruce Clarke (Fordham University Press, 2015). Michael Ruse's *Gaia Hypothesis* recounts their collaboration through its wider historical and philosophical context. For a scientifically skeptical view of their intellectual work, see Toby Tyrrell, *On Gaia: A Critical Investigation of the Relationship between Life and the Earth* (Princeton University Press, 2013).
52. Ludwik Fleck, *Genesis and Development of a Scientific Fact* (University of Chicago Press, 1979), 100.
53. Bruce Clarke, "Gaia Is Not an Organism: Scenes from the Early Collaboration between Lynn Margulis and James Lovelock," in Sagan, *Lynn Margulis,* 35.
54. Luckhurst, "Pseudoscience," 406.
55. Lovelock, *Gaia*, 6, 23, all emphases added.
56. Lovelock, 137.
57. Lovelock, 138.

58. Lovelock, 138.

59. Margulis, *Symbiotic Planet,* 113.

60. Margulis, 113.

61. Sagan, "Life on a Margulisian Planet," 24.

62. Lovelock, *Gaia,* 139.

63. Lovelock, 140.

64. Lovelock, 139.

65. The ontogenetic model of human consciousness is a theory developed by Erich Jantsch and Conrad Waddington that models human consciousness as a movement through four learning modes: virtual, functional, conscious, and superconscious. The last stage is central to noogenesis and is characterized by superconscious self-regulation of cultural and mankind processes. For more on the superconscious, see Erich Jantsch, "Evolution: Self-Realization through Self Transcendence," in *Evolution and Consciousness: Human Systems in Transition,* ed. Erich Jantsch and Conrad H. Waddington (Addison-Wesley, 1976), 43.

66. Lovelock, *Homage,* 264.

67. Ruse, *Gaia Hypothesis,* 223.

68. David Hess, *Science in the New Age* (University of Wisconsin Press, 1993), 3.

69. Wouter Hanegraaff, *New Age Religion and Western Culture: Esotericism in the Mirror of Secular Thought* (Brill, 1996), 10.

70. Sarah Pike, *New Age and Neopagan Religions in America* (Columbia University Press, 2006), 14.

71. Kaiser, *How the Hippies Saved Physics,* xiii.

72. Kaiser, 262.

73. Kaiser, 109.

74. J. Krishnamurti and David Bohm, *The Limits of Thought* (Routledge, 1999), 110.

75. Gregory Bateson, *Mind and Nature* (Dutton, 1979), 93.

76. Bateson, 37.

77. Peter Russell, *The Awakening Earth: The Global Brain* (Ark Paperbacks, 1982), 37, 78.

78. Russell, 78.

79. Russell, 80.

80. Russell, 85.

81. Russell, 211.

82. Fritjof Capra, *The Turning Point* (Bantam, 1982), 285.

83. Capra, 291.

84. Capra, 296.

85. O. W. Markley, "Human Consciousness in Transformation," in Jantsch and Waddington, *Evolution and Consciousness,* 225.

86. Markley, 225–26.

87. Markley, 225.

88. Stanislaw Lem, *Solaris* (Berkeley Medallion, 1961), 178.

89. Isaac Asimov, *Foundation's Edge* (Doubleday, 1982), 300.
90. Roger Luckhurst, "Catastrophism, American Style: The Fiction of Greg Bear," *Yearbook of English Studies* 37, no. 2 (2007): 217.
91. Cladogenesis is an evolutionary splitting event where a new species branches off from the parent species. This is distinct from the more commonly known anagenesis, which refers to evolution within the original lineage.
92. Greg Bear, *Darwin's Radio* (Ballantine Books, 1999), 84.
93. Bear, 81.
94. Bear, 80.
95. Luckhurst, "Catastrophism," 226.
96. Bear, *Darwin's Radio*, 192.
97. Bear, 169.
98. Bear, 104.
99. Bear, 192.
100. Greg Bear, "When Genes Go Walkabout," *Proceedings of the American Philosophical Society* 148 (2004): 330.
101. Bear, 325.
102. Bear, *Darwin's Radio*, 150.
103. Bear, 180.
104. Greg Bear, *Vitals* (Ballantine Books, 2002), 34.
105. Bear, 205.
106. Scott Gilbert and Jan Sapp, "A Symbiotic View of Life: We Have Never Been Individuals," *Quarterly Review of Biology* 87, no. 4 (2012): 334.
107. Myra Hird, *The Origins of Sociable Life: Evolution after Science Studies* (Palgrave Macmillan, 2009), 1.
108. Hird, 84.
109. Hird, 20.
110. Hird, 84.
111. Margulis and Sagan, *What Is Life?*, 69.
112. Bear, *Vitals*, 152.
113. Bear, 44–46.
114. Bear, 121.
115. Bear, 342.
116. Bear, 343–44.
117. Lynn Margulis, "Gaia Is a Tough Bitch," in *The Third Culture: Beyond the Scientific Revolution*, ed. John Brockman (Simon & Schuster, 1995), 140.
118. Westfahl, *Mechanics of Wonder*, 185.
119. Ruse, *Gaia Hypothesis*, 217.
120. "2001 Amsterdam Declaration on Earth System Science," International Geosphere-Biosphere Programme (IGBP), July 13, 2001, http://www.igbp.net/about/history/2001amsterdamdeclarationonearthsystemscience.4.1b8ae20512db692f2a680001312.html.
121. Bertalanffy, *General System Theory*, 45.

3. Remote Viewing

1. Paul Erickson et al., *How Reason Almost Lost Its Mind: The Strange Career of Cold War Rationality* (University of Chicago Press, 2013), 3.

2. Uri Geller is an Israeli British psychic best known for illusions performed on television. During the 1970s and 1980s, Geller appeared on *The Tonight Show with Johnny Carson, The Merv Griffin Show,* and other programs where he would bend spoons, stop watches, and describe hidden drawings using only the powers of his mind. His abilities attracted interest from New Age and psi community members, including Hal Puthoff (Stanford Research Institute) and Jack Sarfatti (Fundamental Fysiks Group). For more information, see Kaiser, *How the Hippies Saved Physics,* chap. 4.

3. Ingo Swann, *To Kiss the Earth Goodbye* (Hawthorn Books, 1975), 81.

4. Swann, 84.

5. Ingo Swann, *Psychic Literacy and the Coming Psychic Renaissance* (Swann-Ryder Productions, 2018), 90.

6. Ingo Swann, *Your Nostradamus Factor: Accessing Your Innate Ability to See into the Future* (Panta Rei, 1993), 41.

7. Swann, 41.

8. *Kirlian photography* is an umbrella term for photographic techniques that capture electrical coronal discharges. In the 1960s, Professor Thelma Moss of UCLA's Neuropsychic Institute argued that Kirlian photography captured the invisible "auras" of objects and beings. This concept was popularized to the general public in a best-selling book: Lynn Schroeder and Sheila Ostrander, *Psychic Discoveries behind the Iron Curtain* (Bantam, 1970).

9. Puglionesi, *Common Phantoms,* 4.

10. G. N. M. Tyrrell, *Apparitions* (Duckworth, 1943), 149.

11. Swann, *To Kiss,* 6.

12. Swann, 5.

13. Swann, 4.

14. Swann, 6.

15. Swann, 9.

16. After the death of James Hyslop in 1920, strife broke out between the pro-spiritualist (believers in ghosts and mediums) and the pro-science (advocates of laboratory science and medium testing) factions of the ASPR. When spiritualist Frederick Edwards assumed the ASPR presidency in 1923, the latter group left and created the Boston Society for Psychical Research in 1925. For more, see Inglis, *Science and Parascience,* chap. 5.

17. Puglionesi, *Common Phantoms,* 8.

18. Swann, *To Kiss,* 3.

19. Michel Foucault, "Governmentality," in *The Foucault Effect: Studies in Governmentality* (University of Chicago Press, 1991), 92.

20. Foucault, 96.

21. Foucault, 93.

22. Erickson et al., *How Reason,* 5, 12.

23. Velminski, *Homo Sovieticus,* 38.

24. Velminski, 36.

25. Velminski, 37.

26. Velminski, 71.

27. Velminski, 85.

28. Department of Defense (DoD), "Controlled Offensive Behavior—USSR," July 1972, 39, https://www.dia.mil/FOIA/FOIA-Electronic-Reading-Room/FOIA-Reading-Room-Russia/FileId/122008/.

29. DoD, 40.

30. Vasiliev, *Experiments,* 5.

31. DoD, "Controlled Offensive Behavior," 25.

32. DoD, "Controlled Offensive Behavior," 39.

33. Both Hal Puthoff and Russell Targ were laser physicists who dabbled in the California New Age scene, so it was not out of the ordinary for the two engineers to study psi phenomena in the 1970s. Puthoff was also a Scientologist. For more on their connection to the esoteric culture and science of the period, see Kaiser, *How the Hippies Saved Physics,* 69–70.

34. Jim Schnabel, *Remote Viewers: The Secret History of America's Psychic Spies* (Dell, 1997), 97.

35. Schnabel, 59.

36. Swann, *To Kiss,* 58.

37. Swann, 106.

38. Schnabel, *Remote Viewers,* 100.

39. Schnabel, 101.

40. Swann, *To Kiss,* 109.

41. Schnabel, *Remote Viewers,* 102.

42. Ingo Swann, "Notes on Experiments and Research at Stanford Research Institute, December 1972 through August 1973," August 1973, Box 127, Folder 8, Ingo Swann Papers, University of West Georgia Special Collections, University of West Georgia Ingram Library, Carollton, Ga.

43. Ingo Swann, "Coordinate Remote Viewing," May 1, 1986, 1, Box 138, Folder 1, Ingo Swann Papers.

44. Swann, 1.

45. Swann, 2.

46. Mary Craig Sinclair provides precise steps for receiving images telepathically sent by her husband in *Mental Radio.* After sitting in a dimly lit room to minimize distractions, "say to the unconscious mind, 'I want the picture which is on this card, or paper, presented to my consciousness.' . . . Then relax into blankness again and hold blankness for a few moments, then try gently, without straining, to see whatever forms may appear on the void into which you look with closed eyes. Do not

try to conjure up something to see; just wait expectantly and let something come." For more detailed instructions, see Upton Sinclair, *Mental Radio* (Charles C. Thomas, 1930), 116–24.

47. Swann, *To Kiss*, 10.
48. Harold Puthoff and Russell Targ, "Perceptual Augmentation Techniques. Part Two: Research Report," Stanford Research Institute (1975), 4, https://avalonlibrary.net/ Perceptual_Augmentation_Techniques_SRI_1973-1975_%28Remote_Viewing_ Puthoff_%26_Targ%29/CIA-RDP96-00791R000100420002-0_SRI%20-%20PERCEP TUAL%20AUGMENTATION%20TECHNIQUES%20-%20PART%20I%201974-75 -76.pdf.
49. Puthoff and Targ, 6.
50. Schnabel, *Remote Viewers*, 112.
51. Harold Puthoff and Russell Targ, "A Perceptual Channel for Information Transfer over Kilometer Distances: Historical Perspective and Recent Research," *Proceedings of the IEEE* 64, no. 3 (1976): 337.
52. Russell Targ and Harold Puthoff, "Information Transmission under Conditions of Sensory Shielding," *Nature* 251 (1974): 607, emphasis added.
53. Puthoff and Targ, "Perceptual Channel," 350.
54. Puthoff and Targ, 329.
55. Swann, "Coordinate Remote Viewing," 1.
56. Puthoff and Targ, "Perceptual Channel," 330.
57. Swann also developed noncoordinate forms of remote viewing. For example, in 1973, Swann used his remote viewing powers to probe the surface of the planet Jupiter, with his results cross-checked with data from the Pioneer 10, Pioneer 11, Voyager 1, and Voyager 2 satellite flybys. For more information, see Harold Sherman and Ingo Swann, "An Experimental Psychic Probe of the Planet Jupiter" (1973).
58. Targ and Puthoff, "Information Transmission," 607.
59. Over the years, SRI's psychical surveillance programs took on several different names that were based on funding, leadership, and administrative changes, such as Project Scanate, Project Grill Flame, Project Gondola Wish, and Project Stargate. I use *Project Stargate* as an umbrella term for all government projects involving remote viewing.
60. Swann, "Coordinate Remote Viewing."
61. Swann, 5.
62. In his personal notes, Swann includes a brainstorm on the etymology of the word *matrix*, including its origin from the Latin term *mater* (womb); "something within which something else develops"; "the natural material within which something is embedded"; "a place or medium in which something is produced"; "a place of point of origin or growth," "the formative part"; and "everything should have its formative matrix within which and out of which it is formed." See Ingo Swann, "Matrix 1996–1997," n.p., Box 136, Folder 7, Ingo Swann Papers. Although Swann likely arrived at the term *matrix* independently of *Neuromancer*, it is possible that

he liked William Gibson's idea of a consensual digital hallucination and applied it to his psychical realm. It is also possible that he came across Gardner Murphy's description of "the world of the paranormal as a kind of matrix from which proceed impressions which influence the specific psychological events which happen from day to day, and upon which they in turn make some impression." Gardner Murphy, "Field Theory and Survival," *Journal of the American Society for Psychical Research* 39, no. 4 (1945): 203.

63. Swann, "Coordinate Remote Viewing," 5.
64. William James, "The Last Report," in *William James on Psychical Research,* comp. and ed. Gardner Murphy and Robert O. Ballou (Viking, 1960), 324.
65. James, 324.
66. Murphy, "Field Theory," 192.
67. Swann, "Coordinate Remote Viewing," 5.
68. Swann, 5.
69. Swann, 5–6.
70. Swann, 6.
71. Swann, 7.
72. Swann, 7–8.
73. Swann, 8.
74. Swann, 7.
75. Puglionesi, *Common Phantoms,* 116.
76. Puglionesi, 116.
77. Swann, "Coordinate Remote Viewing," 3.
78. Swann, 3.
79. Swann, 3.
80. Swann, 3.
81. Myers, *Human Personality,* 1:12.
82. Ingo Swann, "Letter from Swann to Dr. Dennis Edmondson," November 29, 1991, Box 18, Folder 3, Ingo Swann Papers.
83. Schnabel, *Remote Viewers,* 114.
84. Schnabel, 218–19.
85. Schnabel, 27.
86. Schnabel, 38.
87. Schnabel, 77.
88. Michael Mumford, Andrew Rose, and David Goslin, "An Evaluation of Remote Viewing: Research and Applications," American Institutes for Research, September 29, 1995, 4–11, https://www.cia.gov/readingroom/docs/CIA-RDP96-00791R0002001 80006-4.pdf.
89. Mumford, Rose, and Goslin, 4–8.
90. Schnabel, *Remote Viewers,* xi.
91. Project MKUltra was a CIA project designed to develop drugs and procedures for use in interrogation, psychological torture, and brainwashing. Starting in 1953, the

CIA funded internal and external studies on prisoners, psych ward patients, ethnic minorities, and sex workers, many of whom were never notified they were being experimented on. For more, see Dyck, *Psychedelic Psychiatry,* chap. 2.

92. Ingo Swann, *Star Fire* (Souvenir Press, 1978), 255.
93. Swann, 68.
94. Swann, 75.
95. Swann, 44.
96. Swann, *To Kiss,* 95.
97. Swann, *Star Fire,* 48–49.
98. Swann, 44.
99. Swann, 14.
100. Swann, 14.
101. Swann, 35.
102. Swann, 158.
103. DoD, "Controlled Offensive Behavior," 18, 48.
104. Swann, *Star Fire,* 14.
105. Swann, 31.
106. DoD, "Controlled Offensive Behavior," 17–18.
107. Swann, *Star Fire,* 49–50.
108. Swann, 69.
109. Swann, 68.
110. Swann, 72.
111. Swann, 60.
112. Ido Hartogsohn, *American Trip: Set, Setting, and the Psychedelic Experience in the Twentieth Century* (MIT Press, 2020), 102.
113. Swann, *Star Fire,* 302.
114. In 1991, *Natural ESP* was republished under the new title *Everybody's Guide to Natural ESP* (1991).
115. Ingo Swann, *Everybody's Guide to Natural ESP: Unlocking the Extrasensory Power of Your Mind* (Swann-Ryder Productions, 1991), xiv.
116. Swann, 3.
117. Annie Besant and Charles Leadbeater, *Occult Chemistry: Clairvoyant Observations on the Chemical Elements* (Zuubooks, 1919), 7.
118. Swann, *Everybody's Guide,* 43.
119. Swann, 36.
120. Swann, 59.
121. Swann, 58.
122. Swann, 123.
123. Swann, 123.
124. Swann, 125.
125. Swann, 125.
126. Swann, xv.

127. Ingo Swann, "Proposal for Natural ESP," n.d., Box 142, Folder 2, Ingo Swann Papers.

128. Swann, *Your Nostradamus Factor,* 121.

129. Swann, 3.

130. Swann, 122.

131. Swann, 122.

132. Swann, 123.

133. Swann, 126–27.

134. Swann, 22.

135. Swann, 22.

136. Swann, 22.

137. Swann, 25–26.

138. Swann, 23.

139. Ingo Swann, "Panoramic Consciousness—Prospectus for International School for Applied Psychic Studies," n.d., n.p., Box 143, Folders 1–10, Ingo Swann Papers.

140. Swann.

141. Rhine, *Extra-sensory Perception,* 181.

142. Swann, "Panoramic Consciousness."

143. Swann.

144. Swann.

145. Swann.

146. Swann, *Psychic Literacy,* 8.

147. Swann, 8.

4. On Ghosts and Ghost Vision

1. Leping Zha and Tron McConnell, "Parapsychology in the People's Republic of China, 1979–1989," Central Intelligence Agency, February 13, 1990, 2, https://www.cia.gov/readingroom/document/cia-rdp96-00789r002600290003-0.

2. Zha and McConnell, 2.

3. Zha and McConnell, 9.

4. Zha and McConnell, 8.

5. Zha and McConnell, 10.

6. Zha and McConnell, 11.

7. Comte, *Positive Philosophy,* 28.

8. *Internalist epistemology* refers to the belief that the success of modern science is insured purely through its internal features: the scientific method, objectivity, rationality, mathematics for exposing nature's laws, and so on. In contrast, external epistemology refers to the belief that outside forces—society, politics, history—alone lead to scientific claims. Sandra Harding argues that both extremes are troubling and suggests that the constructionist model of post-Kuhnian science studies—that is, the notion that science and society co-construct scientific knowledge—provides

a better way of understanding knowledge production. For more on internalist and externalist epistemology, see Harding, *Is Science Multicultural?*, chap. 1.

9. Harding, 2.
10. Bronislaw Malinowski, *Magic, Science, and Religion* (Doubleday Anchor, 1948), 17–18.
11. Latour, *We Have Never Been Modern*, 10.
12. My use of *premodern* acknowledges a vexing connotation. The term is supremely useful in identifying early and marginalized arenas of thought while still participating in the problematic rhetorical centrality of the modern. I nevertheless use the term, given its widespread use in science and technology studies.
13. Harding, *Is Science Multicultural?*, 10.
14. Harding, 16.
15. Sandra Harding, *Sciences from Below: Feminisms, Postcolonialities, and Modernities* (Duke University Press, 2008), 138.
16. Harding, 140.
17. Mignolo, "On Pluriversality and Multipolarity," x.
18. Reiter, introduction to *Constructing the Pluriverse*, 2.
19. Reiter, 7.
20. Brogan, *Cultural Haunting*, 5.
21. Brogan, 7.
22. Avery Gordon, *Ghostly Matters: Haunting and the Sociological Imagination* (University of Minnesota Press, 1997), 8.
23. Gordon, 195.
24. David Lloyd, *Under Representation: The Racial Regime of Aesthetics* (Fordham University Press, 2019), 7.
25. Judy, *Sentient Flesh*, 319.
26. Michel Foucault, "Intellectuals and Power," in *Language, Counter-memory, Practice: Selected Essays and Interviews*, ed. Donald F. Blanchard (Cornell University Press, 1977), 209.
27. Harding, *Science Question*, 9.
28. Harding, 21.
29. Harding, 60.
30. Harding, 60.
31. Galvan, *Sympathetic Medium*, 30.
32. Galvan, 16.
33. Galvan, 12.
34. Galvan, 9.
35. Galvan, 13.
36. Galvan, 10.
37. Adam J. Rock, "Introduction: The Medium and the Message," in *Survival Hypothesis: Essays on Mediumship* (McFarland, 2014), 11.
38. Galvan, *Sympathetic Medium*, 30.

39. Galvan, 30.
40. Rock, "Introduction," 11.
41. Gurney, Podmore, and Myers, *Phantasms*, 2:537.
42. Myers, *Human Personality*, 2:74.
43. Ruth Maxey, "'The East Is Where Things Begin': Writing the Ancestral Homeland in Amy Tan and Maxine Hong Kingston," *Orbis Litterarum* 60 (2005): 4.
44. Edward Said, *Orientalism* (Pantheon Books, 1978), 5.
45. Amy Tan, *The Hundred Secret Senses* (Putnam, 1995), 15–16.
46. Tan, 15.
47. Tan, 3.
48. Sheng-Mei Ma, *The Deathly Embrace: Orientalism and Asian American Identity* (University of Minnesota Press, 2000), 116.
49. Ma, 125.
50. Lina Unali, "Americanization and Hybridization in *The Hundred Secret Senses* by Amy Tan," *Hitting Critical Mass* 4, no. 1 (1996): 142.
51. Dominic Zoehrer, "From Fluidum to Prana: Reading Mesmerism through Orientalist Lenses," in *The Occult Nineteenth Century: Roots, Developments, and Impact on the Modern World*, ed. Lukas Pokorny and Franz Winter (Palgrave Macmillan, 2021), 86.
52. Zoehrer, 88–89.
53. Zoehrer, 89.
54. Zoehrer, 93.
55. Zoehrer, 99.
56. Magdelena Kraler, "Tracing Vivekananda's Prana and Akasa: The Yogavasistha and Rama Prasad's Occult Science of Breath," in Pokorny and Winter, *Occult Nineteenth Century*, 378.
57. Robin Yang, "Yinyang Narrative of Reality," in *Chinese Metaphysics and Its Problems* (Cambridge University Press, 2015), 16–17.
58. Yang, 21.
59. Stoic philosopher Posidonius postulated that a "vital force" differentiated living from nonliving things. Henri Bergson modified the term into *élan vital* to signify the self-organization and complexity of living things. For more, see Henri Bergson, *Creative Evolution* (1907), trans. Arthur Mitchell (Holt, 1911).
60. JeeLoo Liu, "In Defense of Chinese Qi-Naturalism," in *Chinese Metaphysics and Its Problems*, ed. Chenyang Li and Franklin Perkins (Cambridge University Press, 2015), 38–39.
61. Liu, 41.
62. Liu, 41.
63. Liu, 41.
64. Liu, 41.
65. Liu, 39.
66. Liu, 48.

67. Liu, 48.

68. Liu, 49.

69. Liu, 49.

70. Myers, *Human Personality*, 2:263.

71. Jiena Sun, "Telling and Enacting Ghost Stories: Narrative and Agency in Amy Tan's *The Hundred Secret Senses*," *Interdisciplinary Literary Studies* 19, no. 3 (2017): 264.

72. Tan, *Hundred Secret Senses*, 18.

73. Tan, 18.

74. According to the meridional theory on which Chinese medicine is based, meridians are a system of metaphysical channels in the body through which qi flows. Superficial meridians connect an organism with the outer environment, principal meridians connect muscles and tissue, distinct meridians nourish deep organs, and curious meridians connect the major channels and serve as energy reservoirs. Each of the twelve regular or principal meridians is associated with an organ, such as the heart, pericardium, lung, spleen, liver, kidney (the yin organs) and stomach, gallbladder, large and small intestine, urinary bladder, and triheater or triple burner (the yang organs). See John C. Longhurst, "Defining Meridians: A Modern Basis of Understanding," *Journal of Acupuncture and Meridian Studies* 3, no. 2 (2010): 67–68.

75. Longhurst, "Defining Meridians," 68.

76. Tan, *Hundred Secret Senses*, 14.

77. Tan, 15.

78. Tan, 28.

79. Tan, 29.

80. Tan, 102.

81. Tan, 211–12.

82. Madam H. P. Blavatsky, *Isis Unveiled* (Bouton, 1877), 1.6.

83. Tan, *Hundred Secret Senses*, 102.

84. Pamela Thurschwell, *Literature, Technology and Magical Thinking, 1880–1920* (Cambridge University Press, 2001), 27.

85. Tan, *Hundred Secret Senses*, 102.

86. Tan, 29.

87. Tan, 205.

88. Tan, 332.

89. Tan, 312.

90. Tan, 316–17.

91. Tan, 317.

92. Tan, 320.

93. Tan, 320.

94. Sun, "Telling," 272.

95. Tan, *Hundred Secret Senses*, 204.

96. Patricia Chu argues the Asian American bildungsroman traditionally involves reconciling the Asian past with the American present identities to produce a new, third identity—the Asian American—who can achieve a sense of wholeness in life. Patricia Chu, "Bildung and the Asian American Bildungsroman," in *The Routledge Companion to Asian American and Pacific Islander Literature,* ed. Rachel Lee (Routledge, 2014), 409.

97. Tan, *Hundred Secret Senses,* 345.

98. Tan, 349.

99. Benzi Zhang, "Reading Amy Tan's Hologram: *The Hundred Secret Senses,*" *International Fiction Review* 31, no. 1–2 (2004): 14, https://journals.lib.unb.ca/index.php/IFR/article/view/7766/8823.

100. The disclosure and acknowledgment of Japan's military prostitution program is complicated. For obvious reasons, the Japanese government refused for decades to acknowledge the dark history of comfort women. The term itself comes from *ianfu,* "a woman who gives ease and consolation," a euphemism designed to disguise the sexual violence inherent in the role. In 1985, Shirota Suzuko published an autobiography describing her miseries as a prostitute for World War II Japanese soldiers. In 1991, Kim Hak-Sun sued the Japanese government for reparations after her World War II sufferings at a so-called comfort station; this opened the door for many more cases demanding formal apologies and compensation. While the Japanese government initially apologized for its wartime prostitution program, right-wing officials and academics have successfully waged antiapology and antiredress campaigns from the mid-1990s onward, arguing that comfort women willingly volunteered for their positions, so the military ought not be held responsible. For more information, see Li Hongxi, "The Extreme Secrecy of the Japanese Army's 'Comfort Women' System," *Chinese Studies in History* 53, no. 1 (2020): 28–40; and Yang Li, "Reflections on Postwar Nationalism: Debates and Challenges in the Japanese Academic Critique of the 'Comfort Women' System," *Chinese Studies in History* 53, no. 1 (2020): 41–55.

101. Tina Chen, *Double Agency: Acts of Impersonation in Asian American Literature and Culture* (Stanford University Press, 2005), 131.

102. Jodi Kim, "Haunting History: Violence, Trauma, and the Politics of Memory in Nora Okja Keller's *Comfort Woman,*" *Hitting Critical Mass* 6, no. 1 (1999): 66.

103. Patricia Chu, "'To Hide Her True Self': Sentimentality and the Search for an Intersubjective Self in Nora Okja Keller's *Comfort Woman,*" in *Asian North American Identities: Beyond the Hyphen,* ed. Eleanor Ty and Donald C. Goellnicht (Indiana University Press, 2004), 72.

104. Janna Odabas, *The Ghosts Within: Literary Imaginations of Asian America* (Deutsche Nationalbibliothek, 2018), 12–14.

105. Bonnie Winsbro, *Supernatural Forces: Belief, Difference, and Power in Contemporary Works by Ethnic Women* (University of Massachusetts Press, 1993), 6.

106. Winsbro, 4.

107. Winsbro, 4.

108. Winsbro, 4.

109. Tina Chen argues that Korean shamanism acts as a response to various forms of oppression. As a proxy for Korean national identity, shamanism emerges over the course of *Comfort Woman* as Akiko Bradley's challenge to colonialism, neocolonialism, patriarchy, Christianity, and American assimilation. For more, see Chen, *Double Agency*, chap. 5.

110. Hahm Pyong-Choon, "Shamanism and the Korean World-view, Family Life Cycle, Society, and Social Life," in *Shamanism: The Spirit World of Korea*, ed. R. Guisso and Chai-shin Yu (Asian Humanities Press, 1988), 60.

111. Chang Chu-Kun, "An Introduction to Korean Shamanism," trans. Yoo Young-Sik, in Guisso and Yu, *Shamanism*, 30.

112. Pyong-Choon, "Shamanism," 61.

113. Pyong-Choon, 72.

114. Pyong-Choon, 74.

115. Pyong-Choon, 75.

116. Pyong-Choon, 76.

117. Pyong-Choon, 76.

118. Laurel Kendall, *Shamans, Housewives, and Other Restless Spirits: Women in Korean Ritual Life* (University of Hawai'i Press, 1985), 55.

119. Kendall, 100.

120. Kendall, 99.

121. Kendall, 99.

122. Kendall, 99.

123. Kendall, 101.

124. Kendall, 102.

125. Kendall, 103.

126. Alan Carter Covell, *Folk Art and Magic: Shamanism in Korea* (Hollym, 1986), 14–15.

127. Kendall, *Shamans*, 21.

128. Kendall, 29.

129. Kendall, 35.

130. Kendall, 25.

131. Chu, "To Hide," 71.

132. Sung Hee Yook, "Mourning Unmourned Deaths: Shamanic Rituals in Nora Okja Keller's *Comfort Woman*," *Feminist Studies in English Literature* 19, no. 3 (2011): 136.

133. Nora Okja Keller, *Comfort Woman* (Viking, 1997), 20.

134. Keller, 21.

135. Keller, 96.

136. Kendall, *Shamans*, 99.

137. Pyong-Choon, "Shamanism," 76.

138. Keller, *Comfort Woman*, 38.

139. Keller, 38.

140. Keller, 38.

141. Keller, 55.

142. Keller, 144.

143. Keller, 35.

144. Pyong-Choon, "Shamanism," 132.

145. Kendall, *Shamans,* 20.

146. Keller, *Comfort Woman,* 6.

147. Kim Kwang-Il, "Kut and the Treatment of Mental Disorder," trans. Suh Kik-on and Im Hye-young, in Guisso and Yu, *Shamanism,* 145.

148. Keller, *Comfort Woman,* 8.

149. The scholarship on *Comfort Woman* and historical violence is copious given the canonical status of Keller's text in Asian American literature. For good starting points, see Kim, "Haunting History"; Chu, "To Hide"; Chen, *Double Agency,* chap. 5; and Silvia Schultermandl, "Writing Rape Trauma, and Transnationality onto the Female Body: Matrilineal Em-body-ment in Nora Okja Keller's *Comfort Woman,*" in Lee, *Routledge Companion to Asian and Pacific Islander Literature,* 403–14.

150. Chen, *Double Agency,* 143.

5. The Multiverse and the Mind

1. Dubey, "Future of Race," 16.

2. Treuer, *Native American Fiction,* 165.

3. The *ethnographic gaze* is a term used to comprehend historical photographs and imagery, particularly those existing outside the cultural framework of the presumed white photographer. The subject of the photographic lens is often aestheticized as some societal ideal and thus romanticized, orientalized, anonymized, etc. Such a gaze enacts traditional hierarchies of Western power/knowledge. For more, see Maria Kakavoulia, *Ethnographica Moralia: Experiments in Interpretive Anthropology* (Fordham University Press, 2008), 213–17.

4. Dubey, "Future of Race," 21.

5. Cathryn Josefina Merla-Watson and B. V. Olguin, *Altermundos: Latin@ Speculative Literature, Film, and Popular Culture* (UCLA Chicano Studies Research Center Press, 2017), 4.

6. Ramón Saldívar, "Historical Fantasy, Speculative Realism, and Postrace Aesthetics in Contemporary American Fiction," *American Literary History* 23, no. 3 (2011): 574.

7. Mignolo, "On Pluriversality and Multipolarity," x.

8. Blavatsky, *Secret Doctrine,* 1.

9. Gurney, Podmore, and Myers, *Phantasms,* 1:lx–lxi.

10. Gurney, Podmore, and Myers, 1:120–21.

11. Myers, *Human Personality,* 1:26.

12. Gurney, Podmore, and Myers, *Phantasms,* 1:271.

13. Gurney, Podmore, and Myers, 1:271.

14. Kandice Chuh, *The Difference Aesthetics Makes* (Duke University Press, 2019), 4.

15. Judy, *Sentient Flesh*, 350.

16. Judy, 353.

17. Judy, 349.

18. Judy, 13.

19. Judy, 19.

20. Ramón Saldívar, "The Other Side of History, the Other Side of Fiction: Form and Genre in Sesshu Foster's *Atomik Aztex*," in *American Studies as Transnational Practice: Turning toward the Transpacific*, ed. Yuan Shu and Donald E. Pease (Dartmouth College Press, 2015), 159.

21. Saldívar, 160.

22. Saldívar, 158.

23. Sesshu Foster, *Atomik Aztex* (City Lights, 2005), 1.

24. Sesshu Foster, "Prefatory Note to *Atomik Aztex*," *Ameriasia Journal* 27, no. 2 (2001): 1.

25. For the rest of this chapter, to minimize any confusion, I will generally use *Zenzontli* to refer to the military officer in the Mesoamerican-dominated universe and *Zenzon* to refer to the immigrant in a Western-dominated universe. There are overlaps between these characters, however, so his two names will sometimes mingle.

26. Scholarship on *Atomik Aztex* is small but growing. Sascha Pöhlmann interprets the novel as a "cosmographic metafiction" highly effective for its political critique. Stephen Hong Sohn views the novel as a satire of capitalism and socialism. Ramón Saldívar has written about the text as a key instantiation of speculative realist critique of colonialism. See Sascha Pöhlmann, "Cosmographic Metafiction in Sesshu Foster's *Atomik Aztex*," *Amerikastudien/American Studies* 55, no. 2 (2010): 226; Stephen Hong Sohn, *Racial Asymmetries* (New York University Press, 2014), 82; and Saldívar, "Other Side of History," 158.

27. Vincent Malmstrom, *Cycles of the Sun, Mysteries of the Moon: The Calendar in Mesoamerican Civilization* (University of Texas Press, 1997), 14.

28. Elizabeth Boone, *Cycles of Time and Meaning in the Mexican Books of Fate* (University of Texas Press, 2007), 3.

29. Boone, 3.

30. Burr Brundage, *The Fifth Sun: Aztec Gods, Aztec World* (University of Texas Press, 1979), 14.

31. The twenty days in each calendar refer to the twenty divinities: alligator, wind, horse, lizard, snake, death, deer, rabbit, water, dog, monkey, grass, reed, jaguar, eagle, vulture, earthquake, knife, rain, and flower.

32. Boone, *Cycles of Time*, 17.

33. Brundage, *Fifth Sun*, 26.

34. Brundage, 26.

35. David Carrasco, *City of Sacrifice: The Aztec Empire and the Role of Violence in Civilization* (Beacon Press, 1999), 3.

36. Brundage, *Fifth Sun*, 26–27.

37. Harding, *Is Science Multicultural?*, 10.

38. Foster, *Atomik Aztex*, 1.

39. Foster, 5.

40. Foster, 1.

41. David Wallace, *The Emergent Multiverse: Quantum Theory According to the Everett Interpretation* (Oxford University Press, 2012), 44.

42. Foster, *Atomik Aztex*, 3.

43. Kay Read, *Time and Sacrifice in the Aztec Cosmos* (Indiana University Press, 1998), 106.

44. Read, 101.

45. Foster, *Atomik Aztex*, 73.

46. Foster, 131.

47. Judy, *Sentient Flesh*, 392–93.

48. Foster, *Atomik Aztex*, 130–31.

49. Foster, 5.

50. Foster, 3.

51. Read, *Time and Sacrifice*, 100.

52. Read, 110.

53. Read, 112.

54. See Derek Lee, "Postquantum: A Tale for the Time Being, *Atomik Aztex*, and Hacking Modern Space-Time." *MELUS* 45, no. 1 (2020): 15–16.

55. Foster, *Atomik Aztex*, 102.

56. Read, *Time and Sacrifice*, 101.

57. Foster, *Atomik Aztex*, 152.

58. Foster, 152.

59. Foster, 3.

60. Foster, 200.

61. Foster, 165–66.

62. Clifford Abbot, "Floyd Lounsbury," National Academy of Sciences, 2013, 5, https://www.nasonline.org/wp-content/uploads/2024/06/lounsbury-floyd.pdf.

63. For a good introduction to the burgeoning field of Ozeki scholarship, see Shaheem Black, "Fertile Cosmofeminism: Ruth L. Ozeki and Transnational Reproduction," *Meridians* 5, no. 1 (2004): 226–56; Monica Chiu, "Postnational Globalization and (En)gendered Meat Production in Ruth L. Ozeki's *My Year of Meats*," *LIT* 12 (2001): 99–128; Michelle Huang, "Ecologies of Entanglement in the Great Pacific Garbage Patch," *Journal of Asian American Studies* 20, no. 1 (2017): 95–117; Susan McHugh, "Flora, not Fauna: GM Culture and Agriculture," *Literature and Medicine* 26, no. 1 (2007): 25–54; Molly Wallace, "Discomfort Foods: Analogy, Biotechnology, and Risk in Ruth Ozeki's *All Over Creation*," *Arizona Quarterly* 67, no. 4 (2011): 155–81; and

Andrew Wallis; "Towards a Global Eco-consciousness in Ruth Ozeki's *My Year of Meats,*" *Interdisciplinary Studies in Literature and Environment* 20, no. 4 (2013): 837–54.

64. *Shishosetsu,* also known as the I-novel, is a Japanese literary genre that generally uses first-person narration, casual rhetoric, and naturalistic technique to expose the darker elements of society.

65. Schrödinger's cat is a thought experiment devised by Erwin Schrödinger in 1935 to illustrate the paradoxes of the Copenhagen interpretation of quantum mechanics. In Schrödinger's famous thought experiment, a cat is sealed in a box that has a fifty-fifty chance of releasing a poison. Under the Copenhagen interpretation, the cat is technically both dead and alive until an observer opens the box and peers inside, at which point all the quantum possibilities collapse to one (e.g., dead). Everett argues that the cat continues to be both dead and alive because it already exists in two realities that have branched off from the original reality. The cat may appear dead in our reality, but we cannot rule out its living in another reality that we cannot perceive.

66. Hee-Jin Kim, *Dogen Kigen: Mystical Realist* (University of Arizona Press, 1987), 30.

67. Taigen Leighton, *Vision of Awakening Space and Time: Dogen and the Lotus Sutra* (Oxford University Press, 2007), 3.

68. Kim, *Dogen Kigen,* 7.

69. Leighton, *Vision,* 3.

70. Leighton, 9.

71. Kim, *Dogen Kigen,* 104.

72. Kim, 116.

73. Kim, 112.

74. Kim, 120.

75. Kim, 113.

76. Steven Heine, *Existential and Ontological Dimensions of Time In Heidegger and Dogen* (State University of New York Press, 1985), 1.

77. Heine, 1.

78. Heine, 4.

79. Heine, 51.

80. Heine, 55.

81. Kim, *Dogen Kigen,* 140.

82. Leighton, *Vision,* 107.

83. Dogen, *Shobogenzo: Zen Essays by Dogen* (University of Hawai'i Press, 1986), 106.

84. Leighton, *Vision,* 108.

85. Dogen, *Shobogenzo,* 108.

86. Leighton, *Vision,* 145.

87. Kim, *Dogen Kigen,* 111–12.

88. Ruth Ozeki, *A Tale for the Time Being* (Viking, 2013), 70.

89. *Hentai* is a genre of Japanese-influenced pornography that emphasizes perverse sexual desires.

90. Ozeki, *Tale,* 182.

91. Ozeki, 183.

92. Ozeki, 213.

93. Ozeki, 213.

94. Ozeki, 249.

95. Ozeki, 212.

96. Ozeki, 409.

97. *Spin* refers to a form of angular momentum carried by elementary particles like photons, atomic nuclei, and hadrons. It was originally conceived as the rotation of particles around some axis, with the spin obeying the same mathematical laws as other quantized angular momenta. For photons, spin is the quantum-mechanical counterpart to the polarization of light.

98. Albert Einstein, Max Born, and Hedwig Born, *The Born–Einstein Letters: Correspondence between Albert Einstein and Max and Hedwig Born from 1916 to 1955* (Macmillan, 1971), 158.

99. Ozeki, *Tale,* 3.

100. Ozeki, 39.

101. Ozeki, 178.

102. Ozeki, 341.

103. Ozeki, 348.

104. Ozeki, 348.

105. Ozeki, 350.

106. Ozeki, 341.

107. Owen, *Place of Enchantment,* 128.

108. Owen, 128.

109. Gauld, *Founders,* 44, 64.

110. Mignolo, "On Pluriversality and Multipolarity," ix.

6. The Structure of Parascientific Revolutions

1. Daryl Bem, "Feeling the Future: Experimental Evidence for Anomalous Retroactive Influences on Cognition and Affect," *Journal of Personality and Social Psychology* 100, no. 3 (2011): 409.

2. Bem, 419.

3. Bem, 421.

4. Daniel Engber, "Daryl Bem Proved ESP Is Real: Which Means Science Is Broken," *Slate,* May 17, 2017, https://slate.com/health-and-science/2017/06/daryl-bem -proved-esp-is-real-showed-science-is-broken.html.

5. Benedict Carey, "Journal's Paper on ESP Expected to Prompt Outrage," *New York Times,* January 5, 2011, https://www.nytimes.com/2011/01/06/science/06esp.html.

6. Andrew Gelman, "Too Good to Be True," *Slate,* July 24, 2013, https://slate.com/technology/2013/07/statistics-and-psychology-multiple-comparisons-give-spurious-results.html.
7. Engber, "Daryl Bem."
8. Inglis, *Science and Parascience,* 315.
9. Inglis, 315.
10. Collins and Pinch, "Construction," 244–58.
11. Henry James, *The Aspern Papers and the Turn of the Screw* (Penguin, 1984), 169.
12. David Lodge, *Consciousness and the Novel* (Harvard University Press, 2002), 10.
13. Harding, *Science Question,* 62.
14. Fleck, *Genesis,* 30.
15. Fleck, 100.

INDEX

Adams, Bill, 228n56

alchemy, element transfiguration in, 129

Amsterdam Declaration on Earth System Science (2001), 96

analytical overlay: degrading of remote viewing, 116, 133; military intelligence and, 116

anamnesis (loss of forgetfulness), 52

ancestors: Korean spirits, 166, 168, 173; perpetuity of, 202. *See also* spirit mediums

Anderson, Carl David, 43

antimatter: and concept of reality, 45; negative charge of, 43

antiprotons, discovery of, 43

antivaccination movement, quasi-scientific information in, 5–6

Aristotle, 152

art: effect of quantum mechanics on, 42; science and, 3

Asimov, Isaac, 13; *Foundation's Edge,* 86

Asprem, Egil, 5, 15

Astounding Science Fiction (magazine), 13

astral projection, 112, 117; Swann's, 104–5, 110

atomism, 38; logical, 70, 232n14

auras, photography of, 242n8

autopoiesis, of living systems, 74–75, 77

Aztecs: intellectual life of, 185; New Fire Ceremony, 186–87, 194. *See also* calendar, Aztec; chronology, Aztec; science, Aztec

Bacon, Francis, 143

bacteria: behavioral neurotoxins of, 93; dictating of human existence, 91; Gaian, 91–94; global superorganism of, 86; proprioceptic, 79; superiority over human systems, 94

Barnes, David, 61

Bateson, Gregory, 82

Bear, Greg, 86–96; apocalyptic fiction of, 30; "Blood Music," 86; influences on, 90; planetary mind concept, 89; reimagination of the paranormal, 95; rhizomatic connections in, 86; use of systems theory, 89. See also *Darwin's Radio*; *Vitals*

Bekhterev, V. M., 126; telepathy experiments, 39

Bem, Daryl: ESP experiments, 31–32, 211–13; "Feeling the Future," 211–12

Benford, Gregory: *Timescape,* 64

Bergson, Henri: on élan vital, 249n59

Bertalanffy, Ludwig von, 30; *General Systems Theory,* 73–74; on gestalt

properties, 78; influence on New Age, 82; on the occult, 96–97; organismic biology of, 73; systems theory of, 73–74, 81, 83; on unexplainable gestalts, 114; on vitalism, 74
Besant, Annie, 27, 216; *Occult Chemistry*, 129
Bester, Alfred: *Demolished Man*, 13
Bilaniuk, O. M. P., 44; on rest mass, 47
biological sciences: gendered language in, 237n1; telekinesis and, 69
biology, female: pathologized, 27
biology, systems: quantum mechanics and, 140
biomedicine, superiority over spiritual healing, 1
Blavatsky, Helena, 27; on *fluidum*, 152; Orientalism of, 158, 179; regression of humanity, 57; Theosophical Society, 8
bodhisattvas, 200
Bohm, David, 82
Bohr, Niels: on complementarity, 42
Boone, Elizabeth, 185
Borges, Jorge Luis: "On Exactitude in Science," 115
Boston Society for Psychical Research, 227n43, 242n16
Boyle, Robert, 143
brain: interaction with psitrons, 48, 64, 66; reception of external signals, 53, 126; subatomic stimuli of, 41. *See also* dendrons; mind; neurons
Bramble, John, 25
Brhadaranyaka Upanishad, prana in, 152
Broad, C. D.: "The Notion of Precognition," 37
Brogan, Kathleen: on cultural haunting, 146
Brundage, Burr, 186
Bryan, Frances, 119

Buddha-nature: ontological unity of, 199; thusness in, 199, 201. *See also* Zen Buddhism

calendar, Aztec, 184; divinities in, 254n31; intellectual life in, 185; month and year, 186 (fig.); ontological control through, 192; organization of daily life, 187; "round," 185–86; sacred (*tonalpohualli*) and secular (*xiuhpohualli*), 185, 189, 190, 195; as unifying structure, 185. *See also* chronology, Aztec; time, Aztec
calendrical science, Mesoamerican, 184–87, 196
Campbell, John, 95; *Astounding Science Fiction*, 13; on power of literature, 23; psionics of, 13–14
Cannon, Walter: homeostasis theory of, 239n38
Capra, Fritjof, 82, 85; influence on Bear, 91; on posthuman Gaia, 86; role in parascientific production, 70; *The Tao of Physics*, 83; *The Turning Point*, 83–84, 94; use of systems theory, 84
Carington, Whately, 213; *Thought Transference*, 38–39
Carrasco, David, 186
Carter, Jimmy, 119
cartography, clairvoyance and, 30
cause and effect: in Korean shamanism, 165; Western, 206
Chamberlain, Owen, 43
Chandler, Nahum, 19, 147, 161
Chandogya Upanishad, prana in, 152
Chen, Tina, 163, 172, 252n109
Chestnut, Charles: *The Conjure Woman*, 146
China: EFHB studies, 139, 155; ESP studies, 139–40; medicine, 139, 156–57, 250n74; nuclear testing, 119; in Tan's fiction, 150. *See also* qi, Chinese

chronology: the present in, 199; temporal control through, 184-85; Zen, 196, 204
chronology, Aztec, 192; in *Atomik Aztek*, 177, 184-87, 190-92, 194, 208; infinite worlds of, 208; sacrifice in, 186; as science, 187; transcending Western time, 190. *See also* calendar, Aztec; time, Aztec
Chu, Patricia, 163, 169, 251n96
Chuh, Kandice, 177; *The Difference Aesthetics Makes*, 181
CIA: "An Evaluation of Remote Viewing," 119-20; monitoring of Chinese parapsychology, 139; paranormal technology for, 109; Project MKUltra, 120, 245n91
cladogenesis: in *Darwin's Radio*, 90; evolutionary, 87, 241n91
clairvoyance, 10, 29; cartography and, 30, 103; for concrete objects, 117; in *Darwin's Radio*, 90; in espionage, 100; hybridized with cartography, 7; military, 136; national security value, 118; in Project Scanate, 111; resilience of, 5; stages of, 117; Swann's models for, 101-2; as technology, 118; traveling, 112; using geographic coordinates, 109-10; in *Watchmen*, 60. *See also* remote viewing; telepathy
Clarke, Bruce, 78, 238n28
climate change denialism, 6
cognition, physicalist model of, 29
Cold War: geopolitical strategy in, 100; paranormal science in, 30, 137, 218; psychical research in, 101, 118-19; psychic statecraft of, 106-8; scientific progress during, 229n75; surveillance in, 7, 101, 106, 137. *See also* governmentality, Cold War
Collins, H. M., 213
colonialism, racist literature of, 176. *See also* decolonialism

Comfort Woman (Nora Ojka Keller), 31, 162-64, 167-72; American protagonist of, 173; angry spirits of, 167, 168, 170; bad death in, 167-68; canonical status of, 253n149; challenge to colonialism, 252n109; coequal world views in, 204; counterdiscursive text of, 173; cultural haunting in, 163, 167; epistemology of, 142, 172, 173; ethnoscience in, 169-70; funeral rites in, 172; ghost seer of, 141-42, 171; ghosts in, 163-64, 167-72; mother-daughter relationship of, 171; phenomenological world of, 170; scholarship on, 253n149; spirit medium of, 163; spiritual ceremonies (*kuts*) in, 170-71; time in, 169; *yŏngsan* in, 167, 168
comfort women (Japanese Army program): reparations for, 251n100; violence against, 162, 251n100. See also *Comfort Woman*
communication: with the dead, 3; knowledge deficit model of, 28, 231n114; across multiverse, 207; with non-Western ghosts, 7, 31, 141, 170, 171; spectral, 29; through viruses, 88. *See also* clairvoyance; telepathy
Comte, Auguste: metanarrative of science, 6; positivism, 143
consciousness: bacterial, 97; beyond death, 149, 192; Earth's, 67; evolution of, 80; Gaian, 70, 80, 86, 95, 128; global, 29, 67, 68, 82; learning modes of, 240n65; in *Natural ESP,* 130 (fig.); New Age, 82; ontogenetic model of, 240n65; paranormal model of, 3; psitron's model of, 50; renewal in parascience, 207; shared, 135; in systems theory, 75, 84; theorizations of, 39; transcendent possibilities of, 178; Western conception of, 177-78
Coover, John Edgar, 228n48
Cornell University, ESP experiments at, 31-32

counterdiscourse: against dominant
powers, 230n92; Foucauldian, 20
Covell, Alan, 166
Cowan, Clyde, 43
Coyne, Jerry, 2
Crowley, Aleister, 205; Thelema of, 8
culture, mass: the paranormal and, 68,
179, 219, 228n56
culture, Western: exclusion of colonized
people, 147; living and dead in, 154;
psitrons in, 64. *See also*
multiculturalism
Curie, Marie, 129
cybernetics: self-regulation through, 76;
systems theory and, 238n28

Daisyworld (computer simulation), 68,
237n5
Darwin's Radio (Greg Bear), 70, 86,
87–89; Anthropocene in, 89; cladogenesis in, 90; clairvoyance in, 90;
evolutionary processor in, 93; Gaia in,
88–91; global superorganism of, 87;
hostile Earth in, 95; human genome in,
87–88, 90; paranormal mind in, 88;
phylogenetic memories in, 90; systems
theory in, 88; telekinesis in, 95; viruses
in, 87, 88, 95
Dasein (existence), 181, 182
dead, the: communication with, 3; minds
of, 36; subliminal selves of, 10. *See also*
ancestors
death: bad, 167–68; consciousness
beyond, 149, 192; humors (*sangmun*),
166; Korean rites of, 168
decolonialism, 182; approaches to the
paranormal, 145–46; of pluriverse, 208;
scholarship on, 161; in speculative
ethnic fiction, 178; for STS, 217; of
Western science, 144
Delanda, Manuel: *A Thousand Years of
Nonlinear History*, 66

Demiurge, Platonic, 71
dendrons: interaction with psychons,
40–41; "mental will" of, 40; psitrons
striking, 48; size of, 233n32
Deng Xiaoping, 140
Department of Defense (DOD), US:
"Controlled Offensive Behavior—
USSR," 107–8, 125, 126
Descartes, René, mechanistic theories of,
71–72
Dick, Philip K., 23, 86; belief in tachyons,
53–54; citizen science experiments, 53;
The Divine Invasion, 54; Exegesis
diary, 53–54, 55, 56; influence of, 64;
"Minority Report," 52; psychedelic
drug use, 52; psychic experiences of,
52–54; as theorist of mind, 55; 2-3-74
experiences, 53–54, 235n72; *The
Transmigration of Timothy Archer*, 54;
Ubik, psi in, 52; vitamin regimen,
52–53, 57, 115. See also *VALIS*
Dirac, Paul, 49; on antielectrons, 43
disorder, spontaneous, 74
DNA: as global system, 89; intelligent,
89–90
Dobbs, Adrian, 29, 37, 184; career of, 46;
on EEG use, 49–50; "The Feasibility of
a Physical Theory of ESP," 46; imaginary mass calculations, 66; interpretation of time, 47; on precasts, 56; on
precognition, 66; psitron theory, 34,
36, 40, 46–52, 57, 61, 132; and quantum mechanics, 47; reconceptualization of mind, 49; in SPR, 46; on
tachyons, 66; theory of consciousness,
36; "Time and Extrasensory Perception," 46; on the unconscious, 62; use
of dualist-interactionism, 48; use of
psychon theory, 46
Dogen, 203; on ancestors, 202; *Shobogenzo*, 198; on time, 200–201, 204;
zazen teachings, 198

Doyle, Arthur Conan, 5; SPR membership, 11
Drohan, Christopher, 60
dualist-interactionism, 35, 40; Dobb's use of, 48
Dubey, Madhu, 176
Duke University Parapsychology Laboratory, 12, 14, 228n48

Earth: agency of, 67–68; bacterial domination of, 92; in classical philosophy, 69, 70–71; ethical stewardship of, 96; gendered rhetoric of, 237n1; higher consciousness of, 67; as living entity, 30; in Plato, 71; self-regulation by, 75–77; as superorganism, 71. *See also* Gaia
Eburne, Jonathan, 19; *Outsider Theory,* 6
Eccles, John: Dobb's use of, 48; on mind–body interaction, 40–41; *The Neurophysiological Basis of Mind,* 40–41; on precognition, 66; on psychons, 233n28; theory of telepathy, 41–42
Eddington, Arthur, 43; on mind control, 233n32
Edmondson, Dennis, 118
$E = mc^2$: c in, 44; subjectivity of, 42
EFHB (exceptional functions of the human body), Chinese, 155; electromagnetic, 139
Einstein, Albert: Annus Mirabilis papers, 42; on quantum entanglement, 203
electroencephalogram (EEG), in psitron theory, 49–50, 65–66
Eliot, T. S., 4
Engber, Daniel: "Daryl Bem Proved ESP Is Real," 212
Enlightenment: epistemic progress following, 5; mechanism of, 72
environmental science, interdisciplinary, 96

epistemology: anomalous, 214; decolonial, 31, 208; external, 247n8; internalist, 143, 247n8; mosaic, 145, 161, 173, 177, 217
epistemology, heterodox: processes of, 18–19
epistemology, scientific: Kuhn's theory of, 15, 16, 28; multiparadigmatic, 144
Erdrich, Louise: *Tracks,* 146
Esalen Institute, New Age at, 82
esoterica, scientific discourse in, 72
espionage, telepathic, 99–100. *See also* surveillance, Cold War
Estabrooks, George Hoben, 228n48
Eternalism, in *Watchmen,* 60–61
ethnoscience, 145, 231n105; decolonization of paranormal mind, 208; in multicultural fiction, 27; past and future in, 177; qi in, 154; of speculative ethnic fiction, 178; temporal, 178; women mediums in, 174
eukaryotes: communication networks of, 92; symbiosis with prokaryotes, 77, 78, 238n45
Everett, Hugh, 197, 208, 256n65; "'Relative State' Formulation of Quantum Mechanics," 189
Everything Everywhere All at Once (2022), parallel realities in, 189
evolution, phylogenetic, 158
extrasensory perception (ESP): academic experiments in, 12–14; ASPR's, 105; Bem's experiments in, 31–32, 211–13; Chinese government study of, 139; core, 129, 131; for detection of psitrons, 49–50; discrediting of, 14; practical application of, 131; reproducible, 211; skepticism of, 135. *See also* Rhine, J. B.; senses

facts: social production of, 28; STS prioritization of, 216

Falen, Douglas: *African Science*, 26

Faraday cages, 39

Father Earth (Gaian figure), 86

Faulkner, William: *As I Lay Dying*, 188

Feinberg, Gerald: "Possibility of Faster-than-Light Particles," 44; on tachyons, 65

Feyerabend, Paul, 24, 214; *Against Method*, 25

fiction: paranormal, 4, 137; role in para-scientification, 35, 52, 58, 215

fiction, Black: opposition to Western tradition, 182

fiction, ethnic: American, 146; the "authentic" in, 176; canonical, 178; disproved reality of, 164; non-Western science in, 146; the paranormal in, 27, 30, 143, 156, 201, 217; parascientification through, 26; pluriversal, 217; spectral figures of, 164; spirit mediums in, 140–41; telepathy in, 31; in twenty-first century, 181

fiction, speculative: cultural influence of, 85; interaction with science, 69; production of the pseudoscientific, 35; psitrons in, 51–52, 58, 64

fiction, speculative ethnic, 31; epistemological politics of, 176, 179; ethnoscience of, 178; as new literary phenomenology, 207–9; non-Western science and, 176; the paranormal in, 178, 209; the pluriverse in, 177; poesis in, 183; political impact of, 177; qi-cosmology in, 182; race in, 176; resistance in, 183; time in, 179

fiction, Victorian: paranormal mind in, 142

571 Institute, Chinese military, 139–40

Fleck, Ludwik, 16, 217–18; *The Genesis and Development of a Scientific Fact*, 78

fluidum, prana and, 152

Forel, Auguste: "Hypnotism, or Suggestions and Psychotherapy," 38; psychome theory, 47

Foster, Sesshu: alternative realities in, 183–84, 193–94, 196; *Atomik Aztek*, 31, 187–95; Aztec calendar in, 184, 195, 200; Aztec science in, 184, 192; capitalism in, 184; celebration of the paranormal and, 209; chronology in, 177, 184–87, 190–92, 194, 208; colonization in, 184; critique of colonialism, 254n26; doppelgängers in, 177, 190; epistemological agenda, 195; ethnoscience in, 190, 195–96; hematic technology of, 193; human sacrifice in, 184, 194; multiple consciousness in, 188; multiverse of, 194–95, 208; paranormal mind in, 192–93, 195, 196, 208; phenomenological design of, 189; psi mechanics of, 191; scholarship on, 254n26; social justice in, 195, 208; space-time in, 177, 184, 188–95; speculative realism of, 177, 183; technology in, 192; telepathy in, 183, 184, 188–92, 194–95; Western science in, 178, 189

Foucault, Michel, 93; on governmentality, 101, 106; on power/knowledge, 181, 195

Fox sisters, in Occult Revival, 8, 21

Freud, Sigmund: "Dreams and Occultism," 158; "Project for a Scientific Psychology," 38; on repression, 2; "The Uncanny," 2

Fuller, Steve, 229n75

Fundamental Fysiks Group (FFG), 81–82

future-seeing. *See* precognition

Gaia: agency of, 78; anthropocentric, 90; bacterial, 91–95; causing psychosis, 93; as Daisyworld, 68; in *Darwin's Radio*, 88–91; emergent properties of, 79, 84; enabling by new sciences, 70;

enforcement of supreme will, 90; as global superorganism, 87; higher mental processes of, 80, 83; ideational space of, 69; manipulation of physical reality, 70, 78; mentation of, 84; paranormal mind of, 30, 69; as planetary mind, 68; posthuman, 69, 84, 86; power of life and death, 86, 93; proprioceptive system of, 79, 80; self-awareness of, 79; self-maintaining, 89; superconsciousness of, 80; telekinetic agency, 68, 69, 86–87, 89–90, 215; teleological ability of, 80; telepathic, 91; in *Vitals*, 91–95, 184

Gaia (mythological goddess), 237n1

Gaia hypothesis, 29–30, 67–70; acceptance of, 96; autopoiesis in, 77; Bertalanffy's systems and, 81; Cartesian logic and, 69; collective unconscious in, 128; critics of, 76, 96; cultural acceptance of, 96; cultural power of, 81; decentering of the human, 70; disciplines influenced by, 69; enabling by new sciences, 70; epistemic groundwork for, 69–70; exploration of posthumanism, 69; humanity in, 70, 80, 83, 84, 86, 89, 91, 93; human-nature relations in, 71; illogic of, 96–97; intelligence in, 79; literary promulgation of, 68; mainstreaming of, 96; mankind in, 79–80; marginalized ideas of, 68; New Age and, 81–85; oceans in, 76–77; paranormal consciousness in, 70; political utopianism of, 80; rational/irrational thinking in, 78; reception of, 78, 80–81; in science fiction, 85–86; supporters of, 68; symbiosis in, 77; systems theory and, 30, 70, 78; teleology of, 76–77

Galileo, mechanism of, 72

Galvan, Jill, 148; on mediums, 149

García, Cristina: *Dreaming in Cuban*, 141

gaze, ethnographic, 176, 253n3

Geller, Uri, 103, 242n2

Gelman, Andrew, 212

genome, human: in *Darwin's Radio*, 87–88, 90; distributed mind of, 88

geodesy, in remote viewing, 110

Ghostbusters (film), 103

ghost particles, 43–44, 46, 58

ghosts: belief in, 3; lived experience of, 146–47; of murder victims, 166; social contexts of, 142, 146; in Western literature, 146

ghosts, non-Western, 149; ancestral, 166, 168, 173; in Asian American fiction, 147, 218; in *Comfort Woman*, 163–64, 167–72; communication by, 7, 31, 141, 170, 171; connections to social history, 173; constitutive of material reality, 142; cultural haunting by, 146, 163, 167; dwelling with the living, 168; in Eastern metaphysics, 31; in *The Hundred Secret Senses*, 151, 157, 162, 273; inhabiting shamans' bodies, 170–71; material effects of, 173; as metonymies, 142; Orientalist view of, 142; in qi, 7, 154, 155, 157, 213; range of behaviors, 169, 172; recognition of the living, 169; sexual partners, 169; spirit hypothesis of, 157; summoning rituals for, 170–71; as trauma symbols, 163, 167–68, 169; *yŏngsan* (angry), 167, 168, 170. *See also* shamanism, Korean

ghost seers: gender politics of, 173; Korean, 141–42, 167, 171; as literary archetypes, 141; revealing of qi, 155. *See also* mediums

ghost vision, 142, 147; ethnoscientific definition of, 171; as telepathic image transfer, 149

Gibbons, Dave: *Watchmen*, 29, 35

Gibson, William, 245n62; *Neuromancer*, 114

Gieryn, Thomas, 28; *Cultural Boundaries of Science,* 20
Gilbert, Scott, 92
Golgi stain, 233n16
governmentality, Cold War: parapsychology in, 106–8, 113, 215, 218; psychical research in, 118–19; psychoenergetics in, 107; weaponization of mind in, 126, 136. *See also* Cold War
governmentality, Foucauldian, 101, 106
Great Tōhoku earthquake (2011), 175
growth, exponential: law of, 238n19
Gurney, Edmund, 149, 181

Hacking, Ian, 229n75
hallucination: consensual digital, 245n62; Global south, 183; telepathic, 149
Hammid, Hella, 111
Hancock, Graham, 1
Hansel, C. E. M., 14
Harding, Sandra, 27, 144–45, 231n105; on Aztec chronology, 187; on constructionist science studies, 247n8; on women in science, 216
Hari, Syamala: "Eccles's Psychons Could Be Zero-Energy Tachyons," 65–66
Harvard University, Richard Hodgson Memorial Fund, 228n48
Hastings, Arthur, 50
healing: alternative, 152; spiritual, 1
healing, qi, 156; yin-yang imbalance in, 157
Heidegger, Martin, 181
Heine, Steven, 199–200
Heisenberg, Werner: uncertainty principle, 40, 42, 233n32
hentai (pornography), 201, 257n89
Hermetic Order of the Golden Dawn, 8
heterodoxy, scientific: cultural consumption of, 21; damage from, 214; epistemological processes of, 18–19;

hybridizations of, 19; ideological platforms of, 24; inability to account for, 28; marginalized ideas in, 32, 137; new audiences for, 66; persistence of, 17; in science fiction, 85. *See also* parascience; pseudoscience; science, alternative
Hird, Myra: *The Origins of Sociable Life,* 92
history, networked awareness of, 80
Hogan, Ernest: *High Aztech,* 183
holism, New Age, 85
homeostasis, 76, 239n38
Hopkins, Pauline: *Of One Blood,* 218
Hubbard, L. Ron, 13
humanity: bacterial symbiosis of, 91–92; within Earth superorganism, 80; in Gaia hypothesis, 70, 80, 83, 84, 86, 89, 91, 93; limits of reality on, 135; microbes' hostility to, 93–95; openness to mysteries, 36; psychic potential of, 121; regression of, 57; self-reflective ability, 82; shared consciousness of, 135. *See also* genome, human
human sacrifice, Aztec, 186, 191, 194; perpetuation of reality, 187
Hundred Secret Senses, The (Amy Tan), 31, 149–52, 156–62; American protagonist of, 173; Boxer Rebellion in, 150, 160; Chinese metaphysics, 152; counterdiscursive text of, 173; epistemologies of, 156, 159, 161, 173; esotericism in, 150; ghosts in, 151, 157, 162, 273; interpretation of reality, 151, 152, 159; the paranormal in, 150; parascientific reading of, 152; pluriverse of, 159, 161; psychic healer of, 156–57; qi in, 152, 160, 200; qi-naturalism in, 157–59, 161–62; reincarnation narrative, 150, 159–61; retrocognition in, 157–60; secret senses in, 158, 159, 160; spirit mediums in, 141; subject formation in, 161; Western/non-Western equality

in, 161, 172; yin eyes in, 151, 156–57, 160–61, 167, 204; yin-yang trope, 151, 156–57
Hu Qiaomu, 140
hylozoism: banishment of, 72; Greek theories of, 71, 78; supernatural, 86–87; technoutopianism and, 83
Hyman, Ray, 212
Hyslop, James, 242n16

ideas: "bad," 2, 6, 34; marginalized, 20, 32, 68, 137
ideograms, from remote viewing, 116, 118, 123, 133
individualism, Cartesian, 86; versus systemic needs, 88
Inglis, Brian, 11, 15, 127; on threat of psi, 213
intentionality, scientizing of, 74
International School for Applied Psychic Studies (ISAPS, proposed), 134, 135
internet, paranormal discourse in, 219, 225n10
Iran hostage crisis, remote viewing during, 119
ixtli (locus of perception), Aztec, 192, 194, 195

James, Henry: The Turn of the Screw, 213
James, William, 11, 12; on metaphysics, 39; "mother-sea" theory, 114, 128, 155; "A Plea for Psychology as a Natural Science," 37
Jantsch, Erich, 240n65; Evolution and Consciousness, 80
Japanese Imperial Army, comfort women program, 162, 251n100. See also Keller, Nora Ojka
Jemisin, N. K.: Broken Earth trilogy, 86
Joyce, James, 4; psychical interests of, 11, 227n40

Joy Luck Club, The (Amy Tan), 147; Asian American tropes of, 149–50
Ju Ching, Master, 198
Judy, R. A., 20, 147, 177; on ethnic speculation, 181; on parasemiosis, 190; on poesis, 182, 208
Jung, Carl: on collective unconsciousness, 84, 114

Kaiser, David, 45; How the Hippies Saved Physics, 81
Kaspirovsky, Anatoly Mikhailovich, 107
Kazhinsky, Bernard, 107, 126
Keller, Nora Okja, 149; mediation of incommensurate sphere, 174. See also Comfort Woman
Kellogg, Chester, 14
Kendall, Laurel, 166, 168; on summoning rituals, 170
Kennedy, John: ESP hypotheses, 14
Kim, Hee-Jin, 198, 199
Kim, Jodi, 163
Kim Hak-Sun, 251n100
Kingston, Maxine Hong: The Woman Warrior, 147
knowledge: colonization of, 144; as ellipse, 32; flowing from science, 232n114; networked, 114; pluriverse of, 27, 208; power and, 181; rational/irrational, 32; Western modes of, 176
knowledge, alternative: non-Western modes of, 173, 217; parascientific authorization of, 24; transmission of, 26
knowledge, scientific: co-construction by society, 247n8; cultural politics of, 2; Eurocentric, 182–83
knowledge, traditional: as valid science, 31; Western modernity and
knowledge production: paradigms of, 17; parascientific, 19, 20–21; pseudoscientific, 21, 28; in STS, 17

Koestler, Arthur, 55; on ghost particles, 43–44, 58; on metaphysics, 45–46; on psychical research, 45; *The Roots of Coincidence,* 43–44, 54

Korea, colonial history of, 163. *See also* shamanism, Korean

Kripal, Jeffrey, 3, 24, 25; on popular culture, 228n56

Kuhn, Thomas, 23, 217, 229n73; on "extraordinary science," 49; on intellectual closure, 65; *The Structure of Scientific Revolutions,* 16–17, 24, 229n75

Kumar, Satish, 1

Kurzweil, Ray, 75

Kuttner, Henry: *Mutant,* 13

Lai, Larissa, 182; *Salt Fish Girl,* 147

Lakatos, Imre, 28

Latour, Bruno, 21, 86, 207; on modernity, 17, 151; and mythology of science, 28, 217; on science/culture co-constitution, 72; *We Have Never Been Modern,* 144

Leadbeater, Charles: *Occult Chemistry,* 129

Leary, Timothy: on psychonauts, 127–28

Lee, Chang Rae: *A Gesture Life,* 147

Lei, Sean Hsiang-lin: *Neither Donkey nor Horse,* 26

Leighton, Taigen Dan, 198, 200–201

Lem, Stanislaw: *Solaris,* 85–86

Lewis, Pepper, 82, 85; *Gaia speaks,* 67, 68; spiritualism of, 84

lichen, symbiotic relationships of, 77, 239n46

life: cosmic outcoming of, 76; scalable, 76; systems approach to, 77, 78; traditional characteristics of, 76

light, speed of: psitrons', 47; time and, 44. *See also* space-time

Linzmeyer, A. J., 12

literacy, psychic, 136

literature: in dissemination of the paranormal, 22–25, 172; effect of quantum mechanics on, 233n36; engagement with mind, 228n56; ethnic realism in, 26; hypothetical particles in, 35; as ideological amplifier, 64; interpretation of science, 22–23; the occult in, 11; outmoded science in, 23; paranormal mind in, 7, 68, 142, 215; as psychical technology, 102, 120–36; psychological sciences and, 230n101; in reimagining psi, 29; role in parascientification, 20, 22–24, 51, 143, 172; shaping of scientific epistemology, 23; STS approaches to, 140; unproven concepts in, 214–15. *See also* fiction; science fiction

literature, Asian American: bildungsromans of, 161, 251n96; ghosts in, 147; spirit medium trope in, 30–31, 141; the unexplainable in, 143. *See also* ghost seers; shamanism, Korean; spirit mediums

literature, mass-market: production of pseudoscience, 28

literature, psychical: amplification of ESP, 128; epistemological power of, 121; generated by Project Stargate, 120; as technology, 102, 120–36

Littlefield, Melissa, 37, 38

Liu, Aimee: *Face,* 147

Liu, JeeLoo, 153, 155

Lloyd, David, 147; *Under Representation,* 27

Lodge, David: *Consciousness and the Novel,* 215

longevity, cellular mechanisms of, 91

Longhurst, John, 157

Lounsbury, Floyd, 196

Lovelock, James, 29, 30, 184; collaboration with Margulis, 76, 78, 239n51;

Daisyworld program of, 237n5; *Gaia: A New Look at Life on Earth*, 79; Gaia hypothesis of, 68, 70, 75, 76–81, 84, 95; *Homage to Gaia*, 80–81; influence on Bear, 90; NASA work, 75; use of systems theory, 75

Luckhurst, Roger, 11, 86; on *Darwin's Radio*, 88; on science fiction, 35

Ma, Sheng-Mei, 151

Macartney, Tim "Mac," 1

Maimonides Medical Center, Dream Laboratory of, 14

Malmstrom, Vincent, 185

manuals: the paranormal in, 4; psychic, 30; Swann's, 100, 113, 123

marginalized, the: in parascience, 24, 217; self-expression of, 147

Margulis, Lynn, 29, 30; collaboration with Lovelock, 76, 78, 239n51; Gaia hypothesis of, 68, 70, 75, 76, 95; influence on Bear, 91; "On the Origin of Mitosing Cells," 77; posthuman Gaia of, 84, 86; *Symbiotic Planet*, 78, 79; *What Is Life?*, 78

Markley, O. W.: "Human Consciousness in Transformation," 84

mass: imaginary, 35, 44, 65, 66; rest, 47

mathematics, immutable constants of, 1

Matrix, The (film), 114

Matrix, the (metaphysical space), 113–18; access to locations, 117; corruption of, 133; data contained in, 114–15; etymology of, 244n62; signal lines to, 115–18; three dimensionality of, 115

matter: exotic, 45, 233n34; intelligence of, 75

Maturana, Humberto, 30; on autopoiesis, 74

Mauskopf, Seymour, 5; on the supernatural, 9

Maxey, Ruth, 150

McConnell, Tron: on Chinese parapsychology, 139

McCulloch, Warren, 238n28; on psychons, 38

McDonough, Kelly: *Indigenous Science and Technology*, 26

McDougal, William, 12

McMoneagle, Joe, 119; *The Stargate Chronicles*, 120

McVaugh, Michael: on the supernatural, 9

mechanism: Galileo's, 72; scientific ideology of, 73; teleology of, 74

mechanism, Cartesian: cultural hegemony of, 71–72; the microscopic in, 72

medicine, Chinese: meridians in, 139, 156–57, 250n74; qi in, 156–57

meditation: Buddhist, 198; scientific interest in, 84

mediums: astral projection by, 104; in psychical research, 63; psychons of, 41; in Western occultism, 202. *See also* ghost seers; shamanism, Korean; spirit mediums

mediums, female, 27, 166–67, 173; in ethnoscience, 174; in patriarchy, 142; power of, 148; subliminal self of, 148

mediumship: living agent hypothesis, 149; non-Western, 166–67, 173–74; politics of, 149; subliminal self in, 10, 148; survival hypothesis, 149; technophilic audience of, 21; Western, 147

memory, phylogenetic, 97

Mental Radio (Upton Sinclair): drawing techniques in, 131; reception of images in, 243n46; telepathic image transmissions, 12, 111

mentation. *See* thought

meridians, in Chinese medicine, 139, 156–57, 250n74

Merla-Watson, Cathryn, 176

Mesmer, Franz Anton: *fluidum* theory of, 152

Mesoamerica: circular time of, 177, 185. *See also* calendrical science, Mesoamerican

metaphysics: condoning of the supernatural, 45–46; intentionality in, 74; of mind–brain interactions, 40; reintroduction into science, 69–70; scientific discourse in, 72; signal lines to the physical, 115

metaphysics, Eastern: Chinese, 152; communication with ghosts in, 31; Zen, 197

"'Meta'-relativity" (Bilaniuk, Deshpande, and Sudarshan), 44

metonymy, in supernatural literary analysis, 142

Micale, Mark, 7; on SPR, 10

microbiology: planetary consciousness and, 69; systems, 87

microontology, societal, 92

Midgley, Mary, 80–81, 92; on holism, 72; on human–nature relations, 71–72

Mignolo, Walter, 144; on pluriverse, 145, 154, 173, 177, 217; universe of, 178; on Western parochialism, 151

Mikhailovich, Vladimir, 107

Mills, George Harper, 25

mind: autonomy of, 40; as Cold War battlefront, 106–8; electromagnetic, 107; geohuman amplification of, 80; interconnectivity of, 198; interpretation of signal line data, 116; prioritizing over body, 41; in self-regulating systems, 82; SPR study of, 9–10; in systems theory, 79; weaponization of, 108; in world soul, 95; Zen view of, 198–99, 201, 202, 205–6

mind, Gaian: biological approaches to, 96; as systems phenomenon, 86

mind, global, 80, 91; biochemical mechanism of, 95; psychic renaissance for, 136; in *Star Fire,* 128

mind–brain interactions: metaphysics of, 40; in neurology, 35

mitochondria, bacterial origins of, 77

modernism, literary, 142; the occult in, 4, 226n12; of Occult Revival, 8; phenomenological primacy of, 25–26; quantum physics and, 195; spiritualism in, 25

modernity, Western: ancestral knowing and, 141; loss of secret senses, 158; post-Enlightenment, 18; power of STS in, 216; the premodern and, 144, 248n12

Moore, Alan, 86; popularity of, 64. See also *Watchmen*

Morrisson, Mark, 7

Morse, Samuel, 21

Moss, Thelma, 242n8

multiculturalism: scientific, 145; supernaturalism in, 146; weak, 181

Murphy, Gardner, 14–15, 228n48; field theory of, 114, 128; on the Matrix, 245n62; on paradoxes, 37

Myers, Frederic W. H., 15, 105; *Human Personality and Its Survival of Bodily Death,* 10, 15, 36, 180, 196; immaterialism of, 50; on precognition, 37; on subliminal self, 39, 50, 114, 132, 180; on telepathy, 179–80

Myers, P. Z., 2

mysticism, of Occult Revival, 179

mysticism, Eastern: quantum mechanics and, 81

myth, science and, 3, 28, 217

Nanahuatl (Aztec god), self-sacrifice of, 186, 187

National Security Agency, use of remote viewers, 119

Natural ESP (Ingo Swann), 102, 121, 128–31; access to ESP in, 129; clairvoyance in, 131; drawing in, 131; ESP core in, 129, 131; goals of, 131; human consciousness in, 130 (fig.); the Matrix in, 131; psi in, 129, 131; second reality of, 129, 130 (fig.), 131; the unconscious in, 129

nature: definition by local culture, 200; mechanistic view of, 1; organismic intention of, 73

Naylor, Gloria: *Mama Day,* 141, 218

necromancers, physicists as, 43

Needleman, Jacob, 80–81

Nelson, Alondra, 176

nervous system: electromagnetic waves and, 107; Gaian, 79; peripheral, 40; planetary, 83; quantum mechanics and, 65

neurons: mass of, 233n32; search for analogies, 38; visualization of, 233n16

neuroquantology, 65

neutrinos: discovery of, 43, 46, 50; as ghost particles, 43–44; theorization of, 234n39

New Age: antecedents of, 81; in California, 243n33; effect on scientific boundaries, 85; Gaia hypothesis and, 81–85; interconnected being beliefs, 82; memoirs of, 4; non-Western philosophy in, 81; paranormal mind in, 142; physicist participants, 81; principles of, 81; rejection of binaries, 84, 85; in religious studies, 70; scientific impossibilities in, 84; self-awareness in, 81; spiritualism and, 81, 84; use of science, 84; women's agency in, 237n1

Nguyen, Viet Thanh: *The Sympathizer,* 147

noogenesis, 240n65

nuclear testing, Chinese: remote viewing of, 119

objects, parascientific, 21, 206, 215; conditions for possibility, 64–65; continued existence of, 22, 65; cyclical, 24; epistemic exclusion of, 21–22; indestructability of, 66; legitimization of, 39; in literature, 23; new narratives for, 66; as normal/paranormal bridge, 207; reimagination of, 22. *See also* parascience; psitrons; tachyons

occult, the: hucksterism in, 11; literary modernism and, 226n12; multiepistemic access points for, 215; quantum mechanics' naturalizing of, 42–44; spectrum of beliefs in, 18; third eye in, 156; Victorian, 117; women in, 27, 216

Occult Revival, 4; modernism of, 8; mysticism of, 179; scientific aims of, 8

Odabas, Janna: *The Ghosts Within,* 163–64

Olcott, Henry Steel, 152

Orientalism: Blavatsky's, 158, 179; depiction of the supernatural, 141, 142, 147; Said on, 150; in Tan's work, 151, 161–62

oscillators, electromagnetic: effect on human behavior, 125

out-of-body experience, 104–5, 112, 117. *See also* astral projection

Owen, Alex, 7, 8; on ultimate reality, 43

Ozeki, Ruth: astral travel in, 205, 208; Buddhist parasemiosis in, 201; chronology of, 177, 199; East–West dichotomy of, 197; epistemology of, 196, 202; ghosts in, 202–3; glasses trope of, 204, 206; Great Tōhoku earthquake in, 175; knowledge systems of, 197; lucid dreaming in, 203–4, 205; multiple realities of, 197; paranormal mind in, 176, 196–97, 202, 208, 209; parascientific objects of, 175, 206–7; pluriverse of, 208; quantum mechanics in, 197–98, 203; retrocognition in, 204; scholarship on, 255n63; Schrödinger's cat in, 197; as second-order parascientific object, 207; space-time in, 175, 176, 205; *A Tale for the Time Being,* 31; telekinesis in, 206; telepathy in, 17; *uji* in, 206; Western science in, 178, 197; zazen meditation in, 201, 203, 208; Zen Buddhism in, 154, 177, 196–207, 208

para (Greek), 19

paradigms, scientific, 16–17, 229n73; anomalous, 63, 216; intellectual closure of, 66; misalignment with psychical research, 15–16; production of facts in, 28; pseudoscience and, 17; supernatural phenomena and, 174; the unproven thriving outside of, 29

paraliteracy: discursive space of, 20; ghosts of, 147

paranormal, the: agency through, 209; in Aztec calendar, 185; Bear's reimagination of, 95; Chinese, 139–40; in contemporary fiction, 4, 137; in contemporary life, 2–3; cultural interpretations of, 32, 120; as cultural universal, 143; decolonial approaches to, 145–46; dissemination through literature, 22–25, 172; as dynamic abstraction, 129; epistemological theory of, 4; estrangement of, 214; in ethnic fiction, 27, 30, 143, 156, 201, 217; Eurocentric disregard for, 209; expansion of scientific thinking, 217; global propagation of, 8, 217, 218; on the internet, 219; legitimate study of, 6; mass culture and, 68, 179, 219, 228n56; mass marketing of, 3, 25; media exposure of, 3; political use of, 118–19; in post-1945 era, 26; the real and, 176; resilience of, 3–4; resistance to scientific critique, 3; rhetoric of physics, 39; in science fiction, 22–23, 26, 218; scientific dismissal of, 14, 134–35; scientific models of, 10;

Swann's influence on, 101; in training manuals, 4; ubiquity of, 215–16; weaponization of, 136; in Western intellectual history, 207; Western marginalization of, 209, 217

paranormal mind: alternative science explanations for, 35; in *Atomik Aztek,* 192–93, 195, 196, 208; banal experiments on, 131; in Bear's works, 95; capabilities of, 3; characteristics of, 10; cultural development of, 4; cultural study of, 6; declining scientific legitimacy, 6; decolonization of, 208; diversity of, 146; as espionage weapon, 99–100; exclusion from scientific narrative, 15; Gaian, 30, 69; in global cultures, 8; in government surveillance, 108, 137; government use of, 109; hybrid existence of, 65; intellectual progression of, 207; in literature, 7, 68, 142, 215; medium-centric, 133; military-political use of, 105, 120, 136; in multicultural science, 31, 177, 181; in multiethnic literature, 29, 140, 149, 217; non-Western epistemes of, 140, 164; in non-Western science, 4, 140, 162, 209; persistence of, 32, 143, 213–14; political use of, 102; as postmodern phenomenon, 25–26; psychical research and, 8–16; reasons for study of, 5; reproducible experiments for, 212; resilience of, 4; scientific irrelevance of, 16; shared, 136; STS approach to, 176; subaltern status of, 2–3; in subatomic space, 50; subliminal self of, 11; Swann's transformation of, 102; in systems theory, 88; of *A Tale for the Time Being,* 176, 196–97, 202, 208, 209; technological, 100–101, 102; transmutations of, 215; variegation of, 219; Victorian models of, 3; women investigators of, 216–17; Zen, 199, 202

parapsychologists, Swann on, 102
parapsychology: in Cold War governmentality, 106–8, 113, 215, 218; effect
of quantum mechanics on, 45–46;
of mid-twentieth century, 127; as
pseudoscientific, 16; science fiction's
engagement with, 13; scientificity of,
128
parapsychology, Chinese: CIA monitoring of, 139
parascience: architectonics of, 216; in
Asian American supernaturalism, 147;
in circulation of pseudoscience, 215;
coexistence with science, 19; discursive networks of, 18–19; epistemic
space of, 7, 20; epistemological justice
in, 217; the excluded in, 216; extension
of post-Kuhnian science, 217; gender
politics of, 27, 173; interdisciplinary
arena of, 7; the marginalized in, 24,
217; multiple discourses of, 21; pseudoscience and, 19–20, 214, 226n14;
recontextualization through, 24;
renewal of consciousness, 207; science
and, 16–25; scientificocultural third
space of, 19, 29, 214; as site of knowledge production, 19, 20–21; women in
science and, 216. *See also* heterodoxy,
scientific; objects, parascientific;
pseudoscience
parascientification, 21, 215; agents of, 22;
cultural reimagination of, 103; epistemic circulation of, 24; in epistemological economy, 219; in ethnic fiction,
26; through literature, 20, 22–24, 51,
143, 172; in multiple disciplines, 65;
natural processes of, 69; patterns of,
23–24; role of fiction in, 35, 52, 58, 215;
in *VALIS*, 58
Park, Therese: *A Gift of the Emperor*, 147
particles, hypothetical, 216; faster than
light, 34, 35; ghost, 43–44, 46, 58;

immaterial, 35; in literature, 35; of
mentation, 233n28; "meta-," 44; of
thought, 71
particles, subatomic: effect on the supernatural, 42–43; forces acting on,
234n46
Partridge, Christopher, 18, 20, 25
Pauli, Wolfgang, 234n39
pedagogy, paranormal, 134–35
phantasms, 149
Phantasms of the Living (Gurney,
Podmore, and Myers), 10, 179
philosophy, classical, 152; living earth in,
69, 70–71
photography, Kirlian, 104
photons: emitted by matter, 47; psychons
and, 38
phrenology, scientific origins of, 22
physics: challenging of absolutes, 45;
impossibility, 42–46; indeterminacy
in, 39; occult, 43; particle, 234n46;
and principles of reality, 44–45; theoretical, 42–43. *See also* quantum
mechanics
physiology, human: symbiotic processes
of, 91–92
Pigliucci, Massimo, 18, 20
Pike, Sarah, 81
Pinch, T. J., 213
Plascencia, Salvador, 182
Plato, 152; hylozoism of, 71, 78; *Timaeus,*
71, 97
poesis: as being-in-the-world, 182;
minority, 208; paranormal, 143, 216; in
speculative ethnic fiction, 183
Pöhlmann, Sascha, 254n26
Popper, Karl: on falsifiability, 28, 49,
231n111
Posidonius, on vital force, 249n59
positivism: Comte's, 143; objectification
in, 198
positrons, in psychical research, 43

posthumanism, of Gaia hypothesis, 69, 84, 86

postmechanism, 75

postmodernism, speculative realism in, 177

power: Gaian, 81, 86, 93; knowledge and, 181; in speculative ethnic fiction, 176

power relations, in epistemological economy, 183

prana, Indian: *fluidum* and, 152

precasts, mental: phenomenological access to, 134; predictive power of, 48; in *VALIS*, 55; in *Watchmen*, 61. *See also* precognition

precognition, 10; deep-level, 133; faster-than-light particles in, 35; future of, 64–66; ideational evolution of, 58; intellectual interest in, 133; in literature, 23; nineteenth-century, 65; as physicalist phenomenon, 35, 58; quantum mechanics and, 7, 21, 69; subliminal self in, 36; thought particles of, 29; in *VALIS*, 55, 56–57; in *Watchmen*, 33–34, 60–63, 62 (fig.); in *Your Nostradamus Factor*, 132. *See also* precasts, mental

premodernity: systems theory and, 72, 81; Western modernity and, 144, 248n12

Price, H. H., 46

Price, Patrick, 111

Prigogine, Ilya, 30, 74

primitive, in Western tradition, 181–82

Prince, Walter, 227n43

progress, epistemic: post-Enlightenment, 5

Project Gondola Wish, 100, 244n59

Project Grill Flame, 100, 244n59

Project MKUltra, 120, 245n91

Project Stargate (psychic spy program), 30, 100, 101, 215; credibility of, 120; critics of, 119; as ideological revolution, 135; literature generated by, 120; loss of funding, 137; memoirs of, 137; militaristic ideology of, 125; purported successes of, 119; remote viewers in, 119

prokaryotes: in human body, 92; symbiosis with eukaryotes, 77, 78, 239n45; in *Vitals*, 91

propriocepsis: of bacteria, 79; of Gaia, 69, 80

Proust, Marcel: *À la recherche du temps perdu*, 175, 197; on linear time, 206

pseudo-knowledge, modern circulation of, 6

pseudoscience: as archaic remnant, 17; effect on public discourse, 5; epistemic production of, 27, 29; intellectual history of, 20; in mass-market literature, 28; in modern culture, 19; parapsychology as, 16; parascience and, 19–20, 214, 226n14; persistence of, 2, 17–18, 20, 213–14, 218; place in science studies, 216; rhetorical power of, 19; in science fiction, 35; scientific development and, 5, 18–19, 28–29; scientific paradigms and, 17; theoretical processes of, 20. *See also* heterodoxy, scientific; parascience

psi (the paranormal), 12; alternative views of nature in, 142; Chinese government study of, 139–40, 152; citizen science in, 132; communist use of, 118; digital natives' exploration of, 219; hereditary abilities, 103; in indigenous science, 146; lack of standard theory, 14; literary reimagining of, 29; long-distance, 108; practical use of, 101; scientific authorization of, 22; shift in awareness of, 137; as threat to science, 213; universal ability for, 131

psi-fi, 13–14, 228n56

psionics, 13

psitrons (hypothetical particles), 34, 36, 206, 215; conceptual formation of, 35; cultural presence of, 64; Dobbs's theory of, 34, 36, 40, 46–52, 57, 61, 132; durability in imaginary, 35; electromagnetic characteristics of, 63, 65–66; emitted by future events, 47–48; endurance of, 54, 65, 66; ESP detection of, 49–50; imaginary mass of, 47; information from future, 56; intellectual history of, 35; interaction with brain, 48, 64, 65; laws of physics and, 51; as literary objects, 64; as material objects, 64; model of consciousness, 50; quantum mechanics and, 51, 69; relationship with space-time, 47, 48, 63; in speculative fiction, 51–52, 58, 64; striking dendrons, 48; subatomic, 50–51; superluminal velocity, 47, 64; theory and practice of, 46–52; toxic, 57, 62; in *VALIS*, 52–58, 63, 64; in *Watchmen*, 58–64, 63 (fig.)

psychical research: academic, 228n48; attacks on, 16; Bem's research and, 213; by citizen scientists, 105; in Cold War, 101, 118–19; disciplinary borders of, 219; empiricist, 50; failure to duplicate, 14–15; first wave, 50; hybridizations of, 11; ideological division within, 50; laboratory-based, 127; in later twentieth century, 3–4; mediums in, 63; methods of, 9; of mid-twentieth century, 46; misalignment with scientific paradigms, 15–16; paranormal mind and, 8–16; positivistic science and, 12; positrons in, 43; quantum mechanics and, 45; scientific tactics hindering, 213; secret history of, 137; in *Star Fire*, 126–27; Swann's role in, 105, 136; as technology, 100; threat to modern science, 15–16; twentieth-century, 3–4, 120; Victorian, 95, 179–81; women in, 148–49

psychic weapons, Soviet-era, 101, 106–8, 120

psychoanalysis, Freudian: untestable hypotheses of, 231n111

psychoenergetics (mind energy), Soviet programs, 107, 125–26

psychogenesis, Chinese government study of, 139

psychological sciences: of early twentieth century, 11; literature and, 230n101; philosophical origins of, 37; replication crisis in, 212; Victorian, 4

psychomes (psychic units), 38, 47

psychonauts, 127–28

psychons (hypothetical particles), 37; autonomous, 39; interaction with dendrons, 40–41; as mental will, 233n28; merging with tachyons, 66; photons and, 38; polyvalence of, 39

Puglionesi, Alicia, 12, 27, 104; on drawing, 117

Puthoff, Hal, 242n2, 243n33; *Mind-Reach*, 120; on Project Scanate, 112; psi research of, 108–9; on remote viewing, 112; Swann and, 108, 109, 110, 111

Pyong-Choon, Hahm, 164–65, 168

qi, Chinese: across space-time, 154; communication with ghosts, 7; constitution of yin and yang, 156; etheric substance of, 157; flowing through all things, 139; macro/micro levels, 154; in medicine, 156–57; meridians channeling, 139, 156–57, 250n74; omnipresent, 153; powering psi programs, 152; reincarnation in, 155; seen/unseen reality of, 153; in Tan's work, 27, 31; universality of, 154

Qian Xuesen, 571 Institute, 139–40

qi cosmology: conceptual scheme of, 154; in speculative ethnic fiction, 182

qigong (qi cultivation), 139; science, 140
qigong masters, Chinese government use of, 140
qi-naturalism, 153–54, 176; epistemic validity of, 159; ghosts in, 157; humanity/nature unity in, 165; in *The Hundred Secret Senses*, 157–59; as knowledge system, 142; retrocognition in, 155; scientific framework of, 156; spirit medium of, 154–55; types of qi in, 165
quantum mechanics: the bizarre in, 45; Eastern mysticism and, 81; effect on culture, 42; effect on literature, 195, 233n36; effect on parapsychology, 45–46; many-worlds interpretation, 189, 208; naturalizing of the occult, 42–44; nervous system and, 65; psitrons and, 51, 69; psychical research and, 45; systems biology and, 140
quantum physics: imaginary mass in, 35; neural sciences and, 65; precognition and, 7, 21, 69

race, in speculative ethnic fiction, 176
radio communication, 118; biological, 107
Ramín y Cajal, Santiago, 233n16
rationalism, 32; European, 27; in Gaia hypothesis, 78
Rauschler, Elizabeth, 82
Read, Kay: on Aztec time, 189–90, 194; on Mesoamerican calendar, 192
realism, speculative, 176–77; epistemological justice in, 182–83; social justice and, 181, 182
reality: antimatter and, 45; Aztec, 183–84, 187, 192; defamiliarizations of, 176; enjoined with mind, 199; ethnoscientific understanding of, 170, 173; Gaia's manipulation of, 70, 78; of *The Hundred Secret Senses*, 151, 152, 159; limits on human race, 135; material/

immaterial, 152–53; mind's coterminous with, 205; multiple, 145; parallel, 189; physics and, 44–45; qi as fundamental of, 159; social-specific boundaries of, 134–35
reason, third space of, 32
red disaster (*hongaek*), Korean, 166
reductionism, Cartesian, 84
reincarnation, in *The Hundred Secret Senses*, 150, 159–60; in qi, 155
Reiter, Bernd, 31, 144; on mosaic epistemology, 145, 173, 177, 217
religion, New Age disenchantment with, 81
remote viewing, 100; access to data, 120; analytical overlay degrading, 116, 133; antecedents of, 113; of Chinese nuclear testing, 119; CIA evaluation of, 119–20; coordinate (CRV), 113; drawing in, 116–17, 131; geodesy in, 118; geographic coordinates in, 109–10, 112–13; ideograms from, 116, 118, 123, 133; learning of technique, 113; map produced during, 124 (fig.); military purpose of, 123; nationalistic agenda of, 125; noncoordinate forms of, 244n57; political context of, 113; in Project Scanate, 110–12; in Project Stargate, 119; psi ideas and, 118; as psychoenergetic perception, 112; science of, 108–20; signal lines in, 115–16; subliminal self and, 117; Swann's, 111; theoretical frameworks for, 101; training for, 117; as type of clairvoyance, 112; uncertainty surrounding, 112; as universal talent, 113; websites, 137; will to believe in, 119. *See also* clairvoyance; telepathy
researchers, degrees of freedom for, 212
retrocognition: in *The Hundred Secret Senses*, 157–60; in qi-naturalism, 155
retrocognition (knowledge of past), 36

Rhine, J. B., 12–13, 49, 213; discrediting of, 14; empiricism of, 50; *Extrasensory Perception*, 13; Parapsychology Laboratory of, 12, 14, 113. *See also* extrasensory perception

Richet, Charles, 5; materialism of, 50

Rickels, Laurence, 54

Romains, Jules, 15

romance, supernatural, 3

Rose, Charlie, 120

Royal Society (London), 143, 144

Ruse, Michael, 81; *Gaia Hypothesis*, 239n51

Russell, Bertrand: atomism of, 38, 232n14

Russell, Peter, 85; *The Awakening Earth*, 82, 87; Gaiafield of, 82, 83, 87, 88, 92; influence on Bear, 90; parapsychology experiments, 109

Russia, televised mass hypnosis in, 4

Said, Edward: Orientalism of, 150

Saldívar, Ramón, 177, 195, 208, 254n26; on speculative realism, 176–77, 182

Sapp, Jan, 92

Sarfatti, Jack, 82, 242n2

Schnabel, Jim: *Remote Viewers*, 120

Schrödinger's cat, 197, 256n65

Schrödinger wave mechanics, 43, 66

science: "almost," 18; androgenic, 27; broken, 212; coexistence with parascience, 19; in development of paranormal mind, 7; epistemic priority of, 232n14; epistemic production of, 27; factitious study of, 28; gender politics of, 27; interaction with speculative fiction, 69; interdisciplinary, 27, 96; metaphysics in, 69–70; methodological problems, 212–13; parascience and, 16–25; in perpetuation of pseudoscience, 28–29; postcolonial studies, 145; post-Kuhnian, 18, 28, 217, 247n8; power in, 216; processing

through culture, 23; questioning of validity, 2; reinforcement of magical thinking, 216; rejected ideas in, 7; role in parascientification, 51; social nature of, 5, 144; thought styles in, 217–18

science, alternative: the paranormal in, 35, 68. *See also* heterodoxy, scientific

science, Aztec, 184, 195–96; heart, 191–92, 196, 208

science, indigenous, 26; in ethnic fiction, 146; literature and, 172; within parascience, 201; psi in, 146; telepathy in, 145

science, marginalized, 32; in cultural consciousness, 68–69; in STS, 28

science, multicultural, 208; in literary analysis, 141; paranormal mind in, 31; politics of, 173; Western science in, 27

science, non-Western, 231n105; Asian, 27; literature and, 140; marginalized ideas and, 20; multiethnic literature's use of, 29; paranormal mind in, 4, 140, 142, 162, 209; in twenty-first century STS, 217

science, paranormal: Cold War ideology in, 30, 137, 218

science, positivist: on psychical research, 16

science, Western: decentering of, 178, 217; decolonization of, 144; dominance of, 141, 143–44; Eurocentrism of, 143–44; exclusion of women, 147–48, 216; in multicultural sciences, 27; mythology of, 17, 28, 217; other discourses and, 217; as type of ethnoscience, 29

science and technology studies (STS): approaches to literature, 140; constructionist, 247n8; decolonization of, 217; feminist, 27; knowledge production in, 17; literary, 24; marginal science in, 28; multicultural, 146; on paranormal mind, 176; place of pseudoscience in, 216; prioritization of facts, 216; twenty-first century, 217

science fiction: in cultural consciousness, 7; effect on mass culture, 13; engagement with parapsychology, 13; Gaia hypothesis in, 85–86; golden age of, 218; hypotheses in, 95–96; ideology of, 23; as occult sciences laboratory, 207; the paranormal in, 22–23, 26, 218; pseudoscience in, 35; psitrons in, 64; role in scientificization, 85; scientific imaginary of, 51; socioethical function of, 23; tachyons in, 64; telekinesis and, 29

scientificization: role of science fiction in, 85; of the soul, 10; of the supernatural, 8

seánces, televised, 107

Search for Extraterrestrial Intelligence (SETI), 18, 216

self, subliminal, 15, 50, 214; of the dead, 10; of mediums, 148; in precognition, 36; remote viewing and, 117; survival of body, 39, 192, 195; Swann on, 132; in telepathy, 180; as unconscious, 114

senses: fluidity among, 159; secret, 158; sixth, 12. *See also* extrasensory perception

Serres, Michel, 24, 214; *Hermes* writings, 25; *The Troubadour of Knowledge*, 32

Shackleton, Basil, 48, 49, 55, 234n54

shamanism, Korean, 31, 162–72, 176; cause and effect in, 165; cosmology of, 164; epistemology of, 166; ethnoscientific principles of, 164; female suffering symbols in, 167–68, 169; ghost vision in, 171; humanity/nature unity in, 165; as knowledge system, 142; parascience/literature relationship of, 172; as political resistance, 164;

removal of bad luck, 167; as response to oppression, 252n109; spirit communication via, 141, 166; spirit/humanity coexistence in, 164; summoning rituals (*kuts*), 170–71; supernatural contagion in, 166; time in, 165, 169; women shamans, 166–67. See also *Comfort Woman*; ghosts, non-Western; ghost seers

Shapin, Steven, 78

Sheldrake, Rupert: defenders of, 2; *The Science Delusion,* 1; Tedx talk (2013), 1–3, 4, 17, 213

Shirota Suzuko, 251n100

shishosetsu (I-novel), 256n64

signal lines, 113; access to the unconscious, 116; linking to geophysical knowledge, 116; to the Matrix, 115–18; remote detection of, 115; right/left brain processes of, 116

Sinclair, Mary Craig, 173; *Mental Radio,* 12, 111, 131, 243n46

Sinclair, Upton. See *Mental Radio*

Singularity, the, 75

sixth sense, Rhine's theory of, 12

Skeptical movement, 2, 225n4

skepticism, hampering of precognition, 133

Skinner, B. F., 14

Soal, Samuel G., 105, 235n54

Soal-Shackleton experiments, 48, 234n54

social sciences, methodological problems in, 212

societies, non-Western: approaches to literature, 140

Society for Psychical Research (SPR), 5, 9–11; conflicts within, 11; decline of, 14; goals and methodology of, 9; on the human mind, 9–10; members, 9; *Proceedings,* 15, 66

Society for Psychical Research, American (ASPR): ESP experiments, 105;

spiritualist/scientist conflict in, 11–12, 105, 227n43, 242n15; Swann's work with, 104, 117

Sohn, Stephen Hong, 254n26

soldiers, psychic, 101, 113

Sommer, Andreas, 6

soul, scientificization of, 10

Soviet Union: psychic statecraft of, 101, 106–8, 120, 218; psychoenergetic program, 107, 125–26

space-time: directionless, 205; ethnoscientific principles of, 201; indigenous frameworks of, 208; psitrons' relationship with, 63; qi across, 154; in *A Tale for the Time Being*, 175, 176; telepathy in, 180; Zen, 199–201, 203–5. *See also* time

space-time, Aztec: in *Atomik Aztek*, 177, 184, 188–95; renewal of, 187

spin (quantum mechanics), 203, 257n97

spirit mediums: in Asian American literature, 30–31, 172; clairvoyance of, 141; in ethnic fiction, 140–41; non-Western epistemes of, 141; occult gifts of, 140; as Orientalist caricatures, 141. *See also* ghost seers; mediums; shamanism, Korean

spiritual healing, biomedicine's superiority over, 1

spiritualism: in modernist literature, 25; New Age and, 81, 84; synthesis with scientism, 28; telegraph technology and, 21; Victorian, 8, 39

Squier, Susan: on literary technologies, 22

Stanford, Thomas, 228n48

Stanford Research Institute (SRI), 30; *Coordinate Remote Viewing*, 113–14; discourse of twentieth-century science, 112; military connections of, 108; Project Scanate, 110–12, 119, 244n59; remote viewing programs, 111–12,

244n59; Swann's work with, 101, 109, 110–13

Stanford University, Fellowship in Psychical Research, 228n48

Star Fire (Ingo Swann), 102, 125–28; clairvoyance in, 122–23; conscious/unconscious mind of, 123; geolocation in, 123, 127; geopolitics of, 125; ideograms, 123; as ideological revolution, 135; martial focus of, 123; Matrix-based ideas in, 121–22; mind control in, 126; political ideology of, 125; psi experience in, 122; psychical research in, 126–27; psychonauts of, 128; remote viewing in, 121–23, 127; scientific pedantry in, 127; telekinesis in, 127; vision of world, 122–23; weaponization of mind in, 126; world mind in, 128

Star Trek (2009), parallel reality in, 189

STEM (science, technology, engineering, and mathematics) education, 17; pseudoscientific beliefs and, 28–29

Stoics, hylozoism of, 71

Stranger Things (Netflix series), 137; telepathy in, 99–100

Sturgess, Kylie, 2

subjective experience, systematization of, 85

superheroes: creation myths of, 59; in *Watchmen*, 59–60

supernatural, the: differing scientific paradigms of, 174; naturalization of, 217; scientificization of, 8; social/literary dissemination of, 5; SPR's investigation of, 9; subatomic particles and, 42–43; ubiquity of, 214

supernaturalism: Asian American, 147, 218; scientific authorization of, 42

surveillance, Cold War, 101; paranormal mind in, 108, 137; technology of, 106. *See also* Cold War; espionage, telepathic

Swann, Ingo: ASPR work, 104, 117; astral projection by, 104–5, 110; as Cold War ideologue, 102; ESP guides, 100, 128–31; flower paintings, 104, 127; government work, 101; ideological revolution of, 135; influence of, 120, 136; Korean War service, 103; law of divination, 132–33; literary influence, 101; magnetometer episode, 109, 127; on the Matrix, 115, 121–22, 131, 133, 155, 244n62; memoirs of, 104, 113, 117; models of clairvoyance, 101–2; novels of, 100; parapsychological fiction of, 30; on parapsychologists, 102, 105, 137; and perception of the paranormal, 101; phenomenological theories, 101, 121; probe of Jupiter, 244n57; Project Stargate work, 119, 128; psi experiences, 101, 122, 128, 136; *Psychic Literacy and the Coming Psychic Renaissance*, 121, 136–37; psychic renaissance of, 102, 103, 136–37; remote viewing and, 109–11; role in psychical research, 105; SRI work, 101, 109, 110–13, 123; at Stanford Research Institute, 101; telepathy of, 109; *To Kiss the Earth Goodbye*, 103, 109; training manuals of, 100, 113, 123; United Nations work, 104; use of geographic coordinates, 109–10. See also *Natural ESP*; *Star Fire*; *Your Nostradamus Factor*

Swann family, paranormal minds among, 103

symbiosis, eukaryotic/prokaryotic cells, 77, 78, 239n45

systems theory, 72–75; agency in, 74; authorization of ancient concepts, 97; consciousness in, 75, 84; cybernetics and, 238n28; in *Darwin's Radio*, 88; first-order, 238n28; Gaia hypothesis and, 30, 70, 78; goals of, 74; the holistic in, 74; intelligence in, 79; knowledge paradigm of, 85; mind in, 79; among New Age adherents, 82; paranormal mind in, 88; parascientific, 97; premodern beliefs in, 81; principles of, 73, 238n19; reviving of premodern logic, 72; science/culture co-constitution in, 72; second-order, 74, 238n28; transdisciplinary principles of, 72–73; vitalism of, 75

tachyons (hypothetical particles): Dick's belief in, 53–54; faster-than-light, 44, 64; imaginary mass of, 65; merging with psychons, 66; in science fiction, 64; toxic, 57; in *VALIS*, 55–56; in *Watchmen*, 33, 60–61, 62 (fig.)

Tan, Amy, 149; China in fiction of, 150; mediation of incommensurate sphere, 174; qi cosmology in, 27; self-Orientalism of, 151, 161–62. See also *Hundred Secret Senses, The*; *Joy Luck Club, The*

Targ, Russell, 243n33; *Mind-Reach*, 120; parapsychology experiments, 109; on Project Scanate, 112; psi research of, 108; on remote viewing, 112; and Swann, 108, 110, 111

Tarlaci, Sultan, 65

Taylor, Bayard: *A Visit to India, China, and Japan*, 160

TED website, "The Debate about Rupert Sheldrake's Talk," 2

telegraphy: as electromagnetic, 39; incorporation into spiritualism, 21; probing of consciousness, 217; spiritual, 216; in *Stranger Things*, 99–100. See also clairvoyance; remote viewing

telekinesis, 29; in cultural consciousness, 7; effect of biological sciences on, 69; Gaian, 68, 69, 86–87, 89–90, 215; mechanics of, 97; nineteenth-century, 90; science fiction and, 29; in *Star Fire*,

127; systems theory and, 29; in *A Tale for the Time Being,* 206; viral, 95; in *Watchmen,* 59-60, 60 (fig.)

telepathemes (thought particles), 39, 46

telepathy, 10, 29, 177; among abiotic entities, 2; in *Atomik Aztek,* 183, 184, 188-92, 194-95; in comics, 3; early psychical research on, 179-81; Eccles's theory of, 41-42; as electromagnetic communication, 126; in ethnic fiction, 31; marginality in Western science, 226n14; in native sciences, 145; persistence of, 3, 17; in real time, 180; scientific origins of, 22; Soviet, 107; in space-time, 180; species-wide, 1-2; subliminal self in, 180. *See also* clairvoyance; remote viewing

teyloia (existence beyond death), Aztec, 192

Thales of Miletus, on living earth, 71, 78

Theosophical Society, 8

Thiher, Allen, 42

Thompson, J. J., 5

Thompson, Tade: Wormwood Trilogy, 218

thought: conversion to material action, 42; electromagnetic emissions of, 107; particles of, 71

time: agency over, 200; circular, 31, 177, 185; classical views of, 199; in Korean shamanism, 165, 169; linear, 199, 206; meditation on, 201; Mesoamerican, 177, 185; psitrons' regression through, 47, 48; in speculative ethnic fiction, 179; and speed of light, 44; in *Watchmen,* 61-62. *See also* space-time

time, Aztec, 185-86; circular, 185, 189-90; control over, 186; sequential ages of, 189-90. *See also* chronology, Aztec

Tiptree, James, Jr.: *Up the Walls of the World,* 125

Toltecs: reality of, 192; sacrifice by, 191

Troland, Leonard Thompson, 228n48

Tylor, Edward Burnett, 17

Tyrrell, G. N. M., 104

uji (being-time), 200-201, 203; in *A Tale for the Time Being,* 206; unity of, 205

Ullman, Montague: Dream Laboratory of, 14

Unali, Lina, 151

unconscious, the: global, 82; ideograms from, 118; links of, 135; pre-Freudian theory of, 10; signal line access to, 116; Swann's models for, 101-2; as technology, 101

unconscious, the, collective, 81, 84; in Gaia hypothesis, 128; Jungian, 118; maximization of, 136

unexplainable, the: in Asian American literature, 143; gestalts of, 114; non-Western scientific epistemes of, 174

United States, telepathic espionage programs, 100

universe: geocentric, 229n73; occluded, 43; occult, 35

VALIS (Philip K. Dick), 29, 35; brain-site stimulations, 55; Dobb's work in, 54; false reality in, 54; human regression in, 57; incarnation of Jesus in, 54; narrator of, 54, 55-57; neo-Christian Gnosticism of, 58; parascientification in, 58; the pastoral in, 57; precasts in, 55; precognition in, 55, 56-57; psitrons in, 52-58, 63, 64; tachyons in, 55-56; 2-3-74 in, 54, 55; "Vast Active Living Intelligence System," 54

van Vogt, A. E.: *Slan,* 13

Varela, Francisco: on autopoiesis, 74

Vasiliev, Leonid L., 108; *Experiments in Mental Suggestion,* 39

Velminski, Wladimir, 106-7, 125-26; *Homo Sovieticus,* 218

Vint, Sherryl: *Bodies of Tomorrow,* 23; on science fiction, 22–23
viruses: as agents of will, 89; communication through, 88; in *Darwin's Radio,* 87, 88; as psychical agents, 89; telekinetic, 95
vitalism, 74, 249n59; in Bear's works, 87; qi and, 153; in systems theory, 75
Vitals (Greg Bear), 70, 90–95; annihilation of humanity in, 95; bacterial superorganism of, 86, 91–95; bacterial telepathy in, 91, 93–94; eternal life in, 91, 93, 94–95; Gaia in, 91–95, 184; global mind in, 91; hostile microbes in, 94–95; microbiome in, 91–92; neurotoxins in, 93; prokaryotes in, 91
Vivekananda, on alternative healing, 152
von Neumann, John, 238n28

Waddington, Conrad, 80, 240n65
Warcollier, René, 105
Watchmen (Alan Moore), 29, 35, 52; clairvoyance, 60; decision-making in, 63; Eternalism in, 60–61; experience of time in, 61–62; precasts in, 61; precognition in, 33–34, 60–63, 62 (fig.); psitrons in, 58–64, 63 (fig.), 184; scholarship on, 236n92; superheroes of, 59–60; tachyons in, 33, 60–61, 62 (fig.); telekinesis in, 59–60, 62 (fig.)
Watson, Andrew: Daisyworld program of, 237n5
Weinbaum, Stanley G.: "The Ideal," 38
wholeness: Newtonian, 45; in systems theory, 82
Wiener, Norbert, 76, 238n28
will: mental, 41; viruses as agents of, 89
Winsbro, Bonnie: *Supernatural Forces,* 164
women: comfort, 162; exclusion from Western science, 147–48, 216; neural

biology of, 148–49, 173; in occult science, 27, 216; in psychical research, 148–49; as receptors of information, 146, 148; shamans, 166–67, 216; Victorian, 148. *See also* ghost seers
Woolf, Virginia: *Mrs. Dalloway,* 11
world, psychical: mirroring physical world, 115, 180. *See also* Gaia
world soul: collective agency of, 95; human mind within, 95; in precognition, 36; premodern beliefs in, 81

Yang, Robin, 153
Yeats, W. B., 4
Yeoh, Michelle, 189
yin-yang: imbalance in, 157; symbol, 153
Your Nostradamus Factor (Ingo Swann), 102, 121, 132–34; applied psi in, 132; as how-to manual, 132; ideograms in, 133; as literary technology, 134; the Matrix in, 133; mind-dynamic presence of, 132; nature of time in, 132; precognition in, 132; psi in, 133, 134
Yu, Charles, 182

Zai, Zhang, 154, 155; on ghosts, 157
Zen Buddhism: as alternative chronology, 196; Buddha-nature in, 199; explanation of natural world, 198; mental connection in, 208; metaphysics of, 197; paranormal mind in, 199, 202; Soto school of, 198; space-time in, 199–201, 203–5; in speculative realism, 178; in *A Tale for the Time Being,* 154, 177, 196–207, 208; versus Western phenomenology, 199
Zener cards, 12, 13 (fig.)
Zha, Leping, 139
Zhang, Benzi, 161
Zoehrer, Dominic, 152

DEREK LEE is assistant professor of English and Z. Smith Reynolds Foundation Fellow at Wake Forest University.

Printed and bound by CPI Group (UK) Ltd, Croydon, CR0 4YY

17/08/2025

14719517-0001